微藻碳酸酐酶生物地球化学作用
Biogeochemical Action of Microalgal Carbonic Anhydrase

吴沿友　李海涛　谢腾祥 等 著

Wu Yanyou　Li Haitao　Xie Tengxiang et al.

科学出版社

北　京

内 容 简 介

本书以微藻的碳酸酐酶为研究对象，通过添加和未添加乙酰唑胺到实验系统，研究了微藻的几种碳酸酐酶同工酶基因表达对碳酸氢根离子和 pH 的响应，阐明了微藻碳酸酐酶胞外酶在环境响应中的开关作用。通过对微藻利用无机碳途径的定量，研究了重碳酸盐、pH 以及直接碳酸氢根离子途径对微藻碳汇的影响，阐明了微藻碳酸酐酶对碳源、碳汇的调节作用。通过对碳酸盐岩的微藻生物溶蚀作用的定量以及微藻对碳酸钙碳源利用份额的定量，研究了二氧化碳的供给、酸碱度以及无机碳利用途径对微藻碳酸盐岩溶蚀和碳酸盐碳源的利用影响，阐明了微藻碳酸酐酶在碳酸盐岩的溶蚀过程中的"催化"作用。通过大量元素、微量元素以及几种污染物对微藻碳酸酐酶影响的研究，阐明了无机元素对微藻碳酸酐酶影响的"激素"效应。最后，以岩溶湖泊微藻为例，研究了岩溶湖泊微藻种类组成、岩溶湖泊无机碳变化与碳酸酐酶的关系，估算了不同湖泊微藻无机碳碳源份额和无机碳利用途径份额，评价了岩溶湖泊微藻的碳汇能力，阐明了岩溶湖泊微藻碳酸酐酶对碳汇的制约能力。

本书可作为大中专院校和科研单位从事海洋科学、海洋与湖沼学、生物地球化学以及生态学等广大科研工作者和研究生的参考用书，也可作为高年级本科生了解藻类生理生态学、全球变化生态学以及生物地球化学等领域发展动态的课外资料。

图书在版编目（CIP）数据

微藻碳酸酐酶生物地球化学作用/吴沿友等著. —北京：科学出版社，2015.2
ISBN 978-7-03-043326-8

Ⅰ.①微… Ⅱ.①吴… Ⅲ.①微藻-碳酸酐酶-作用-生物地球化学-研究
Ⅳ.①P593

中国版本图书馆 CIP 数据核字（2015）第 027600 号

责任编辑：孙天任　顾晋饴/责任校对：刘亚琦
责任印制：徐晓晨/封面设计：许　瑞

科 学 出 版 社 出版
北京东黄城根北街 16 号
邮政编码：100717
http://www.sciencep.com

北京摩诚则铭印刷科技有限公司 印刷
科学出版社发行　各地新华书店经销
*

2015 年 2 月第 一 版　　开本：720×1000 1/16
2020 年 4 月第二次印刷　　印张：16 3/4
字数：338 000

定价：129.00 元
（如有印装质量问题，我社负责调换）

序

  地表岩石-土壤-生物-水-大气相互作用带被称为地球关键带（the critical zone）。正确理解关键带中发生的物理、化学和生物过程特征及其对风化壳形成演化，认识其中物质释放-迁移-转化的控制机制是认识关键带的核心。碳酸酐酶能催化二氧化碳与碳酸氢根离子之间的快速转化，它广泛分布于土壤、水体和岩石表层的生物体中，以其特有的快速催化效应，将地表岩石、土壤、水体和大气之间的无机碳紧密地偶联在一起，带动了碳乃至其他元素的生物地球化学循环，加快了各体系之间的物质迁移和转化速率。该书从微观入手，重点研究了在微藻碳酸酐酶参与下的水-CO_2交换、水岩相互作用以及水体无机碳利用的生物学机制和地球化学机制，清晰地诠释了微藻碳酸酐酶如何从基因表达（分子水平）到生理调控（细胞水平），进而导致细胞表面微环境的物理化学变化，最终影响着关键带中的生物地球化学循环的过程。这对认识关键带中的物质循环调控机制有着重要的意义，也为（大）分子水平上的关键带研究提出了建设性意见。

  该书从碳酸酐酶基因表达的角度来探讨影响微藻稳定碳同位素分馏的深层次机制，将分子生物学引入地球化学领域，丰富了地球化学的研究手段。确定了胞外碳酸酐酶对微藻稳定碳同位素的分馏作用，阐明了微藻碳酸酐酶在碳源、碳汇以及生物岩溶过程的调节作用，提出了无机元素对碳酸酐酶影响的"激素"效应假说，评估了岩溶湖泊微藻的各种碳汇能力以及影响碳汇能力的因素。这些研究成果对生物地球化学乃至地球关键带的研究具有重要的理论和实践意义。

  该书充分运用同位素地球化学、植物生理学、化学计量学、分子生物学、环境科学、生态学、生物数学等方向的理论、技术和手段，开展了生物地球化学方面的研究，发展了一个全新的研究视角，是多学科交叉的研究典范。

  该书开发出多种定量方法，定量出微藻碳源份额、无机碳利用途径份额、生物溶蚀作用、岩溶湖泊水体中微藻岩溶碳汇和光合碳汇能力等，为生物地球化学作用的研究从定性走向定量提供了可借鉴的模式。

  该书在室内工作的基础上，开展了岩溶湖泊微藻的研究工作，探讨了碳酸酐酶在调节岩溶湖泊微藻的碳汇组成、响应水环境变化中的作用，开展了碳酸酐酶在碳酸盐岩的溶蚀与沉积作用的研究。将室内获得的理论模型应用到野外水体，是理论与实际相结合的一个典范。

  该书是一部从微观角度入手，研究生物与环境相互作用的深层次机制，并最终解决宏观环境问题的理论专著，在诸多方面颇有创新，并取得了一系列研究成果。相信该书的出版，必将激起生物地球化学及生态环境领域学者更加关注碳酸酐酶乃至其他生物大分子的热潮，推动生物地球化学乃至地球关键带的研究向着微观以及微观与宏观相结合的方向发展，建立多学科交叉的地球化学研究体系。值该书出版之际，我衷心祝愿以吴沿友研究员为首的研究团队在大分子（酶）生物地球化学研究领域取得更大的成就。

中国科学院院士

2014 年 10 月

前　言

碳酸酐酶（carbonic anhydrase，CA，EC 4.2.1.1）是一种含锌的金属酶，催化 CO_2 与 HCO_3^- 之间的可逆转化，调节控制着生物体的碳代谢。生态系统中的碳代谢驱动着碳的生物地球化学循环，进而影响氮、磷、硫等元素的生物地球化学循环。水生生态系统是全球最大的生态系统，它所带来的初级生产力占全球净初级生产力的 50%。微藻在水生生态系统中起到关键的作用，是初级生产力的最主要来源，并广泛受到碳酸酐酶的调节。因此，研究微藻碳酸酐酶生物地球化学作用，将能探明水环境中的无机碳的转运规律，阐明水-岩-气-生系统中无机和有机碳的偶合机制，揭示碳酸盐岩演化、生物成矿以及岩溶碳汇等机制，增加我们对生物地球化学循环中的一些未知过程的理解。同时也为水环境治理提供科学依据。

本书分八章。第一章介绍了碳酸酐酶生物学作用及碳酸酐酶活力的常规测定方法。第二章研究了莱茵衣藻与蛋白核小球藻的碳酸酐酶基因表达，探讨了碳酸酐酶基因表达对 pH、碳酸氢根离子以及氟的响应。第三章探讨了 pH 对微藻稳定碳同位素分馏的影响，确定了胞外碳酸酐酶对稳定碳同位素分馏的影响程度。第四章介绍了微藻无机碳利用途径的定量方法，探讨了微藻碳酸酐酶对碳源、碳汇的调节作用。第五章介绍了微藻与碳酸盐岩交互作用的定量方法，探讨了碳酸酐酶在微藻生物岩溶中的作用。第六章探讨了无机元素对碳酸酐酶影响的"激素"效应，阐明了无机元素与微藻碳酸酐酶的相互作用。第七章介绍了岩溶湖泊微藻种类组成与碳酸酐酶之间的关系，评价了几种岩溶湖泊微藻碳汇能力。第八章展望未来五年内酶生物地球化学研究方面的发展趋势，提出了我们努力的方向。

本书是作者及其科研团队 15 年的相关科研成果的总结。在研究过程中不仅得到了吴征镒院士、欧阳自远院士、袁道先院士、卢耀如院士、刘丛强院士等老一辈科学家的指导和鼓励，同时还得到了国内外众多相关科学家及中国科学院地球化学研究所环境地球化学国家重点实验室、江苏大学现代农业装备与技术教育部重点实验室的同事们的支持和帮助。对此，我们深表谢意！

本书得到了国家重点基础研究发展计划（973 计划）专题"岩溶植被及微藻碳源利用策略及份额估算"（2013CB956701，2013CB956703）、国家自然科学基金"岩溶湖泊水体微藻碳酸酐酶地球化学作用"（40273038）、"微藻碳酸酐酶胞外酶影响下的稳定碳同位素分馏的识别"（40973060）、中国科学院百人计划项目

"喀斯特生态系统的稳定性和适应性"、教育部博士点基金"温室环境制约下的土壤碳酸酐酶的变化研究"（20070299002）、环境地球化学国家重点实验室项目以及江苏省高校优势学科资助项目（苏政办〔2014〕37号）等众多项目的资助。正是有了这些项目的资助，才有了本书的出版，在此，我们表示衷心的感谢。

本书是由吴沿友教授领导的课题组集体撰写完成的。在本研究中，吴沿友教授在学术思想的提升、项目的组织实施、研究方案的设计、技术路线的形成以及学术成果的提炼中起着决定性的作用。各章撰写分工为：第一章，李海涛、吴沿友；第二章，李海涛、吴运东、吴沿友；第三章，吴沿友、徐莹、李海涛；第四章，李海涛、吴沿友；第五章，谢腾祥、吴沿友；第六章，吴沿友、王宝利、李潜、吴运东；第七章，谢腾祥、吴沿友、李海涛、徐莹、王宝利、何梅、陈椽；第八章，吴沿友。此外，课题组的多位在读研究生参与了本书的资料收集和整理。杭红涛、张开艳、饶森等同学对部分章节内容进行了校对。

我们要特别感谢刘丛强院士！这不仅是因为刘丛强院士为本书热情作序以及对我们的长期关怀和支持，更重要的是因为，在碳酸酐酶研究方面，刘丛强院士为我们指明了学术道路。早在2001年时，时任中国科学院地球化学研究所所长的刘丛强院士就敏锐地窥探出碳酸酐酶的生物地球化学作用的重要性，鼓励我们开展这方面的研究，将"红枫湖水体碳酸酐酶及其生物地球化学作用研究"列为中国科学院地球化学研究所创新领域课题。有了这个课题和刘丛强院士等众多科学家的长期支持，本书终于得以问世。

碳酸酐酶生物地球化学作用或者说大分子（酶）生物地球化学作用涉及极其复杂的物理、化学以及生物学过程，内容极其丰富、复杂，本书仅是冰山一角，只能起着抛砖引玉的作用。出版本书，也只希望引来众多科学家的关注和兴趣。希望今后有更多的学者和专家从宏观和微观多角度多层次开展大分子（酶）生物地球化学作用研究，推动地球化学、生态学的发展。

由于著者学术水平有限，时间仓促，疏漏之处在所难免，恳请读者不吝赐教！

作　者

2014年9月29日

目 录

序
前言
第一章 碳酸酐酶的生物学作用 ·· 1
  第一节 碳酸酐酶简介 ··· 2
    一、碳酸酐酶的结构和性质 ··· 2
    二、碳酸酐酶的分布 ··· 2
    三、碳酸酐酶的多样性 ··· 2
  第二节 碳酸酐酶的作用与意义 ··· 4
    一、碳酸酐酶的生理生化作用 ·· 4
    二、影响碳酸酐酶活力的主要因素 ····································· 5
  第三节 碳酸酐酶的抑制剂及应用 ······································ 7
    一、碳酸酐酶抑制剂的种类 ·· 7
    二、乙酰唑胺对碳酸酐酶的抑制作用原理 ····························· 7
    三、碳酸酐酶抑制剂的应用 ·· 7
  第四节 碳酸酐酶活力的常规测定方法 ································· 8
    一、放射免疫分析法 ··· 8
    二、pH 计法 ··· 8
    三、同位素二氧化碳质谱分析法 ······································· 8
    四、mRNA 测定法 ·· 8
    五、比色法 ··· 9
  参考文献 ··· 9
第二章 微藻的碳酸酐酶基因表达及其对环境的响应 ······················ 15
  第一节 莱茵衣藻碳酸酐酶基因的生物信息学 ······················ 17
    一、莱茵衣藻是碳酸酐酶基因的生物信息学研究的模式生物 ········ 17
    二、莱茵衣藻碳酸酐酶基因的生物信息学 ··························· 18
  第二节 莱茵衣藻与蛋白核小球藻碳酸酐酶基因表达 ·············· 20
    一、实验材料与方法 ·· 20
    二、常规 PCR 扩增结果 ·· 27
    三、莱茵衣藻与蛋白核小球藻碳酸酐酶基因的同源性 ··············· 27
    四、碳酸酐酶同工酶基因的亲缘关系及基因表达的差异性 ·········· 28

第三节 微藻碳酸酐酶基因表达对碳酸氢根离子的响应 ······ 29
　　一、碳酸氢钠对碳酸酐酶基因表达的影响 ······ 30
　　二、乙酰唑胺对碳酸酐酶基因表达的影响 ······ 33
　　三、乙酰唑胺与碳酸氢钠共同作用下的碳酸酐酶基因表达 ······ 35
第四节 微藻碳酸酐酶基因表达对pH的响应 ······ 39
　　一、pH对碳酸酐酶基因表达的影响 ······ 39
　　二、乙酰唑胺与pH共同作用下的碳酸酐酶基因表达 ······ 41
第五节 微藻碳酸酐酶基因表达对氟的响应 ······ 45
参考文献 ······ 47

第三章 胞外碳酸酐酶对稳定碳同位素分馏的影响　50
第一节 碳酸酐酶胞外酶对稳定碳同位素分馏作用及其识别基础 ······ 52
第二节 pH对微藻稳定碳同位素分馏的影响 ······ 55
　　一、研究材料和测定方法 ······ 55
　　二、pH对微藻生长的影响 ······ 56
　　三、pH对微藻碳酸酐酶胞外酶活力的影响 ······ 57
　　四、pH对微藻稳定碳同位素分馏的影响 ······ 58
第三节 乙酰唑胺作用下重碳酸盐对微藻稳定碳同位素分馏的影响 ······ 60
　　一、培养条件和处理方法 ······ 60
　　二、乙酰唑胺作用下重碳酸盐对微藻生长的影响 ······ 61
　　三、乙酰唑胺作用下重碳酸盐对微藻碳酸酐酶胞外酶活力的影响 ······ 63
　　四、乙酰唑胺作用下重碳酸盐对微藻稳定碳同位素分馏的影响 ······ 64
第四节 乙酰唑胺对微藻稳定碳同位素分馏的影响 ······ 67
　　一、培养条件和处理方法 ······ 67
　　二、乙酰唑胺对微藻生长的影响 ······ 68
　　三、乙酰唑胺对微藻碳酸酐酶胞外酶活力的影响 ······ 69
　　四、乙酰唑胺对微藻稳定碳同位素分馏的影响 ······ 69
参考文献 ······ 71

第四章 微藻碳酸酐酶对碳源、碳汇的调节作用　75
第一节 碳酸酐酶在无机碳代谢中的作用 ······ 77
　　一、碳酸酐酶与无机碳的转运 ······ 77
　　二、碳酸酐酶与光合作用 ······ 79
第二节 微藻无机碳利用途径的定量方法 ······ 79
　　一、微藻碳同位素测定方法及碳同位素修正模型 ······ 79
　　二、微藻利用不同碳源的份额计算方法 ······ 80
　　三、微藻利用不同形态无机碳的途径份额计算方法 ······ 82

四、直接碳汇和间接碳汇的计算方法 ………………………………… 83
　第三节　无机碳利用途径对微藻碳汇的影响 …………………………… 84
　　　一、重碳酸盐对微藻碳汇的影响 ………………………………………… 84
　　　二、碳酸酐酶胞外酶对微藻碳汇的影响 ………………………………… 89
　　　三、直接碳酸氢根离子利用途径对微藻碳汇的影响 …………………… 97
　　　四、碳酸酐酶胞外酶和直接碳酸氢根离子利用途径对微藻碳汇的复合影响 … 102
　第四节　微藻碳汇组成与"遗失的碳汇" ………………………………… 105
　参考文献 …………………………………………………………………… 107

第五章　微藻碳酸酐酶与生物岩溶作用 …………………………………… 111
　第一节　碳酸酐酶对碳酸盐岩的溶蚀作用 ……………………………… 114
　　　一、碳酸酐酶与碳酸盐岩的溶蚀 ………………………………………… 114
　　　二、碳酸酐酶与碳酸钙的沉积 …………………………………………… 115
　　　三、碳酸酐酶与生物成矿作用 …………………………………………… 116
　第二节　微藻与碳酸盐岩交互作用的定量方法 ………………………… 117
　　　一、碳酸盐岩的微藻生物溶蚀作用定量方法 …………………………… 117
　　　二、微藻对碳酸钙碳源利用份额的定量方法 …………………………… 118
　　　三、方解石生物溶蚀所释放的无机碳在各介质中分配的定量方法 …… 119
　　　四、岩溶碳汇和光合碳汇计算方法 ……………………………………… 120
　第三节　碳酸酐酶在微藻生物溶蚀中的作用 …………………………… 120
　　　一、微藻生物溶蚀作用的种类特征 ……………………………………… 120
　　　二、pH与微藻生物溶蚀作用 …………………………………………… 122
　　　三、碳酸酐酶胞外酶在微藻生物溶蚀中的作用 ………………………… 129
　　　四、阴离子通道在微藻生物溶蚀中的作用 ……………………………… 137
　第四节　碳酸酐酶影响下的微藻对碳酸钙碳源的利用 ………………… 144
　　　一、酸碱度对微藻利用碳酸钙碳源的影响 ……………………………… 145
　　　二、胞外碳酸酐酶对微藻利用碳酸钙碳源的影响 ……………………… 147
　　　三、阴离子通道对微藻利用碳酸钙碳源的影响 ………………………… 151
　第五节　碳酸酐酶影响下的生物岩溶碳汇作用 ………………………… 153
　　　一、酸碱度对生物岩溶碳汇的作用 ……………………………………… 153
　　　二、胞外碳酸酐酶对岩溶碳汇的作用 …………………………………… 156
　　　三、阴离子通道对岩溶碳汇的作用 ……………………………………… 158
　参考文献 …………………………………………………………………… 160

第六章　无机元素与微藻碳酸酐酶的相互作用 …………………………… 165
　第一节　无机元素对碳酸酐酶影响的"激素"效应 …………………… 168
　　　一、无机元素对碳酸酐酶影响的剂量效应 ……………………………… 168

二、"激素"对碳酸酐酶的影响 …………………………………… 169
三、无机元素对碳酸酐酶作用的"激素"效应的可能机制 ……… 170
第二节 微藻碳酸酐酶对氮磷钾吸收的影响 ………………………… 171
一、植物离子吸收动力学 …………………………………………… 171
二、胞外碳酸酐酶抑制剂对微藻氮磷钾吸收的影响 ……………… 171
三、微藻碳酸酐酶影响无机元素吸收的可能机制 ………………… 175
第三节 无机元素对微藻碳酸酐酶的影响 …………………………… 176
一、氟对微藻碳酸酐酶的影响 ……………………………………… 176
二、重金属污染物对微藻碳酸酐酶的影响 ………………………… 185
参考文献 …………………………………………………………………… 190

第七章 岩溶湖泊中微藻碳酸酐酶的生物地球化学作用 ……………… 198
第一节 碳酸酐酶与岩溶水环境中的碳循环 ………………………… 200
第二节 岩溶湖泊水环境特征 ………………………………………… 204
一、岩溶湖泊水文及物理特征 ……………………………………… 205
二、岩溶湖泊水化学特征 …………………………………………… 206
三、岩溶湖泊微藻的生长特征 ……………………………………… 211
第三节 岩溶湖泊微藻种类组成与碳酸酐酶——以 2002~2003 年为例
………………………………………………………………………… 214
一、微藻种群组成和碳酸酐酶变化特征 …………………………… 214
二、碳酸酐酶活力与微藻的组成和种类的相关性 ………………… 216
三、岩溶湖泊微藻种类与碳酸酐酶协同变化特征 ………………… 217
第四节 岩溶湖泊无机碳的变化和微藻碳酸酐酶 …………………… 222
一、岩溶湖泊无机碳浓度及碳同位素组成的变化特征 …………… 222
二、湖泊水体中的碳酸氢根离子浓度与藻体 $\delta^{13}C$ 的关系 ……… 224
第五节 岩溶湖泊微藻碳汇能力 ……………………………………… 227
一、岩溶湖泊微藻的生物量的变化 ………………………………… 227
二、岩溶湖泊微藻无机碳碳源份额和无机碳利用途径 …………… 228
三、乙酰唑胺与碳酸氢钠共同作用下的岩溶湖泊微藻碳汇 ……… 230
四、岩溶湖泊微藻碳汇能力的评价 ………………………………… 233
参考文献 …………………………………………………………………… 237

第八章 展望 ……………………………………………………………… 246

Contents

Preface

Foreword

Chapter 1 Biological function of carbonic anhydrase ················ 1
  1.1 Introduction to carbonic anhydrase ················ 2
    1.1.1 Structure and properties of carbonic anhydrase ················ 2
    1.1.2 Distribution of carbonic anhydrase ················ 2
    1.1.3 Diversity of carbonic anhydrase ················ 2
  1.2 Function and meaning of carbonic anhydrase ················ 4
    1.2.1 Physiological and biochemical function of carbonic anhydrase ················ 4
    1.2.2 The main factors influencing the carbonic anhydrase activity ················ 5
  1.3 Carbonic anhydrase inhibitor and application ················ 7
    1.3.1 Carbonic anhydrase inhibitor species ················ 7
    1.3.2 Principle of the inhibition on carbonic anhydrase by acetazolamide ················ 7
    1.3.3 Application of carbonic anhydrase inhibitor ················ 7
  1.4 Determination on the activity of carbonic anhydrase ················ 8
    1.4.1 Radioimmunoassay ················ 8
    1.4.2 pH meter method ················ 8
    1.4.3 Isotopes of carbon dioxide mass spectrometry ················ 8
    1.4.4 mRNA assay ················ 8
    1.4.5 Colorimetry ················ 9
  Reference ················ 9

Chapter 2 The gene expression of carbonic anhydrase in microalgae and their response to the environment ················ 16
  2.1 Bioinformatics of carbonic anhydrase genes in *Chlamydomonas reinhardtii* ················ 17
    2.1.1 *Chlamydomonas reinhardtii*, a model for the research on bioinformatics of carbonic anhydrase genes ················ 17
    2.1.2 Bioinformatics of carbonic anhydrase genes in *Chlamydomonas reinhardtii* ················ 18
  2.2 The gene expression of carbonic anhydrase in *Chlamydomonas*

    reinhardtii and *Chlorella pyrenoidosa* ·· 20
   2.2.1 The experimental materials and methods ······························· 20
   2.2.2 The results of conventional PCR amplification ······················ 27
   2.2.3 The homology of carbonic anhydrase gene in *Chlamydomonas*
     reinhardtii and *Chlorella pyrenoidosa* ·································· 27
   2.2.4 The genetic relationship among carbonic anhydrase isozyme genes and
     the differences of gene expression ·· 28
  2.3 The gene expression of carbonic anhydrase in microalgae in response
   to bicarbonate ·· 29
   2.3.1 The influence of sodium bicarbonate on carbonic anhydrase gene expres-
     sion ·· 30
   2.3.2 The influence of acetazolamide on carbonic anhydrase gene expression
     ·· 33
   2.3.3 Carbonic anhydrase gene expression under the action of acetazolamide in
     combination with sodium bicarbonate ······································ 35
  2.4 The gene expression of carbonic anhydrase in microalgae in response
   to pH ··· 39
   2.4.1 The influence of pH on carbonic anhydrase gene expression ············ 39
   2.4.2 Carbonic anhydrase gene expression under the action of acetazolamide in
     combination with pH ··· 41
  2.5 The gene expression of carbonic anhydrase in microalgae in response
   to fluoride ·· 45
  References ·· 47
**Chapter 3 The effect of extracellular carbonic anhydrase on stable carbon
     isotope fraction** ··· 51
  3.1 The role and fundamental of extracellular carbonic anhydrase on
   stable carbon isotope fraction ··· 52
  3.2 The influence of pH on stable carbon isotope fraction ··············· 55
   3.2.1 Materials and methods ··· 55
   3.2.2 The influence of pH on microalgal growth ······························ 56
   3.2.3 The influence of pH on the activity of extracellular carbonic anhydrase in
     microalgal ··· 57
   3.2.4 The influence of pH on stable carbon isotope fraction in microalgal ··· 58
  3.3 The effect of bicarbonate on microalgal stable carbon isotope
   fraction under acetazolamide ··· 60

3.3.1　Culture conditions and methods ········· 60
3.3.2　The effect of bicarbonate on microalgal growth under acetazolamide ··· 61
3.3.3　The effect of bicarbonate on the activity of extracellular carbon anhydrase under acetazolamide ········· 63
3.3.4　The effect of bicarbonate on microalgal stable carbon isotope fraction under acetazolamide ········· 64

3.4　The influence of acetazolamide on microalgae stable carbon isotope fraction ········· 67
3.4.1　Culture conditions and methods ········· 67
3.4.2　The influence of acetazolamide on microalgal growth ········· 68
3.4.3　The influence of acetazolamide on the activity of extracellular carbonic anhydrase ········· 69
3.4.4　The influence of acetazolamide on microalgal stable carbon isotope fraction ········· 69

References ········· 71

Chapter 4　The regulation on carbon source and carbon sequestration by microalgal carbonic anhydrase ········· 76

4.1　The role of carbonic anhydrase in inorganic carbon metabolism ··· 77
4.1.1　Carbonic anhydrase versus inorganic carbon transport ········· 77
4.1.2　Carbonic anhydrase versus photosynthesis ········· 79

4.2　Quantitative method of inorganic carbon utilization pathways in microalgae ········· 79
4.2.1　The determination on stable carbon isotope and the correction model of stable carbon isotope in microalgae ········· 79
4.2.2　The calculation on the proportion of inorganic carbon sources used by microalgae ········· 80
4.2.3　The calculation on the proportion of inorganic carbon utilization pathways by microalgae ········· 82
4.2.4　The calculation on the direct and indirect carbon sequestrations ········· 83

4.3　The impact on carbon sequestrations by inorganic carbon utilization pathways in microalgae ········· 84
4.3.1　The influence of bicarbonate on carbon sequestrations in microalgae ··· 84
4.3.2　The influence of extracellular carbonic anhydrase on carbon sequestrations in microalgae ········· 89
4.3.3　The influence of direct bicarbonate utilization pathway on carbon seques-

    tration in microalgae ··· 97
  4.3.4 The compound influence of extracellular carbonic anhydrase and direct bicarbonate utilization pathway on carbon sequestrations in microalgae ··· 102
4.4 The composition of microalgal carbon sequestrations versus "missing carbon sink" ··· 105
References ·· 107

Chapter 5 Microalgal carbonic anhydrase versus the Bio-Karst ················ 112

5.1 The role of carbonic anhydrase in the dissolution of carbonate rock ··· 114
  5.1.1 Carbonic anhydrase versus the dissolution of carbonate rock ············ 114
  5.1.2 Carbonic anhydrase versus the precipitation of calcium carbonate ······ 115
  5.1.3 Carbonic anhydrase versus bio-mineralization ································ 116
5.2 The methods of quantifying the interactive action between microalgae and carbonate rock ··· 117
  5.2.1 The methods of quantifying the bio-dissolution of carbonate rock ······ 117
  5.2.2 The methods of quantifying the utilization on the carbon in calcium carbonate by microalgae ··· 118
  5.2.3 The methods of quantifying the distribution in the inorganic carbon releasing from calcite's bio-dissolution in different mediums ··············· 119
  5.2.4 The methods of calculation on the karst carbon sink and photosynthetic carbon sink ··· 120
5.3 The role of carbonic anhydrase in the microalgal bio-dissolution of carbonate rock ··· 120
  5.3.1 The type of microalgal bio-dissolution ·· 120
  5.3.2 The pH versus microalgal bio-dissolution ··· 122
  5.3.3 The role of extracellular carbonic anhydrase on microalgal bio-dissolution ·· 129
  5.3.4 The role of anion channel system on microalgae's bio-dissolution ······ 137
5.4 The utilization on the carbon in calcium carbonate by microalgae under the influence of carbonic anhydrase ·· 144
  5.4.1 The effect of pH on the utilization of the carbon in calcium carbonate by microalgae ··· 145
  5.4.2 The effect of extracellular carbonic anhydrase upon the utilization on the carbon in calcium carbonate by microalgae ······································· 147
  5.4.3 The effect of anion channel system upon the utilization on the carbon in

calcium carbonate by microalgae ⋯⋯⋯⋯⋯⋯⋯⋯⋯⋯⋯⋯⋯⋯⋯⋯⋯⋯ 151
5.5 The bio-karstic carbon sink under the influence of carbonic anhydrase ⋯⋯⋯⋯⋯⋯⋯⋯⋯⋯⋯⋯⋯⋯⋯⋯⋯⋯⋯⋯⋯⋯⋯⋯⋯⋯⋯⋯⋯⋯⋯⋯⋯⋯⋯⋯ 153
    5.5.1 The role of pH on the bio-karstic carbon sink ⋯⋯⋯⋯⋯⋯⋯⋯ 153
    5.5.2 The role of extracellular carbonic anhydrase on the bio-karstic carbon sink ⋯⋯⋯⋯⋯⋯⋯⋯⋯⋯⋯⋯⋯⋯⋯⋯⋯⋯⋯⋯⋯⋯⋯⋯⋯⋯⋯⋯⋯⋯⋯ 156
    5.5.3 The role of anion channel system on the bio-karstic carbon sink ⋯⋯⋯ 158
References ⋯⋯⋯⋯⋯⋯⋯⋯⋯⋯⋯⋯⋯⋯⋯⋯⋯⋯⋯⋯⋯⋯⋯⋯⋯⋯⋯⋯⋯⋯⋯ 160

Chapter 6 The interaction between inorganic elements and carbonic anhydrase in microalgae ⋯⋯⋯⋯⋯⋯⋯⋯⋯⋯⋯⋯⋯⋯⋯⋯⋯⋯⋯⋯⋯⋯⋯⋯⋯ 166
6.1 The hormone effect of inorganic elements on carbonic anhydrase ⋯⋯ 168
    6.1.1 The dosage effect of inorganic elements on carbonic anhydrase ⋯⋯⋯⋯ 168
    6.1.2 The influence of hormone on carbonic anhydrase ⋯⋯⋯⋯⋯⋯⋯⋯ 169
    6.1.3 The possible mechanism on the hormone effect of inorganic elements on carbonic anhydrase ⋯⋯⋯⋯⋯⋯⋯⋯⋯⋯⋯⋯⋯⋯⋯⋯⋯⋯⋯⋯⋯⋯ 170
6.2 Influence of carbonic anhydrase on nitrogen, phosphorus and potassium uptake by microalgae ⋯⋯⋯⋯⋯⋯⋯⋯⋯⋯⋯⋯⋯⋯⋯⋯⋯⋯⋯⋯⋯⋯⋯ 171
    6.2.1 Ion uptake kinetics in plants ⋯⋯⋯⋯⋯⋯⋯⋯⋯⋯⋯⋯⋯⋯⋯⋯ 171
    6.2.2 The influence of extracellular carbonic anhydrase inhibitor on nitrogen, phosphorus and potassium uptake by microalgae ⋯⋯⋯⋯⋯⋯⋯⋯ 171
    6.2.3 The possible mechanism on the influence of carbonic anhydrase upon inorganic elements uptake by microalgae ⋯⋯⋯⋯⋯⋯⋯⋯⋯⋯⋯⋯⋯⋯ 175
6.3 Influence of inorganic elements on carbonic anhydrase ⋯⋯⋯⋯⋯⋯ 176
    6.3.1 Influence of fluorine on carbonic anhydrase in microalgae ⋯⋯⋯⋯⋯ 176
    6.3.2 Influence of heavy metal pollutant on carbonic anhydrase in microalgae ⋯⋯⋯⋯⋯⋯⋯⋯⋯⋯⋯⋯⋯⋯⋯⋯⋯⋯⋯⋯⋯⋯⋯⋯⋯⋯⋯⋯⋯⋯⋯⋯⋯⋯ 185
References ⋯⋯⋯⋯⋯⋯⋯⋯⋯⋯⋯⋯⋯⋯⋯⋯⋯⋯⋯⋯⋯⋯⋯⋯⋯⋯⋯⋯⋯⋯⋯ 190

Chapter 7 Biogeochemical action of microalgal carbonic anhydrase in karst lakes ⋯⋯⋯⋯⋯⋯⋯⋯⋯⋯⋯⋯⋯⋯⋯⋯⋯⋯⋯⋯⋯⋯⋯⋯⋯⋯⋯⋯⋯⋯⋯⋯⋯⋯⋯ 199
7.1 Carbonic anhydrase versus carbon cycle in karst water environment ⋯⋯ 200
7.2 Water environment characteristics in karst lakes ⋯⋯⋯⋯⋯⋯⋯⋯⋯ 204
    7.2.1 Hydrological and physical characteristics in karst lakes ⋯⋯⋯⋯⋯ 205
    7.2.2 Hydrochemical characteristics in karst lakes ⋯⋯⋯⋯⋯⋯⋯⋯⋯ 206

    7.2.3 Growth characteristics of microalgae in karst lakes ·················· 211
  7.3 Microalgal Composition and carbonic anhydrase in karst lakes ··· 214
    7.3.1 Characteristics of the microalgae species composition and carbonic anhydrase ··· 214
    7.3.2 Correlation between carbonic anhydrase activity and the microalgal composition and species ··· 216
    7.3.3 Covariance relations between microalgal species and carbonic anhydrase 2002~2003 as an example ·· 217
  7.4 Variation of inorganic carbon versus microalgal carbonic anhydrase in karst lakes ·· 222
    7.4.1 Variation characteristics of concentration and carbon isotopic composition of inorganic carbon in karst lakes ·················· 222
    7.4.2 Relationship between the concentration of bicarbonate and microalgal $\delta^{13}C$ of in karst lakes ·· 224
  7.5 Carbon sink ability of microalgae in karst lakes ···················· 227
    7.5.1 Variation of microalgae biomass in karst lakes ················· 227
    7.5.2 Proportion and pathway of the utilization of inorganic carbon by microalgae in karst lakes ··· 228
    7.5.3 Carbon sink of microalgae in karst lakes under the treatment with acetazolamide and sodium carbonate ·· 230
    7.5.4 Evaluation on the carbon sink of microalgae in karst lakes ············ 233
  References ··· 237
Chapter 8 Prospects ·· 249

第一章　碳酸酐酶的生物学作用

摘　　要

碳酸酐酶(carbonic anhydrase，CA，EC 4.2.1.1)是一种含锌的金属酶，催化 CO_2 与 HCO_3^- 之间的可逆转化。碳酸酐酶在生物界分布广泛，种类众多，它的活性受环境条件的影响很大，主要影响因素有：水分、光照、环境 pH、CO_2 浓度、HCO_3^- 浓度等环境条件。碳酸酐酶在光合作用、呼吸作用、维持生物体内酸碱平衡及离子交换等方面都具有非常重要的生理功能。

碳酸酐酶的分析方法主要有：pH 计法、放射免疫分析法、同位素二氧化碳质谱分析法、mRNA 测定法、比色法等，其中，以改进的 pH 计法最常用。

碳酸酐酶的抑制剂主要有乙酰唑胺(acetazolamide，AZ)和乙氧苯并噻唑磺胺(ethoxyzolamide，EZ)，它们在碳酸酐酶的研究中具有非常重要的意义。本书中我们利用乙酰唑胺来研究碳酸酐酶胞外酶的一些生物地球化学过程。

Chapter 1　Biological function of carbonic anhydrase

Abstract

Carbonic anhydrase (CA，EC4.2.1.1)，a zinc-containing metalloenzyme, catalyzes the reversible interconversion between bicarbonate (HCO_3^-) and carbon dioxide (CO_2). CA is ubiquitously found in all kindoms of life and it has multiple isoforms. The activities of CAs are varied largely at different conditions, such as moisture, light, pH, CO_2 concentration and HCO_3^- concentration etc. CA plays many physiological functions in photosynthesis, respiration, pH homeostasis, and ion transport etc.

The main methods to determine CA activity are as follows: pH meter method, radioimmunoassay, isotopes of carbon dioxide mass spectrometry, mRNA assay, colorimetry. Among these methods, the pH meter method with slight modifications was used commonly.

The acetazolamide (AZ) and ethoxyzolamide (EZ) are two inhibitors, which have very important significance in the study of CA. In this book, AZ was used to study some biogeochemical process of extracellular carbonic anhydrase.

第一节 碳酸酐酶简介

一、碳酸酐酶的结构和性质

碳酸酐酶(carbonic anhydrase，CA)(EC4.2.1.1)是一种含 Zn 的金属酶，分子质量约为 30000Da。它的结构含一条卷曲的蛋白质链和一个 Zn^{2+}，其中，Zn^{2+} 处于变形四面体的配位环境。

CA 广泛存在于动物和植物中，甚至在有些微生物中也含有该酶(Smith and Ferry，2000)，到目前为止，已报道出 80 多种含锌的金属酶，位居各种金属酶的首位。它可以催化 CO_2 和 HCO_3^- 之间的相互转化(方程(1-1))。在无 CA 的条件下，平衡需要 1min；而在有 CA 催化的条件下，平衡只需要 10^{-6}s(Khalifah，1971)。

$$HCO_3^- + H^+ \longleftrightarrow CO_2 + H_2O \tag{1-1}$$

碳酸酐酶在光合作用、离子交换、钙化作用、维持生物体内酸碱平衡、无机碳转运等一系列的生理过程中都有十分重要的生理功能(Badger and Price，1994；Gilmour and Perry，2009)。

二、碳酸酐酶的分布

自从 1933 年首次从牛红细胞中纯化得到第一个碳酸酐酶以来(Meldrum and Roughton，1933)，碳酸酐酶的研究就受到了生物学家的广泛关注，陆续在包括人类在内的哺乳动物、植物和单细胞绿藻中发现了多种多样的碳酸酐酶，而且在有些微生物中也含有该酶，甚至在古细菌界和细菌界等原核生物中也发现了碳酸酐酶(Smith and Ferry，2000)。

三、碳酸酐酶的多样性

碳酸酐酶分布广泛，种类多样，根据其在细胞中的位置，碳酸酐酶可分为质膜上和质膜内两大类，据此可以分为胞外碳酸酐酶和胞内碳酸酐酶。胞外碳酸酐酶分布在质膜上及细胞周质空间；胞内碳酸酐酶分布在质膜内，根据其在各个细胞器中的功能又可分为叶绿体碳酸酐酶、线粒体碳酸酐酶、细胞质碳酸酐酶等。

胞外碳酸酐酶(也称碳酸酐酶胞外酶，CAex)通过金属离子与细胞外表面相连接，催化细胞扩散层中的 HCO_3^- 迅速水解生成游离的 CO_2，从而保证了 CO_2

的快速供应。在微藻中，CAex还能直接引起无机碳的浓缩（CCM机制），并具有区室化特征。在生物的碳代谢过程中，碳酸酐酶胞外酶的存在与否能够显著影响生物体的生化反应速率，一般情况下，CAex的活性是不足的，而且在不同微藻中的差异悬殊，相差可达几十倍、甚至几百倍（Colman et al.，2002）。

叶绿体碳酸酐酶被认为是最重要的碳酸酐酶，进入生物体中的一部分HCO_3^-在细胞内经叶绿体膜蛋白直接把HCO_3^-主动转运到叶绿体内，最后经叶绿体碳酸酐酶转化为CO_2供核酮糖-1，5二磷酸羧化/加氧酶（rubisco）固定（Badger and Price，1994；Park et al.，1999；Sültemeyer，1998）。

线粒体碳酸酐酶的催化代谢方向与叶绿体碳酸酐酶的相反，但对生物的生命活动同样重要，它在生物呼吸代谢中具有重要意义。线粒体碳酸酐酶催化生物的呼吸代谢，为生物提供充足的能量，并把呼吸代谢中生成的CO_2催化为HCO_3^-快速生成，并再次转运到叶绿体基质中，被叶绿体碳酸酐酶催化为CO_2，再次为光合作用提供原料。

根据氨基酸序列和晶体结构分析，碳酸酐酶可以分为以下α、β、γ、δ、ε、ζ六类。

α类碳酸酐酶的研究较多，同时它也是活性较大的一类碳酸酐酶，分子质量约为30kDa，在Zn原子周围具有三个组氨酸（Moroney et al.，2001）。它分布广泛，在脊椎动物（Meldrum and Roughton，1933）、藻类（Fujiwara et al.，1990；Fukuzawa et al.，1990）、高等植物（Moroney et al.，2001；Tuskan et al.，2006）、真菌（Li et al.，2009；Elleuche and Pöggeler，2010）等都有发现。

β类碳酸酐酶的研究历史较短，直到1990年才陆续有所报道。最早是在高等植物中发现的（Burnell et al.，1990a；Fawcett et al.，1990），接下来在蓝细菌中（Fukuzawa et al.，1992）、微藻（Eriksson et al.，1996）以及细菌界和古细菌界等原核生物中被发现（Gotz et al.，1999；Smith and Ferry，1999，Supuran and Scozzafava，2007）。β类碳酸酐酶分子质量较大，一般为100~200kDa，由23~25kDa亚基构成二聚体或多聚体。β类碳酸酐酶的晶体结构显示Zn^{2+}与2个保守的半胱氨酸和1个保守的组氨酸连接在一起（Bracey et al.，1994；Rowlett et al.，1994）。

γ类碳酸酐酶最早于1994年在古细菌 *Methanosarcina thermophila* 中发现的（Alber and Ferry，1994）。接下来，在真菌和植物中发现了编码γ结构蛋白质的碳酸酐酶基因（Newman，1994）。γ类碳酸酐酶具有左手平行的β-螺旋折叠结构，它是包含三个Zn原子的同型三聚体。目前认为：它是在30亿年至40亿年前进化来的，分子质量约为69kDa。最新研究显示：γ类碳酸酐酶在植物和微藻中的线粒体电子传递链中起重要作用（Klodmann et al.，2010；Sunderhaus et al.，2006）。

以上是三类常见的碳酸酐酶结构，此外，还有三种不常见的碳酸酐酶结构，它们分别为：在硅藻中发现的 δ 类碳酸酐酶(Roberts et al.，1997)；在细菌中发现的 ε 类碳酸酐酶(So and Espie，2005；So et al.，2004)；在海洋原核生物中发现的 ζ 类碳酸酐酶。ε 类碳酸酐酶的活性中心与 β 类碳酸酐酶的相似，也是由 2 个半胱氨酸和 1 个组氨酸构成(So et al.，2004)。ζ 类碳酸酐酶也与 β 类碳酸酐酶相似(Lane and Morel，2000；Park et al.，2007)，不同的是 Zn 金属中心可以由 Cd 或 Co 替换(Lane and Morel，2000；Lane et al.，2005；Park et al.，2007)。

目前，在莱茵衣藻中发现的碳酸酐酶有 α 类、β 类和 γ 类(Lindskog，1997；Moroney et al.，2011)。虽然所有的碳酸酐酶中都含有 Zn 离子，而且它们的催化机制都是相似的(Moroney et al.，2001)，但是，不同结构的碳酸酐酶之间没有显著的序列一致性，而且都是独立进化的。因此，碳酸酐酶是催化功能趋同进化的极好例子。

第二节 碳酸酐酶的作用与意义

一、碳酸酐酶的生理生化作用

碳酸酐酶催化 CO_2 与 HCO_3^- 之间的可逆转化，在脱水作用时，HCO_3^- 在活性中心与 $Zn-H_2O$ 反应，从而使反应快速向右进行(方程(1-1))；而在水合作用时，CO_2 在活性中心与 $Zn-OH$ 反应，从而使反应快速向左进行(方程(1-1))，总之，碳酸酐酶催化 CO_2 与 HCO_3^- 之间的相互转化，最终促进生物的生长发育。

碳酸酐酶在光合作用、离子交换、钙化作用、维持生物体内酸碱平衡、无机碳转运等一系列的生理过程中都有着十分重要的生理功能(Badger and Price，1994；Gilmour and Perry，2009)。

在光合作用过程中，碳酸酐酶通过催化 CO_2 与 HCO_3^- 之间的相互转化来降低 CO_2 在叶肉细胞中的扩散阻力，加快 CO_2 的供应，为羧化作用提供充足的底物，并参与调节光合作用，最终加快光合作用的进程(Haglund et al.，1992a)。

在无机碳的运移过程中，以微藻为代表的水生生物在碳酸酐酶的参与下形成了一种在细胞内提高 CO_2 浓度的机制——无机碳浓缩机制(carbon-concentrating mechanisms，CCM 机制)(Badger et al.，1998；Badger and Price，1992；Bozzo and Colman，2000；Colman et al.，2002；Giordano et al.，2005)。一般来说，具有 CCM 机制的微藻体内 CO_2 浓度是环境中的 5～75 倍(Badger et al.，1998)。依靠 CCM 机制，微藻可以在水体 CO_2 浓度较低的条件下达到较高的光合速率。

在离子交换方面，根据碳酸酐酶的催化方程(方程(1-1))，它有两种关键的离子，分别为阴离子(HCO_3^-)和阳离子(H^+)，植物生长所需的 K^+、Na^+、

Ca^{2+}、Mg^{2+}等阳离子，很容易通过离子交换进入生物体；而 PO_4^{3-}、NO_3^-等阴离子，也可以通过离子交换的方式而快速被生物利用。

在钙化作用方面，有碳酸酐酶参与的情况下，钙化沉积速度加快，如鸟蛋壳的形成，软体动物的贝壳等。在禽类的产蛋过程中，碳酸酐酶在蛋壳的形成中具有重要的催化作用，它能够加速蛋壳膜上碳酸钙晶体的生长和积聚（Fernandez et al.，2004）。

在维持酸碱平衡方面，碳酸酐酶在 HCO_3^- 的脱水过程中，释放 OH^-，有利于抑制环境酸化；而在 CO_2 的水合过程中，释放出 H^+，抑制环境的 pH 过高。总之，在生物碳酸酐酶的参与下，有利于维持环境 pH 的稳定。

二、影响碳酸酐酶活力的主要因素

碳酸酐酶是一种诱导酶，其活性大小受环境影响很大，主要有：水分、pH、光照、CO_2浓度、HCO_3^-浓度等因素，尤其是环境中CO_2浓度的下降，诱导碳酸酐酶活性升高的报道最多（Badger et al.，2002；Burkhardt et al.，2001；Moroney and Somanchi，1999；陈雄文和高坤山，2003；Price，2011；Hofmann et al.，2013）。

1. 水分与碳酸酐酶活性

水是光合作用的原料，没有水就不能进行光合作用。CA 作为光合作用的重要调节酶，也会因水分不同而表现出不同的活性，以适应不同的水环境，满足光合作用的要求。在水分干旱逆境条件下，CA 活性的变化与光合速率的变化保持一致，CA 可能影响整个光合作用过程（Downton and Slatyer，1972）。水稻旗叶在全展后可逆衰退阶段经历土壤干旱处理时，如干旱程度不十分严重，气孔导度等光合参数都会相应下降，但 CA 活性却因为逆境诱导而有较大幅度的上升，这促使了叶肉细胞气孔导度的上升和CO_2有效供给的增加，在一定程度上补偿由于气孔关闭而引起的光合速率下降。倒 5 叶期由于处于不可逆衰退阶段，其光合机构已经衰退，干旱处理不能诱导 CA 活性增加，反而引起大幅度下降。说明在水分逆境条件下 CA 活性对光合速率有调节作用，植物不同品种在水分逆境条件下仍能维持较高的光合速率可能是其广适性的原因，因此 CA 活性对水分逆境的响应程度可能是品种适应性强弱的指标之一（戴新宾等，2000）。脱水和聚乙二醇诱导的渗透胁迫对拟南芥的碳酸酐酶基因表达具有显著影响（Wu et al.，2012）。在水分胁迫下，橄榄树的碳酸酐酶基因表达下调，复水后该基因表达上调（Perez-Martin et al.，2014）。

2. pH 与碳酸酐酶活性

pH 对植物的酶促反应有调节作用。细胞中每一种酶都有最适的 pH 和最适

pH 的微环境。即使是 pH 的微小变化也会导致酶促反应速度的改变。酸性条件下生长的椭圆小球藻的 CA 的活力明显比在碱性条件下生长的 CA 的活力小(Rotatore and Colman, 1991); Janette 和 John 认为造成酸性 pH 时 CA 活性明显低于碱性 pH 时的原因,是 pH 影响无机碳存在形式:酸性条件下,无机碳主要以 CO_2 形式存在,它有足够的量进入细胞催化 HCO_3^- 迅速转化成 CO_2,碱性条件下无机碳主要以 HCO_3^- 存在,必须依靠 CA 催化 HCO_3^- 迅速转化成 CO_2,以满足机体的需要(Janette and John, 1994)。对莱氏衣藻在不同 pH 下胞外碳酸酐酶活性变化的观测表明,在 pH 7.2 时诱导的酶活性最高,而 pH 5.5 时诱导的酶活性明显低于 pH 7.2 和 pH 9.0 时诱导的酶活性(陈雄文和戴新宾,2000)。对甲藻而言,pH 7.5 时对胞外碳酸酐酶的诱导效果最好(戴芳芳等,2011)。此外,碱性条件还可能直接影响植物的生长,以致影响机体 CA 的表达(Williams and Colman, 1996)。

3. 光强对碳酸酐酶活性的影响

CA 是参与 CO_2 传导而进入羧化位点的重要光合酶,其活性高低对光合作用有较大影响。光是影响光合作用的重要环境因子之一。在 C_3 植物中,CA 可通过促进 CO_2 的液相扩散调节光合作用(Burnell et al., 1990b)。CA 催化 CO_2 转变为 DIC 及 DIC 转变为光合羧化部位所需的 CO_2。CA 活性的变化受光强的调节,与 Rubisco 羧化速率的变化相似,光诱导 Rubisco 活性和羧化速率增高的同时出现 CA 活性相应增高(Haglund et al., 1992b)。因为羧化作用增强使光合羧化位点的 CO_2 分压下降,有利于促进 CA 催化反应进程(Rawat and Moroney, 1995)。生长在强光下的 *Gracilaria tenuistipitata* 的 DIC 传输速率增高(Mercado et al., 2000)。同时,CA 活性的增高又有利于增大 Rubisco 对 CO_2 的亲和力。已证明水稻 CA 活性增高可降低 CO_2 固定的 K_m 值,提高光合 CO_2 固定能力;CA 活性增高也被视为水稻减轻在强光下出现光抑制的手段之一(季本华等,1997)。植物通过调节 CA 活性而适应光环境的变化。

4. 无机、有机阴离子及其他因素与碳酸酐酶

活性 CA 作为一种含 Zn 金属酶,被 Zn 专性活化,Zn^{2+} 对 CA 的活性具有调节作用,金属离子如 Cd^{2+}、Co^{2+} 也影响 CA 的活力。Na 也影响 CA 的活力:植株缺 Na,叶片中的 CA 的活力是对照的两倍(Brownell et al., 1991)。

无机和有机阴离子能强烈抑制 CA 活力(Hatch, 1991; Johansson and Forsman, 1994);氮的缺乏会降低 CA 的活力,同时,可通过增加硝酸盐而慢慢恢复到原活力(Burnell et al., 1990b);HCO_3^-、醋酸盐能调节该酶的活性(Merrett et al., 1996);甘氨酸、葡萄糖和其他有机碳对 CA 的诱导有抑制作用(Umino and

Shiraiwa，1991)。

CA 可以与卤素离子、羧酸根、酚、醇、咪唑、羧酸酰胺、硫酰胺及 SCN^- 等结合，它们会抑制 CA 的催化活性(Jönsson et al.，1993)。

植物激素也严重影响酶的诱导和酶的活力。研究表明，细胞分裂素可明显促进 CA 的 mRNA 的表达(Sugiharto et al.，1992)，除了 ABA 外，IAA、GA_3、KIN 以及 HBR(2,8-homobrassinolide)都不同程度地增强了 CA 的活力(Hayat et al.，2001)，温度和氧化-还原电位等对 CA 酶的活力和诱导也有影响(Moubarak-Milad and Stemler，1994)。

第三节 碳酸酐酶的抑制剂及应用

一、碳酸酐酶抑制剂的种类

乙酰唑胺(acetazolamide，AZ)是含 1，3，4-噻二唑环的杂环磺酰胺类碳酸酐酶胞外酶抑制剂，由于其不能跨过细胞膜，因此，其作用底物只能是碳酸酐酶胞外酶(CAex)，且对 CAex 有较好的抑制作用(Moroney et al.，1985；Williams and Turpin，1987；肖忠海等，2008)。

乙氧苯并噻唑磺胺(ethoxyzolamide，EZ)是另一类碳酸酐酶的抑制剂，EZ 由于具有细胞膜穿透性，所以它对胞外碳酸酐酶和胞内碳酸酐酶均能产生抑制作用。

二、乙酰唑胺对碳酸酐酶的抑制作用原理

AZ 对碳酸酐酶的抑制作用是通过其磺酰胺基团中的 N 与碳酸酐酶活性中心的 Zn^{2+} 相连接，取代与 Zn^{2+} 相连接的水分子与 Thr-199 的—OH 形成氢键。基团中的一个氧原子与 Thr-199 的—NH 形成氢键，另一个氧原子朝向 Zn^{2+} 位置(图 1-1)。总之，AZ 中的 $(R)SO_2NH_2$ 基团是发挥主要作用的官能团(Lindskog，1997；曾广智等，2006)。

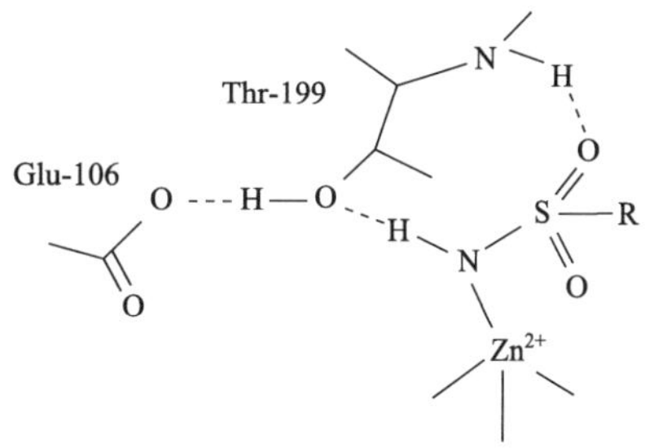

三、碳酸酐酶抑制剂的应用

通过向培养液中添加 AZ，来抑制微藻碳酸酐酶胞外酶的活性，进而可以准确分析微藻碳酸酐酶胞外酶对稳定碳同位素分

图 1-1 AZ 对 CA 抑制作用的结合位点
图片来源：曾广智等，2006
Fig. 1-1 The inhibition of AZ to the active site of CA from Zeng Guangzhi et al.，2006

馏的贡献,并进一步阐述间接利用碳酸氢根离子能力对微藻碳汇的影响。

通过向培养液中添加 EZ,达到完全抑制微藻的各种碳酸酐酶活性的目的,从而分析所有碳酸酐酶对稳定碳同位素分馏的贡献,再与之前的 AZ 处理相结合,进而区分除胞外碳酸酐酶以外的总的胞内碳酸酐酶的贡献。

第四节 碳酸酐酶活力的常规测定方法

一、放射免疫分析法

此法一般采用皮内注射的方式把放射性标记物^{125}I 标记于活体动物身上(常选用兔子),经过一定的实验周期,最后使用专业的放射免疫计数器来检测计数(吴婵群和江一民,1995)。此法的最大缺点是有放射性污染、且操作过程复杂,测定周期过长等,因此,不便于在常规实验室进行推广使用。

二、pH 计法

Wilbur 和 Anderson(1948)建立了电化学法来测定碳酸酐酶活力的方法,奠定了碳酸酐酶的测定方法体系。并在此基础上,学者们对碳酸酐酶活力的传统电化学测定方法进行了不断改进,通过配置一定浓度的巴比妥钠缓冲液来监测 pH 下降一个单位(pH 从 8.3 下降到 7.3)所需要的时间。尤其是近年来,本书作者研究团队对传统电化学方法进行了较大改进,此方法引入了电化学工作站来采集碳酸酐酶反应体系中的电位变化过程曲线(采用自制的锑微电极),并在最后的数据处理方面使用 Origin 软件将匀速变化的一段数据进行线性拟合,最终使碳酸酐酶的测定精度得到了较大提高(Wu et al.,2011;施倩倩等,2010)。

三、同位素二氧化碳质谱分析法

此法是通过比较同一条件下,NaH^{14}CO$_3$ 在有碳酸酐酶参与下^{14}CO$_2$ 的生成速率与 NaH^{14}CO$_3$ 在无碳酸酐酶参与下^{14}CO$_2$ 的生成速率,通过比较两者的差值来计算碳酸酐酶的活性大小(Hirano et al.,1980)。

四、mRNA 测定法

此法是通过提取生物样品的 RNA,并设计特异的引物,从分子生物学的角度来检测不同碳酸酐酶基因表达的差异,并由此来推测碳酸酐酶的活性大小。此法的优点是可以区分不同碳酸酐酶基因表达的差异,从分子水平上来阐述引起碳酸酐酶活性大小的机理。缺点主要有:由于物种基因的差异性,每一个物种都需要设计相应的引物,且操作过程复杂,试剂成本高,对检测者的要求高,重现性

五、比色法

HCO_3^- 的含量受 pH 的影响较大，通过复合指示剂显色，它在 546nm 处有吸收峰，其 HCO_3^- 含量与吸光度成正比（董理等，2007）。在一定的环境条件下，样品中的 HCO_3^- 和 CA 活性呈反比，由此可以计算碳酸酐酶的活性。

另一种方法是碳酸酐酶催化醋酸对硝基酚生成黄色对硝基苯酚，在 405nm 附近有吸收峰，在一定的反应时间内，生成的对硝基酚吸光度变化速率与碳酸酐酶活性呈正比，由此来推算碳酸酐酶的活性大小（胡云良等，2004）。

参 考 文 献

陈雄文，戴新宾.2000.pH 值和氮素对莱氏衣藻胞外碳酸酐酶活性的影响.南京农业大学学报，23(1)：27-29.

陈雄文，高坤山.2003.CO_2 浓度对中肋骨条藻的光合无机碳吸收和胞外碳酸酐酶活性的影响.科学通报，48(21)：2275-2279.

戴芳芳，周成旭，严小军.2011.pH 及光照对两种赤潮甲藻种群生长和胞外碳酸酐酶活性的影响.海洋环境科学，30(5)：694-698.

戴新宾，翟虎渠，张红生，张荣铣.2000.土壤干旱对水稻叶片光合速率和碳酸酐酶活性的影响.植物生理学报，26(2)：133-136.

董理，安伟奇，刘虹久.2007.血浆（清）HCO_3^- 碳酸酐酶比色法的特点及临床应用.中国实验诊断学，11(9)：1255-1256.

胡云良，徐晓杰，丁红香，郑静，陆永绥.2004.碳酸酐酶自动生化分析法测定血清锌.中国卫生检验杂志，13(6)：684-685.

季本华，李霞，焦德茂.1997.光抑制条件下水稻叶内碳酸酐酶的适应调节及其生理作用（英文）.中国水稻科学，11(4)：238-240.

施倩倩，吴沿友，朱咏莉，宋艳娇.2010.构树与桑树叶片的碳酸酐酶胞外酶活力比较.安徽农业科学，38(16)：8376-8377.

吴婵群，江一民.1995.碳酸酐酶放射免疫分析法的建立.苏州医学院学报，15(1)：62-63.

肖忠海，王林，汪海.2008.碳酸酐酶抑制剂乙酰唑胺的临床应用进展.中国新药杂志，17(16)：1390-1394.

曾广智，黄火强，谭宁华，嵇长久，潘蓄林.2006.碳酸酐酶Ⅱ及其抑制剂研究进展.云南植物研究，28(5)：543-552.

Alber B E, Ferry J G. 1994. A carbonic anhydrase from the archaeon methanosarcina-thermophila. Proceedings of the National Academy of Sciences of the United States of America, 91(15): 6909-6913.

Badger M R Andrews T J, Whitney S, et al. 1998. The diversity and coevolution of rubisco, plastids, pyrenoids, and chloroplast-based CO_2-concentrating mechanisms in algae.

Canadian Journal of Botany, 76(6): 1052-1071.

Badger M R, Hanson D, Price G D. 2002. Evolution and diversity of CO_2 concentrating mechanisms in cyanobacteria. Functional Plant Biology, 29(3): 161-173.

Badger M R, Price G D. 1992. The CO_2 concentrating mechanism in cyanobacteria and microalgae. Physiologia Plantarum, 84(4): 606-615.

Badger M R, Price G D. 1994. The role of carbonic anhydrase in photosynthesis. Annual Review of Plant Biology, 45(1): 369-392.

Bozzo G G, Colman B. 2000. The induction of inorganic carbon transport and external carbonic anhydrase in *Chlamydomonas reinhardtii* is regulated by external CO_2 concentration. Plant Cell and Environment, 23(10): 1137-1144.

Bracey M H, Christiansen J, Tovar P, Cramer S P, Bartlett S G. 1995. Spinach carbonic anhydrase investigation of the zinc binding ligands by site directed mutagenesis, elemental analysis and exafs. Journal of Cellular Biochemistry, 19(A): 151.

Brownell P F, Bielig L M, Grof C P L. 1991. Increased carbonic anhydrase activity in leaves of sodium-deficient C_4 plants. Functional Plant Biology, 18(6): 589-592.

Burkhardt S, Amoroso G, Riebesell U, Sultemeyer D. 2001. CO_2 and HCO_3^- uptake in marine diatoms acclimated to different CO_2 concentrations. Limnology and Oceanography, 46: 1378-1391.

Burnell J N, Gibbs M J, Mason J G. 1990a. Spinach chloroplastic carbonic anhydrase nucleotide sequence analysis of cDNA. Plant Physiology, 92(1): 37-40.

Burnell J N, Suzuki I, Sugiyama T. 1990b. Light induction and the effect of nitrogen status upon the activity of carbonic anhydrase in maize leaves. Plant Physiology, 94(1): 384-387.

Colman B, Huertas I E, Bhatti S, Dason J S. 2002. The diversity of inorganic carbon acquisition mechanisms in eukaryotic microalgae. Functional Plant Biology, 29(3): 261-270.

Downton J, Slatyer R O. 1972. Temperature dependence of photosynthesis in cotton. Plant Physiology, 50(4): 518-522.

Elleuche S, Pöggeler S. 2010. Carbonic anhydrases in fungi. Microbiology, 156(1): 23-29.

Eriksson M, Karlsson J, Ramazanov Z, Gardeström P, Samuelsson G. 1996. Discovery of an algal mitochondrial carbonic anhydrase: molecular cloning and characterization of a low-CO_2-induced polypeptide in *Chlamydomonas reinhardtii*. Proceedings of the National Academy of Sciences of the United States of America, 93(21): 12031-12034.

Fawcett T W, Browse J A, Volokita M, Bartlett S G. 1990. Spinach carbonic anhydrase primary structure dedeced from the sequenve of cDNA clone. Journal of Biological Chemistry, 265(10): 5414-5417.

Fernandez M, Passalacqua K, Arias J, Arias J. 2004. Partial biomimetic reconstitution of avian eggshell formation. Journal of Structural Biology, 148(1): 1-10.

Fett J P, Coleman J R. 1994. Regulation of periplasmic carbonic anhydrase expression in *Chlamydomonas reinhardtii* by acetate and pH. Plant Physiology, 106(1): 103-108.

Fujiwara S, Fukuzawa H, Tachiki A, Miyachi S. 1990. Structure and differential expression of 2 genes encoding carbonic anhydrase in *Chlamydomonas reinhardtii*. Proceedings of the National Academy of Sciences of the United States of America, 87(24): 9779-9783.

Fukuzawa H, Fujiwara S, Yamamoto Y, Dionisiosese M L, Miyachi S. 1990. cDNA clone, sequence and expression of carbonic anhydrase in *Chlamydomonas reinhardtii* regulation by environmental CO_2 concentration. Proceedings of the National Academy of Sciences of the United States of America, 87(11): 4383-4387.

Fukuzawa H, Suzuki E, Komukai Y, Miyachi S. 1992. A gene homologous to chloroplast carbonic anhydrase (ICFA) is essential to photosynthetic carbon dioxide fixation by synechococcus PCC7942. Proceedings of the National Academy of Sciences of the United States of America, 89(10): 4437-4441.

Gilmour K M, Perry S F. 2009. Carbonic anhydrase and acid-base regulation in fish. Journal of Experimental Biology, 212(11): 1647-1661.

Giordano M, Beardall J, Raven J A. 2005. CO_2 concentrating mechanisms in algae: mechanisms, environmental modulation and evolution. Annual Review Plant Biology, 56: 99-131.

Gotz R, Gnann A, Zimmermann F K. 1999. Deletion of the carbonic anhydrase-like gene NCE103 of the yeast Saccharomyces cerevisiae causes an oxygen-sensitive growth defect. Yeast, 15(10A): 855-864.

Haglund K, Björk M, Ramazanov Z, García-Reina G, Pedersén M. 1992. Role of carbonic anhydrase in photosynthesis and inorganic-carbon assimilation in the red alga. *Gracilaria tenuistipitata*. Planta, 187(2): 275-281.

Haglund K, Ramazanov Z, Mtolera M, Pedersén M. 1992b. Role of external carbonic anhydrase in light-dependent alkalization by *Fucus serratus* L. and *Laminaria saccharina* (L.) Lamour. (Phaeophyta). Planta, 188(1): 1-6.

Hatch M D. 1991. Carbonic anhydrase assay: Strong inhibition of the leaf enzyme by CO_2 in certain buffers. Analytical Biochemistry, 192(1): 85-89.

Hayat S, Ahmad A, Mobin M, Fariduddin Q, Azam Z M. 2001. Carbonic anhydrase, photosynthesis, and seed yield in mustard plants treated with phytohormones. Photosynthetica, 39(1): 111-114.

Hirano S, Asou H, Noda Y, Shibata I. 1980. Radiologic assay for carbonic anhydrase. Analytical Biochemistry, 106: 427-431.

Hofmann L C, Straub S, Bischof K. 2013. Elevated CO_2 levels affect the activity of nitrate reductase and carbonic anhydrase in the calcifying Rhodophyte *Corallina officinalis*. Journal of Experimental Botany, 64(4): 899-908.

Janette P F, John R C. 1994. Regulation of periplasmic carbnic anhydrase expression in *Chlamydomonas reinhardtii* by acctate an pH. Plant Physiology, 106(1): 103-108.

Jönsson B M, Håkansson K, Liljas A. 1993. The structure of human carbonic anhydrase II in complex with bromide and azide. Federation of European Biochemical Societies letters,

322(2): 186-190.

Johansson I M, Forsman C. 1994. Solvent hydrogen isotope effects and anion inhibition of CO_2 hydration catalysed by carbonic anhydrase from *Pisum sativum*. European Journal of Biochemistry, 224(3): 901-907.

Khalifah R G. 1971. The carbon dioxide hydration activity of carbonic anhydrase: Ⅰ. Stop-flow kinetic studies on native human isoenzymes B and C. Journal of Biological Chemistry, 246(8): 2561-2573.

Klodmann J, Sunderhaus S, Nimtz M, Jänsch L, Braun H P. 2010. Internal architecture of mitochondrial complex I from *Arabidopsis thaliana*. The Plant Cell Online, 22(3): 797-810.

Lane T W, Morel F M M. 2000. Regulation of carbonic anhydrase expression by zinc, cobalt, and carbon dioxide in the marine diatom *Thalassiosira weissflogii*. Plant Physiology, 123(1): 345-352.

Lane T W, Saito M A, George G N, Pickering I J, Prince R C, Morel F M M. 2005. A cadmium enzyme from a marine diatom. Nature, 435(7038): 42.

Lindskog S. 1997. Structure and mechanism of carbonic anhydrase. Pharmacology & Therapeutics, 74(1): 1-20.

Li W, Zhou P P, Jia L P, Yu L J, Li X L, Zhu M. 2009. Limestone dissolution induced by fungal mycelia, acidic materials, and carbonic anhydrase from fungi. Mycopathologia, 167(1): 37-46.

Meldrum N U, Roughton F J. 1933. Carbonic anhydrase. Its preparation and properties. The Journal of Physiology, 80(2): 113-142.

Mercado J M, Carmona R, Niell F X. 2000. Affinity for inorganic carbon of *Gracilaria tenuistipitata* cultured at low and high irradiance. Planta, 210(5): 758-764.

Merrett M J, Nimer N A, Dong L F. 1996. The utilization of bicarbonate ions by the marine microalga *Nannochloropsis oculata* (Droop) Hibberd. Plant, Cell and Environment, 19(4): 478-484.

Moroney J V, Bartlett S G, Samuelsson G. 2001. Carbonic anhydrases in plants and algae. Plant Cell and Environment, 24(2): 141-153.

Moroney J V, Ma Y, Frey W D, Fusilier K A, Pham T T, Simms T A, Dimario R J, Yang J, Mukherjee B. 2011. The carbonic anhydrase isoforms of *Chlamydomonas reinhardtii*: intracellular location, expression, and physiological roles. Photosynthesis Research, 109(1): 133-149.

Moroney J V, Husic H D, Tolbert N. 1985. Effect of carbonic anhydrase inhibitors on inorganic carbon accumulation by *Chlamydomonas reinhardtii*. Plant Physiology, 79(1): 177-183.

Moroney J V, Somanchi A. 1999. How do algae concentrate CO_2 to increase the efficiency of photosynthetic carbon fixation? Plant Physiology, 119(1): 9-16.

Moubarak-Milad M, Stemler A. 1994. Oxidation-reduction potential dependence of photosystem II carbonic anhydrase in maize thylakoids. Biochemistry, 33(14): 4432-4438.

Newman E A. 1994. A physiological messure of carbonic anhydrase in muller cells. Glia, 11(4): 291-299.

Park H, Song B, Morel F M M. 2007. Diversity of the cadmium-containing carbonic anhydrase in marine diatoms and natural waters. Environmental Microbiology, 9(2): 403-413.

Park Y I, Karlsson J, Rojdestvenski I, et al. 1999. Role of a novel photosystem II-associated carbonic anhydrase in photosynthetic carbon assimilation in *Chlamydomonas reinhardtii*. FEBS Letters, 444(1): 102-105.

Perez-Martin A, Michelazzo C, Torres-Ruiz J M, et al. 2014. Regulation of photosynthesis and stomatal and mesophyll conductance under water stress and recovery in olive trees: correlation with gene expression of carbonic anhydrase and aquaporins. Journal of Experimental Botany, doi: 10.1093/jxb/eru160.

Price G D. 2011. Inorganic carbon transporters of the cyanobacterial CO_2 concentrating mechanism. Photosynthesis Research, 109(1-3): 47-57.

Rawat M, Moroney J V. 1995. The regulation of carbonic anhydrase and ribulose-1,5-bisphosphate carboxylase/oxygenase activase by light and CO_2 in *Chlamydomonas reinhardtii*. Plant Physiology, 109(3): 937-944.

Roberts S B, Lane T W, Morel F M M. 1997. Carbonic anhydrase in the marine diatom *Thalassiosira weissflogii* (Bacillariophyceae). Journal of Phycology, 33(5): 845-850.

Rotatore C, Colman B. 1991. The acquisition and accumulation of inorganic carbon by the unicellular green alga *Chlorella ellipsoidea*. Plant, Cell and Environment, 14(4): 377-382.

Rowlett R S, Chance M R, Writ M D, et al. 1994. Kinetic and structural characterization of spinach carbonic anhydrase. Biochemistry, 33(47): 13967-13976.

Sültemeyer D. 1998. Carbonic anhydrase in eukaryotic algae: characterization, regulation, and possible function during photosynthesis. Canadian Journal of Botany, 76(6): 962-972.

Smith K S, Ferry J G. 1999. A plant-type (beta-class) carbonic anhydrase in the thermophilic methanoarchaeon *Methanobacterium thermoautotrophicum*. Journal of Bacteriology, 181(20): 6247-6253.

Smith K S, Ferry J G. 2000. Prokaryotic carbonic anhydrases. Fems Microbiology Reviews, 24(4): 335-366.

So A K C, Espie G S. 2005. Cyanobacterial carbonic anhydrases. Canadian Journal of Botany-Revue Canadienne De Botanique, 83(7): 721-734.

So A K C, Espie G S, Williams E B, et al. 2004. A novel evolutionary lineage of carbonic anhydrase (epsilon class) is a component of the carboxysome shell. Journal of Bacteriology, 186(3): 623-630.

Sugiharto B, Burnell J N, Sugiyama T. 1992. Cytokinin is required to induce the nitrogen-dependent accumulation of mRNAs for phosphoenolpyruvate carboxylase and carbonic anhydrase in detached maize leaves. Plant Physiology, 100(1): 153-156.

Sunderhaus S, Dudkina N V, Jansch L, et al. 2006. Carbonic anhydrase subunits form a matrix-exposed domain attached to the membrane arm of mitochondrial complex I in plants. Journal of Biological Chemistry, 281(10): 6482-6488.

Supuran C T, Scozzafava A. 2007. Carbonic anhydrases as targets for medicinal chemistry. Bioorganic and Medicinal Chemistry, 15(13): 4336-4350.

Tuskan G A, Difazio S, Jansson S, et al. 2006. The genome of black cottonwood, *Populus trichocarpa*. Science, 313(5793): 1596-1604.

Umino Y, Shiraiwa Y. 1991. Effect of metabolites on carbonic anhydrase induction in *Chlorella regularis*. Journal of Plant Physiology, 139(1): 41-44.

Wilbur K M, Anderson N G. 1948. Electrometric and colorimetric determination of carbonic anhydrase. Journal of Biological Chemistry, 176: 147-154.

Williams T G, Colman B. 1996. The effects of pH and dissolved inorganic carbon on external carbonic anhydrase activity in *Chlorella saccharophila*. Plant, Cell and Environment, 19(4): 485-489.

Williams T G, Turpin D H. 1987. The role of external carbonic anhydrase in inorganic carbon acquisition by *Chlamydomonas reinhardtii* at alkaline pH. Plant Physiology, 83: 92-96.

Wu Y Y, Shi Q Q, Wang K, et al. 2011. An electrochemical approach coupled with Sb microelectrode to determine the activities of carbonic anhydrase in the plant leaves. Future Intelligent Information Systems, 86: 87-94.

Wu Y Y, Vreugdenhil D, Liu C Q, Fu W G. 2012. Expression of carbonic anhydrase genes under dehydration and osmotic stress in *Arabidopsis thaliana* leaves. Advanced Science Letters, 17(1): 261-265.

第二章 微藻的碳酸酐酶基因表达及其对环境的响应

摘　要

莱茵衣藻是碳酸酐酶基因的生物信息学研究的模式生物。编码莱茵衣藻胞外碳酸酐酶(CAH1)、叶绿体碳酸酐酶(CAH3)、线粒体碳酸酐酶(CAH5)、γ碳酸酐酶(CAG1)与蛋白核小球藻相应的编码碳酸酐酶基因具有同源性。低浓度碳酸氢钠能够诱导微藻碳酸酐酶胞外酶活性；高浓度的碳酸氢钠抑制微藻碳酸酐酶胞外酶活性，造成编码 CAH1 基因表达下调。编码 CAH3 基因的表达量对碳酸氢钠不敏感。编码 CAH5 基因表达量随添加碳酸氢钠浓度的增加而急剧下降。编码 CAG1 基因表达量随添加碳酸氢钠浓度的增加而略有上升。胞外碳酸酐酶抑制剂乙酰唑胺(AZ)能使蛋白核小球藻的编码 CAH1 基因表达上调，而只在低浓度碳酸氢钠下才使编码莱茵衣藻的 CAH1 基因表达上调。

在莱茵衣藻中，编码 CAH1 基因在酸性条件下表达量较低；而在碱性条件下，表达量有所上升；编码 CAH3 和 CAH5 的基因表达对 pH 的响应模式为：除 pH 为 5.3 外，低 pH 上调，高 pH 下调其表达。蛋白核小球藻的编码 CAH3 和 CAH5 基因在碱性环境下表达量上调；除了在 pH5.3 外，编码 CAH1 的基因表达受 pH 的影响较小。在 AZ 的作用下，四种碳酸酐酶的基因表达对 pH 的响应模式为：莱茵衣藻为低 pH 上调，高 pH 下调；蛋白核小球藻为除 pH 为 5.3 外，随着 pH 的升高基因表达量增高。

莱茵衣藻 4 种碳酸酐酶基因在低氟浓度下相对表达量低，在高氟浓度下，编码 CAH1 表达量随着氟浓度的升高而增大，最大表达量出现在 100mmol/L 的氟处理中；而编码 CAH3、CAH5 和 CAG1 的最大表达量出现在 10mmol/L 的氟处理中。

研究微藻不同类型的碳酸酐酶基因表达对无机碳、pH 和污染物的响应，可为揭示碳酸酐酶的生物地球化学作用提供分子基础。

Chapter 2 The gene expression of carbonic anhydrase in microalgae and their response to the environment

Abstract

Chlamydomonas reinhardtii is a model using for the research on bioinformatics of carbonic anhydrase genes. The gene encoded for extracellular carbonic anhydrase (CAH1), chloroplastic carbonic anhydrase (CAH3), mitochondrial, carbonic anhydrase (CAH5), and gamma carbonic anhydrase (CAG1) in *Chlamydomonas reinhardtii* are homologous to the corresponding that in *Chlorella pyrenoidosa*, respectively. Low concentration of sodium bicarbonate can induce algal extracellular carbonic anhydrase activity, and high concentration of sodium bicarbonate inhibits algal extracellular carbonic anhydrase activity, causing the encoded for CAH1 gene expression down-regulation. However, the encoded for CAH3 gene expression is not sensitive to sodium bicarbonate. The encoded for CAH5 gene expression decreased sharply with the increase of adding sodium bicarbonate concentration. In Contrast, the encoded for CAG1 gene expression was slightly increasing with the increase of adding sodium bicarbonate concentration. Acetazolamide (AZ), an extracellular carbonic anhydrase inhibitor, had up-regulated the CAH1 gene expression in *Chlorella pyrenoidosa* at all of the treatments, but had up-regulated that in *Chlamydomonas reinhardtii* only at low concentrations of sodium bicarbonate.

For *Chlamydomonas reinhardtii*, the encoded for CAH1 gene expression under acidic condition was low; and that under alkaline condition increased; while the mode of the encoded for CAH3 and CAH5 gene expression in response to pH was as following: the gene expression under low pH was up-regulation, and that under high pH was down-regulation without regard to the pH of 5.3. For *Chlorella pyrenoidosa*, the encoded for CAH3 and CAH5 gene expression was up-regulation under alkaline conditions; the encoded for CAH1 gene expression was little affected by pH, without regard to the pH of 5.3. Under the action of acetazolamide, the mode of the encoded for the four carbonic anhydrase gene expression in response to pH was as following: the gene expression in

Chlamydomonas reinhardtii under low pH was up-regulated, and that under high pH was down-regulated. The gene expression in *Chlorella pyrenoidosa* increased with the increase of pH, without regard to the pH of 5.3.

The relative expression of the encoded for the four carbonic anhydrase genes in *Chlamydomonas reinhardtii* was small under low fluoride concentrations. The encoded for the CAH1 gene expression under high fluoride concentration increased concentration of fluoride, and the maximum expression appeared at the treatment of 100 mmol/L fluoride; and the maximum expression of the encoded for CAH3, CAH5 and CAG1 appeared at the treatment of 10 mmol/L fluoride.

The expression of the encoded for different types of carbonic anhydrase genes in response to the inorganic carbon sources, pH and pollutants revealed the carbonic anhydrase molecular mechanism of biogeochemical action.

自1933年人类从牛红细胞中纯化得到第一个碳酸酐酶以来（Meldrum and Roughton，1933），碳酸酐酶的研究就受到了生物学家的广泛关注，科学家们相继在动物和植物中发现了碳酸酐酶，甚至在有些微生物中也发现了该酶的存在（Smith and Ferry，2000）。随着分子生物学的发展，越来越多的物种基因组信息测序完成，特别是莱茵衣藻的碳酸酐酶基因组的公布（Merchant et al.，2007），给我们利用现代分子生物学手段来研究莱茵衣藻的碳酸酐酶基因表达调控的内在机制提供了可能，并将为微藻碳酸酐酶的深入研究带来全新的发展。

实时荧光定量PCR技术是目前受到广泛应用的测定基因表达的新方法，它是通过在PCR反应体系中加入荧光基团，收集整个PCR反应过程中的荧光信号强度，来定量分析扩增产物量的变化，最后通过Ct值或者标准曲线来对起始模板进行定量分析。与普通PCR相比，定量PCR具有特异性强，自动化程度高等优点。利用实时荧光定量PCR技术可以便捷地研究外界环境对微藻碳酸酐酶基因表达的影响。

第一节 莱茵衣藻碳酸酐酶基因的生物信息学

一、莱茵衣藻是碳酸酐酶基因的生物信息学研究的模式生物

日益增多的莱茵衣藻的研究，使得莱茵衣藻（*Chlamydomonas reinhardtii*）成为了模式生物（Lewin，1957，Sager and Granick，1953）。莱茵衣藻也被选择为衣藻基因组计划（包括全基因组计划及EST计划）。现在，全球100多家著名实验室已建立了衣藻实验生物学研究体系及多个研究主题和相应的遗传系统，奠定了莱茵衣藻的学术地位（Quarmby，2000）。

莱茵衣藻是一种真核单细胞绿藻，素有"绿色酵母"之称（Baymann et al.，1999）。莱茵衣藻与酵母有许多共同之处，如世代交替时间短，可以在平板上形成单克隆，也能进行液体培养，以单倍体与二倍体两种形式生长，能对其配子的发育过程进行四分子分析。显然，在酵母已经奠定了其模式生物的角色以后，再选择莱茵衣藻来作为植物的模式生物，可谓应运而生，因为许多细胞学问题不可能在酵母中实现，如光合成（photosynthesis）途径、趋光性（phototaxis）与感光（light perception）、鞭毛（flagella）、中心粒（centrioles）、基体（basal bodies）与轴丝（axoneme）、植物细胞周期的调控与细胞识别、叶绿体的生物（chloroplast biogenesis）发生及其遗传（inheritance）、植物细胞的凋亡、叶绿体的转化，以及叶绿体 DNA 的损伤与修复等（Harris et al.，1989）。

衣藻具有以下特点（Boynton et al.，1988）：①培养条件简单，周期短；②已完成基因组测序，序列结构比较清楚；③核转化与叶绿体转化方法成熟；④具有多种类型的突变株，可针对不同的筛选方法选取不同的宿主突变株；⑤可进行核、叶绿体和线粒体转化。它不但能在核基因组表达真核基因，还能在叶绿体里表达原核基因。

衣藻的生物学特性决定了其能够在短时间内获得大量遗传稳定的转基因后代，为进行遗传学分析提供了便利条件（范国昌和钱凯先，1998）。衣藻具有核染色体、叶绿体和线粒体 3 套基因组。它的核基因组大约为 100mb，具有 GC 含量高（约 62%）、密码子高度偏爱等特点（Rochaix，1978）；叶绿体基因组是约 200kb 的环状 DNA，多拷贝，具有母系遗传的特点（Grant and Chiang，1980）；线粒体基因组是线性 DNA 分子，大约 15.8kb，具有父系遗传的特点（Michaelis et al.，1990），与其他植物和藻类相比，衣藻的线粒体基因很简单，并已完全测序，发现仅有数个与呼吸相关的编码基因（Lumbreras and Purton，1998）。衣藻也是少数 3 套基因组均能进行遗传转化的植物之一（Hippler et al.，1998）。

二、莱茵衣藻碳酸酐酶基因的生物信息学

目前，Genebank 中共收录有 12 种碳酸酐酶基因，分别为：属于 α 家族的 CAH1、CAH2 和 CAH3；属于 β 家族的 CAH4、CAH5、CAH6、CAH7、CAH8 和 CAH9；以及属于 γ 家族的 CAG1、CAG2 和 CAG3。三个家族种类在莱茵衣藻中均有分布，说明莱茵衣藻非常适合作为研究碳酸酐酶基因的模式生物。

（一）莱茵衣藻碳酸酐酶的同源性

从 Genebank 中得到 12 种莱茵衣藻碳酸酐酶氨基酸序列，用 GENEDOC 软件进行同源性分析，结果表明，莱茵衣藻碳酸酐酶氨基酸序列相似性极低，只有

起始氨基酸高度保守(除CAH5),另外3个相对保守位点为70位的缬氨酸和145位、191位的亮氨酸。

(二)莱茵衣藻碳酸酐酶在进化上的关系

将 Genebank 中得到的序列用 MEGA 软件进行序列进化树分析,得到如图 2-1 的进化树。

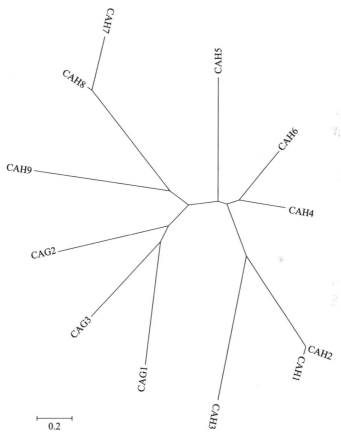

图 2-1 莱茵衣藻碳酸酐酶进化分析
注:进化树使用 MEGA 生成

Fig. 2-1 Phylogenetic tree of carbonic anhydrase in *Chlamydomonas reinhardtii*
The phylogenetic tree was reconstructed by MEGA software

从该进化树可以看出,α 家族碳酸酐酶三个基因 CAH1,CAH2,CAH3 位于同一分支上,并且 CAH1 和 CAH2 遗传距离很短,说明这两个基因可能是由同一个基因发生突变而形成。γ 家族 CAG1、CAG2 和 CAG3 虽然位于同一分支上,但遗传距离较远。β 家族基因种类较多,从图中可看出除 CAH4 和 CAH6 之外遗传距离均较远,表明这些基因可能分别是独立进化的。

第二节　莱茵衣藻与蛋白核小球藻碳酸酐酶基因表达

一、实验材料与方法

(一) 实验材料

微藻藻种购自中国科学院水生生物研究所淡水藻种库(Freshwater Algae Culture Collection of the Institute of Hydrobiology,FACHB)。选用的微藻有两种：莱茵衣藻(*Chlamydomonas reinhardtii*,FACHB-479)和蛋白核小球藻(*Chlorella pyrenoidosa*,FACHB-5)。培养地点在中国科学院地球化学研究所环境地球化学国家重点实验室的人工温室。

(二) 培养方法

微藻培养条件如下,光照强度：$150\mu mol/(m^2 \cdot s)$；光暗周期：12h/12h；白天温度：$25.0℃ \pm 1.0℃$,夜间温度：$22.0℃ \pm 1.0℃$。通过添加少量 NaOH 或 HCl 调节初始 pH 都在 8.00 ± 0.05,为确保藻种保持旺盛的生长状态,不断进行继代培养。严格控制各个处理开始时的微藻接种量,以便降低微藻细胞浓度差异所带来的影响。

具体操作方法如下：微藻培养所需的各种器皿和培养液都采用湿热灭菌法(121℃,30min),微藻培养的所有操作均在无菌条件下进行。微藻一般培养在 250mL 的三角瓶中,50mL 体系,采用无菌封口膜透气培养。使用时,离心收集微藻(1600g,10min),通过测定藻种的光密度值(OD_{650}),使每次处理开始时的微藻接种量接近,以便消除各个处理的藻密度差异所带来的影响。实验处理一般平行培养 5 瓶,且在实验处理过程中,不断随机调换位置,以消除培养室微环境差异对微藻的影响。

(三) RNA 提取

采用 TRIzol Reagent (Invitrogen,California,USA)试剂提取微藻总 RNA。所用耗材、操作环境等,都要保证无 RNA 酶,所有的开盖操作,全部在超净工作台的酒精灯火焰保护下进行。根据试剂使用说明书,具体操作流程如下：

(1) 离心收集约 100mg 微藻,放到 1.5mL EP 管中,加入 1mL Trizol 试剂。

(2) 用取样器吸打数次,盖盖手摇,室温静置 5min。

(3) 加 0.2mL 氯仿，盖盖手摇 15s，确保混匀彻底，室温静置 2~3min。

(4) 离心，12 000g，4℃，15min。

(5) 分装 450μL 异丙醇于 1.5mL EP 管，放入 4℃冰箱冷藏备用。

(6) 配制 75% 的乙醇，放入 4℃冰箱冷藏备用。

(7) 从第 4 步离心后的 EP 管中，用 1mL 取样器小心吸取上层无色水相，移入预冷的装有 0.45mL 异丙醇的 EP 管中，一般吸取 0.40mL，上下颠倒几次，室温静置 10min。

(8) 离心，12 000g，4℃，10min。离心后，管底部可见微量 RNA 沉淀。

(9) 弃上清，加 1mL 75% 乙醇，手摇振荡，使其充分悬浮。

(10) 离心，7500g，4℃，5min。

(11) 弃上清，置于超净台酒精灯火焰保护处，风干 5~10min（小心不要太干燥，去除残留酒精即可，以免过干造成难溶）。

(12) 沉淀溶于 25μL DEPC 水，小心吸打数次，确保沉淀充分溶解。

(13) 促溶（55~60℃水浴 5min），接下来，轻微离心，分装 2μL 于另一新的 EP 管中，以备测其 OD 值。

(14) -70℃保存备用。

取上面第 13 步分装的 2μL RNA 样品，加入 198μL DEPC 水，使用微量比色皿（200μL），测 OD260/OD280，达到 1.8~2.0 的，为合格样品。

RNA 浓度换算公式如下：

$$\text{RNA (ng/μL)} = \text{OD}_{260} \times 稀释倍数 \times 40 \qquad (2\text{-}1)$$

（四）引物设计

根据美国国立生物技术信息中心（NCBI）上面公布的莱茵衣藻的碳酸酐酶基因序列，运用 Primer Premier 5.0 设计合成目标引物。并把设计好的引物序列委托生工生物工程（上海）有限公司（上海生工）合成，采用变性聚丙烯酰胺凝胶电泳（PAGE）纯化方法。根据实验要求，我们设计了两套引物，分别适用于普通 PCR 和实时荧光定量 PCR，详细信息见表 2-1 和表 2-2。

（五）普通 PCR 反应

普通 PCR 反应在 ABI 2700 仪器上进行，选用 SuperScript® III One-Step RT-PCR System with Platinum® Taq (Invitrogen, California, USA) 试剂，采用一步法进行，具体配制方法见表 2-3。操作步骤参照试剂说明书进行，具体操作步骤如下。

表 2-1 莱茵衣藻的基因及引物信息表——普通 PCR
Tab. 2-1 The gene information about *Chlamydomonas reinhardtii* and primer for PCR

目的基因	登录号	片段长度/bp	引物/(F/R)	引物序列
$CAH_1(\alpha)$	XM_001692239.1	245	F	5'-cgagacatttgcgtacaggg-3'
			R	5'-tcgacatggctatgggatt-3
$CAH_3(\alpha)$	XM_001696692.1	232	F	5'-gtgacatcgctacctccca-3'
			R	5'-cctttatcgctcattgtattgt-3'
$CAH_5(\beta)$	XM_001700718.1	189	F	5'-cggtgtttgacatggtgg-3'
			R	5'-tgttacaagagttggaagggt-3'
$CAG_1(\gamma)$	XM_001703185.1	223	F	5'-gcctgctgctacctgctaa-3'
			R	5'-gcctgcggagaacttcat-3'

表 2-2 莱茵衣藻的基因及引物信息表——实时荧光定量 PCR
Tab. 2-2 The gene information about *Chlamydomonas reinhardtii* and primer for real-time quantitative PCR

目的基因	登录号	片段长度/bp	引物/(F/R)	引物序列
a$TUB_2(\beta)$	XM_001693945.1	111	F	5'-gcggccttggcagtttt-3'
			R	5'-tctcacaatccgcctccac-3
$CAH_1(\alpha)$	XM_001692239.1	113	F	5'-tcgcacatgccgatcaa-3
			R	5'-tctcctgtcctccatctccaa-3'
$CAH_3(\alpha)$	XM_001696692.1	105	F	5'-ggggtggactggtttgtgtt-3'
			R	5'-cgtgttggtggcgtatgtct-3'
$CAH_5(\beta)$	XM_001700718.1	133	F	5'-actgtagcctacgggcgtttt-3
			R	5'-ccaagccaaacatccacca-3
$CAG_1(\gamma)$	XM_001703185.1	146	F	5'-cgaggcggagtacctcaaa-3'
			R	5'-cctctgctatctccttcaccatc-3'

a. TUB_2 是看家基因。
a. TUB_2 the housekeeping gene.

表 2-3 PCR 反应体系
Tab. 2-3 Reaction system for PCR

试剂	用量
2×Reaction Mix	12.5μL
RNA 样品	不高于 0.5μg
上游引物(10μmol/L)	0.5μL
下游引物(10μmol/L)	0.5μL
SuperScript® III RT/Platinum® Taq mix	1μL
去 RNA 水	终体积到 25.0μL

反应条件：

第一步：合成 cDNA，一个循环：50℃，25min；

第二步：变性，一个循环：94℃，2min；

第三步：PCR 扩增，40 个循环：94℃，15s；53.5～56℃，30s；68℃，25s；

第四步：延伸，一个循环：68℃，5min；

最后，反应获得的 PCR 产物在－20℃冰箱保存备用。

（六）电泳监测

取 5.0μL PCR 产物于 1.5% 的琼脂糖凝胶电泳，由凝胶成像系统（BINTA, Shanghai, China）拍照，分析结果；并把 PCR 产物测序，进一步验证扩增产物的特异性。

测序工作委托国家基因组北方测序中心完成，选用 ABI 3730xl 序列分析仪，并把测序结果在 NCBI 进行在线 BLAST 分析（http://www.ncbi.nlm.nih.gov/）。

（七）实时荧光定量 PCR

实时荧光定量 PCR 技术用于 RNA 基因表达分析常用的有：一步法和两步法两种（周莉娟和郑伟文，2004）。本实验采用两步法进行，分别为：cDNA 制备和实时荧光定量 PCR。

1. cDNA 的制备

提取的总 RNA 中常常混有基因组 DNA，而且，基因组 DNA 可以直接作为 PCR 模板进行扩增，这将造成解析结果不准确。为消除提取的 RNA 中所含有的少量 DNA 污染，常用的处理方法主要有：①设计引物时，目的片段避开基因组 DNA 扩增区域；②使用 DNase 处理，首先除去基因组 DNA，然后再进行反转录反应。本实验使用 DNase I 处理以除去总 RNA 中的基因组 DNA 污染。

表 2-4 基因组 DNA 的去除反应体系

Tab. 2-4 The reaction system of removal genomic DNA

试剂	用量
5gDNA Eraser Buffer	4.0μL
gDNA Eraser, 1.0μL	2.0μL
RNA 样品	不高于 2μg
去 RNA 水	终体积到 20.0μL

cDNA 合成选用大连宝生物公司的反转录试剂盒（Takara Code ：DRR047A），根据试剂使用说明书，在 0.2mL 的 PCR 管里面配制，详见表 2-4、表 2-5 具体操作过程分为以下两步进行：

第一步：基因组 DNA 的去除反应。

配置好的反应液在 PCR 仪上进行，42℃，2min。

第二步：反转录反应（cDNA 的合成）

表 2-5 cDNA 合成反应体系

Tab. 2-5 The reaction system of cDNA synthesis

试剂	用量
5×PrimeScript® Buffer 2(for real time)	8.0μL
PrimeScript® RT Enzyme Mix 1	2.0μL
RT Primer Mix	2.0μL
第一步的反应液	20.0μL
去 RNA 水	终体积到 40.0 μL

配置好的反应液在 PCR 仪上进行，37℃，15min；85℃，5s。

最后，反转录获得的高纯 cDNA 在 −20℃ 冰箱保存备用。

2. 实时荧光定量 PCR

目前常用的实时定量 PCR 有两种实验方法，分别是探针法和荧光染料法（赵焕英和包金风，2007）。染料法以其经济实惠、适用广泛等优点，得到了广泛应用。本实验采用荧光染料法。

分析方法方面，实时荧光定量 PCR 最常用的两种分析方法是绝对定量和相对定量。绝对定量通过构建质粒标准品，制作标准曲线，计算起始模板的拷贝数；相对定量则是通过比较目的基因与看家基因之间的 CT 值差异，进而计算基因表达的变化。本实验采用荧光染料法进行相对定量分析，每次反应均设置阴性对照、阳性对照及看家基因，反应结束后通过目的基因和看家基因的 CT 值计算求得。

实验在 ABI Step-One 实时荧光定量 PCR 仪器（applied biosystems，California，USA）上面进行。耗材选用 ABI 公司特制的光学八连管，采用 20μL 反应体系（表 2-6）。定量 PCR 试剂购自大连宝生物公司 SYBR® Premix Ex Taq™ II (Perfect Real Time)（Takara, Dalian, China）。依据试剂使用说明书，具体操作流程如下。

表 2-6 实时荧光定量 PCR 反应体系
Tab. 2-6 Reaction system for real time quantitive PCR

试剂	用量
SYBR® Premix Ex Taq™ II (2×)	10.0μL
上游引物/(10μmol/L)	0.5～0.8μL
下游引物/(10μmol/L)	0.5～0.8μL
ROX Reference Dye (50×)	0.4μL
DNA 模板	2.0μL
灭菌蒸馏水	终体积到 20.0μL

定量 PCR 反应条件如下:

第一步:变性,一个循环:95℃,30s;

第二步:PCR 扩增,40 个循环:95℃,5s;62℃,30s(此时收集荧光信号);

第三步:溶解曲线,60～95℃,连续收集荧光信号。

每个样品做 3 个重复,为防止实验器皿、操作过程、引物及试剂所带来的各种潜在污染,每批实验都要做阴性对照。阴性对照的做法参照以上操作步骤,唯一差别就是用灭菌蒸馏水代替样品模板。

3. 定量 PCR 数据质量控制

(1) 扩增的特异性-溶解曲线

根据染料法荧光定量 PCR 的特点,每批定量 PCR 反应之后,都需要进行溶解曲线分析,以便验证非特异性扩增。本实验的各个反应所获得的溶解曲线都是典型的单峰曲线,而且解链温度都在 85℃附近,符合定量 PCR 质量要求。各个基因的扩增曲线如图 2-2 所示。

(2) 扩增效率

选择一个基因表达量较高的模板,按 1:4 进行梯度稀释,进行定量 PCR 扩增效率分析,最后,获得系统自动计算的扩增效率。各个基因的扩增效率详见表 2-7,都达到了进行比较 CT 法的要求(90%～110%)。因此,本研究都采用了比较 CT 法进行实时荧光定量 PCR 分析。

相对定量 PCR 最后获得的 CT 值数据,一般采用 $2^{-\triangle\triangle CT}$ 法进行计算(Livak and Schmittgen, 2001)。

$$\triangle CT = CT_{目的基因} - CT_{看家基因} \qquad (2-2)$$

$$\triangle\triangle CT = \triangle CT_{处理} - \triangle CT_{对照} \qquad (2-3)$$

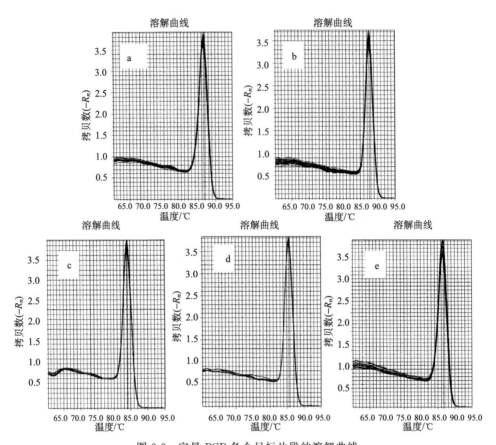

图 2-2 定量 PCR 各个目标片段的溶解曲线

Fig. 2-2 The melt curve of amplified fragments in Real-Time PCR

a. Tubulin2; b. CAH1; c. CAH3; d. CAH5; e. CAG1

表 2-7 定量 PCR 扩增效率表

Tab. 2-7 Amplification efficiency in Real-Time PCR

目的基因 (target)	斜率 (slope)	相关系数 (R^2)	扩增效率 (eff)/%
CAH1	−3.377	0.999	97.74
CAH3	−3.239	0.998	103.59
CAH5	−3.326	0.999	98.85
CAG1	−3.136	0.996	108.42
TUB2	−3.257	0.999	102.79

二、常规 PCR 扩增结果

由图 2-3 可见，经常规 PCR 扩增后，经 1.5% 琼脂糖凝胶电泳分析可见与目的片段一致的特异性条带，表明可以进行荧光定量 PCR 扩增。

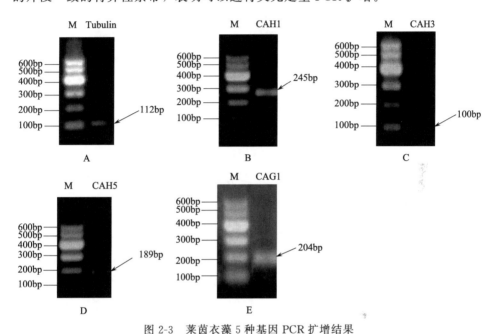

图 2-3　莱茵衣藻 5 种基因 PCR 扩增结果

Fig. 2-3　Electrophoresis results on five types of genes in *Chlamydomonas reinhardtii* amplified by PCR

A. Tubulin；B. CAH1；C. CAH3；D. CAH5；E. CAG1

三、莱茵衣藻与蛋白核小球藻碳酸酐酶基因的同源性

经过 PCR 反应，凝胶电泳检测，测序分析，最后都成功获得了莱茵衣藻的碳酸酐酶基因目标片段。同时，本研究还发现，应用莱茵衣藻的这四对碳酸酐酶引物，在蛋白核小球藻中也都成功扩增出来了相同大小的目标片段（图 2-4）。通过测序分析，也都得到了与莱茵衣藻的相应碳酸酐酶基因相同的目标片段序列，并在 NCBI 中进行 Blast 分析，都显示与莱茵衣藻的碳酸酐酶基因同源（The max ident≥98%）。由此可得出以下结论：本研究所选取的四个碳酸酐酶基因在莱茵衣藻和蛋白核小球藻中具有同源性。目前为止，我们已经向 NCBI 提交了小球藻的 CAG1 和 CAH3 等两个碳酸酐酶基因片段，登录号分别为 JX035902 和 JX426073。

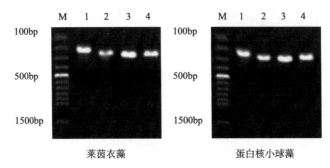

图 2-4 PCR 电泳照片

Fig. 2-4 Electrophoresis photographs of the PCR

M: Marker; 1: CAH1 (245bp); 2: CAH3 (232bp); 3: CAH5 (189bp); 4: CAG1 (223bp)

四、碳酸酐酶同工酶基因的亲缘关系及基因表达的差异性

我们使用分子遗传基因进化分析（molecular evolutionary genetics analysis, MEGA）软件，将本研究所选取的莱茵衣藻的四个碳酸酐酶基因进行序列进化分析，结果显示：这四个碳酸酐酶基因的亲缘关系很近，可以分成两组，其中，CAH1 与 CAH3 的亲缘关系更近，其次是 CAH5 与 CAG1 的亲缘关系也很近（图 2-5）。

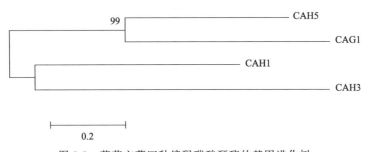

图 2-5 莱茵衣藻四种编码碳酸酐酶的基因进化树

Fig. 2-5 Phylogenetic tree of encoded for the four carbonic anhydrase genes in *Chlamydomonas reinhardtii*

在对实时荧光定量 PCR 数据处理中，为了消除由于核酸浓度差异所带来的影响，一般通过基因表达相对稳定的看家基因来校正。看家基因又叫管家基因或者持家基因。理想的看家基因在各种实验条件下，它都能相对稳定地表达（Suzuki et al., 2000）。

以看家基因为参照，应用 $2^{-\Delta CT}$ 法，定量计算莱茵衣藻和蛋白核小球藻在模拟自然条件下的碳酸酐酶基因表达量，首先可以看出四个碳酸酐酶基因相对于各自看家基因表达量的高低。从定量 PCR 实验结果来看，我们可以得出以下结论：不论是莱茵衣藻还是蛋白核小球藻，四个碳酸酐酶基因表达差异都很大。总体来看，它们的编码碳酸酐酶胞外酶(CAH1)和线粒体碳酸酐酶(CAH5)的基因表达都很旺盛，编码叶绿体碳酸酐酶(CAH3)的基因表达量居中，而编码 γ 碳酸酐酶(CAG1)的表达量都很低。这一结果跟已有的研究结果相一致（Moroney et al.，2011）。这表明胞内碳酸酐酶基因表达相对稳定，而胞外碳酸酐酶的基因表达则有较大的易变性(图 2-6)。

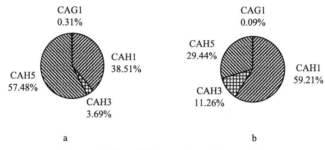

图 2-6 微藻四种编码碳酸酐酶的基因表达
a. 莱茵衣藻；b. 蛋白核小球藻

Fig. 2-6 The gene expressions of the encoded for carbonic anhydrase in algae
a. *Chlamydomonas reinhardtii*；b. *Chlorella pyrenoidosa*

第三节 微藻碳酸酐酶基因表达对碳酸氢根离子的响应

微藻不仅能利用大气中的 CO_2 为原料进行光合作用，还可以通过碳酸酐酶的催化作用，利用环境中的 HCO_3^- 为原料进行光合作用。碳酸氢根离子是喀斯特岩溶湖泊水体中可溶性无机碳的主体，因此，水体中碳酸氢根离子的含量变化对微藻碳酸酐酶基因表达具有决定性影响。本研究通过添加不同浓度的碳酸氢钠来研究碳酸酐酶对碳酸氢根离子的响应。

乙酰唑胺(acetazolamide，AZ)是含 1,3,4-噻二唑环的杂环磺酰胺类碳酸酐酶胞外酶抑制剂(CAex)（Moroney et al.，1985；Williams and Turpin，1987）。通过向实验处理中添加不同浓度的 AZ，来研究微藻碳酸酐酶胞外酶受到抑制的情况下，也就是间接利用碳酸氢根离子途径受阻条件下的微藻碳酸酐酶基因表达的变化。

一、碳酸氢钠对碳酸酐酶基因表达的影响

在培养基中分别添加 0mmol/L、0.5mmol/L、2.0mmol/L、4.0mmol/L、8.0mmol/L 和 16.0mmol/L 的碳酸氢钠对莱茵衣藻和蛋白核小球藻进行培养。3天后,用实时荧光定量 PCR 技术测定它们的基因表达。从本研究获得的编码莱茵衣藻和蛋白核小球藻的四种碳酸酐酶(CAH1,CAH3,CAH5,CAG1)基因表达结果来看,随着添加的碳酸氢钠浓度的增加,它们的基因表达有的上调,有的下调(图 2-7,图 2-8)。总体来看,在同一处理下,两种微藻的四种编码碳酸酐酶基因表达趋势类似。具体来看,编码 CAH1 基因在这两种微藻中,都是在低浓度碳酸氢钠(0~2.00mmol/L)条件下基因表达较高;而在高浓度碳酸氢钠(8.00~16.00mmol/L)条件下,编码 CAH1 基因的表达却受到了较大抑制(图 2-7a,图 2-8a)。以不添加碳酸氢钠条件下的编码 CAH1 基因表达量为对照(100%),在添加高浓度碳酸氢钠(16.00mmol/L)条件下,我们把编码 CAH1 基因所受到的抑制程度进行定量化:莱茵衣藻的表达量为对照的 0.27 倍,而蛋白核小球藻的更低,只有对照的 0.07 倍。

CAH1 位于细胞周质,控制碳酸酐酶胞外酶的活性。已有研究显示,在缺乏无机碳源的情况下,碳酸酐酶胞外酶活性较高,相应地,碳酸酐酶胞外酶基因受到诱导,促使编码 CAH1 基因表达上调(Fujiwara et al.,1990;Fukuzawa et al.,1990;Moroney et al.,2011;Yoshioka et al.,2004)。Fujiwara 等认为不同浓度 CO_2 对编码 CAH1 的基因表达影响明显,在高浓度 CO_2 的条件下,编码 CAH1 基因的表达量几乎为 0(Fujiwara et al.,1990)。与提高 CO_2 浓度对 CAH1 所产生的影响类似,添加的碳酸氢钠浓度增加也会造成编码 CAH1 基因表达量下降。通过本研究,也印证了低浓度碳酸氢钠能够诱导微藻碳酸酐酶胞外酶活性,造成编码 CAH1 表达量上调;而过高浓度的碳酸氢钠却抑制微藻碳酸酐酶胞外酶活性,造成编码 CAH1 表达量下调。

随着添加碳酸氢钠浓度的增加,编码 CAH3 基因在这两种微藻中表达趋势不同,它在莱茵衣藻中呈波浪式缓慢上升的趋势;而在蛋白核小球藻中却呈略微下降的趋势(图 2-7b,图 2-8b)。已有研究显示:CAH3 位于叶绿体的类囊体中,参与光合作用,其主要功能是催化 HCO_3^- 快速脱水形成 CO_2,供 Rubisco 进行同化作用(Duanmu et al.,2009)。一直以来,众多科学家普遍认为编码 CAH3 基因对环境碳源浓度变化不敏感(Karlsson et al.,1998;Markelova et al.,2009)。本研究也有类似的发现:莱茵衣藻的编码 CAH3 基因表达随添加碳酸氢钠浓度的增加略微上调,而蛋白核小球藻的编码 CAH3 基因表达随添加碳酸氢钠浓度的增加却略微下调,总之,它们的编码 CAH3 基因表达受碳酸氢钠浓度的影响都很小,尤其是在高浓度时更是如此。

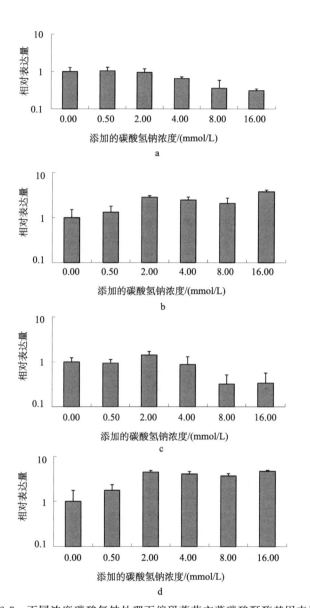

图 2-7 不同浓度碳酸氢钠处理下编码莱茵衣藻碳酸酐酶基因表达
Fig. 2-7 The gene expression of encoded for carbon anhydrase in
Chlamydomonas reinhardtii under different concentration of $NaHCO_3$ treatment
a. CAH1; b. CAH3; c. CAH5; d. CAG1

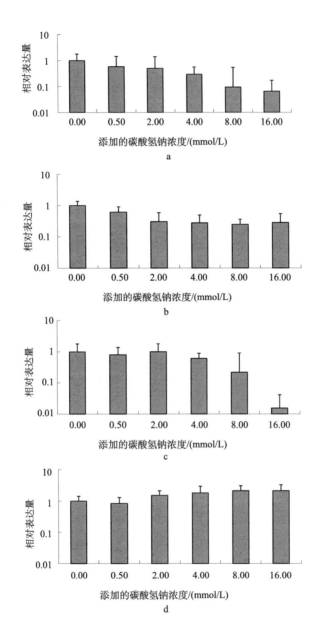

图 2-8 不同浓度碳酸氢钠处理下蛋白核小球编码藻碳酸酐酶基因表达

Fig. 2-8 The gene expression of encoded for carbon anhydrase in *Chlorella pyrenoidosa* under different concentration of $NaHCO_3$ treatment

a. CAH1; b. CAH3; c. CAH5; d. CAG1

与编码 CAH1 的基因表达类似,不论在莱茵衣藻中还是在蛋白核小球藻中,编码 CAH5 的基因表达都是随添加碳酸氢钠浓度的增加而急剧下降(图 2-7c,图 2-8c)。我们以不添加碳酸氢钠条件下的编码 CAH5 基因表达量为对照(100%),在添加高浓度碳酸氢钠(16.00mmol/L)条件下,莱茵衣藻的编码 CAH5 基因表达量为对照组的 0.23 倍,而蛋白核小球藻的编码 CAH5 基因表达量更低,只有对照组的 0.02 倍。

CAH5 参与线粒体电子传递链,在添加低浓度碳酸氢钠条件下,首先造成编码胞外酶 CAH1 基因的表达上调,相应地,此时需要线粒体呼吸作用提供更多的能量和生成 CO_2 供微藻生长所需要,由此带来编码 CAH5 基因表达上调,因此,编码 CAH5 的基因表达具有与 CAH1 相似的趋势,这也与相关研究结果类似(Moroney et al., 2011;Sunderhaus et al., 2006)。

γ 类碳酸酐酶(CAG1)也在线粒体电子传递链中起重要作用(Hewett-Emmett and Tashian,1996;Klodmann et al., 2010;Sunderhaus et al., 2006)。它在莱茵衣藻和蛋白核小球藻中的基因表达趋势相近,随添加碳酸氢钠浓度的增加,两种微藻的编码 CAG1 基因表达都呈上调的趋势(图 2-7d,图 2-8d)。需要说明的是,不论在莱茵衣藻还是蛋白核小球藻中,编码 CAG1 基因的表达量都很小,详见本章第一节,这可能跟其自身的特殊结构和功能有关。

在低浓度碳酸氢钠条件下,编码碳酸酐酶胞外酶(CAH1)基因表达受影响不显著,而高浓度碳酸氢钠则下调编码 CAH1 基因的表达。对比相关研究,不同浓度 CO_2 对编码 CAH1 基因表达的影响明显,低浓度 CO_2 诱导编码 CAH1 基因表达上调,而高浓度 CO_2 下调编码 CAH1 基因表达(Fujiwara et al., 1990;Yoshioka et al., 2004)。本研究通过添加不同浓度的碳酸氢钠所获得的研究结果与其较一致。由此我们认为:水体中缺乏可溶性无机碳能够诱导微藻的编码 CAH1 基因表达量上调,相应地,碳酸酐酶胞外酶活性提高;而可溶性无机碳含量过高则下调编码 CAH1 基因表达,相应地,碳酸酐酶胞外酶活性下降。本研究从机理上阐明了微藻碳酸酐酶胞外酶对不同环境碳源浓度的响应及其在水体缺乏无机碳条件下的适应机制。

二、乙酰唑胺对碳酸酐酶基因表达的影响

在碳酸氢钠处理的培养基中再添加 AZ 对莱茵衣藻和蛋白核小球藻进行培养。3 天后,以不添加 AZ 的碳酸氢钠处理为对照,分别计算添加 10.0mmol/L AZ 对微藻编码碳酸酐酶基因表达的影响。

在不添加碳酸氢钠和添加 0.5mmol/L 碳酸氢钠的条件下,AZ 使莱茵衣藻编码 CAH1 和 CAH3 基因表达量上调;而编码 CAH5 基因表达量却明显下调;编码 CAG1 基因表达量变化较小,详见图 2-9a、图 2-9b。

在添加 2.0mmol/L 碳酸氢钠的条件下，AZ 使莱茵衣藻编码 CAH1 基因表达量上调；编码 CAH3 基因表达量略下调；而编码 CAH5 基因表达量急剧下降；编码 CAG1 基因表达量明显下调，详见图 2-9c。

在添加 4.0mmol/L 碳酸氢钠的条件下，AZ 使莱茵衣藻编码 CAH1 基因表达量下调；编码 CAH3 基因表达量略下调；编码 CAH5 基因表达量变化不大；编码 CAG1 基因表达量下调，详见图 2-9d。

在添加 8.0mmol/L 碳酸氢钠的条件下，AZ 使莱茵衣藻编码 CAH1 基因表达量下调；编码 CAH3 基因表达量略下调；而编码 CAH5 基因表达量急剧下降；编码 CAG1 基因表达量明显下调，详见图 2-9e。

在添加 16.0mmol/L 碳酸氢钠的条件下，AZ 使莱茵衣藻编码 CAH1 基因表达量下调；编码 CAH3 基因表达量略下调；而编码 CAH5 基因表达量急剧下降；编码 CAG1 基因表达量明显下调，详见图 2-9f。

对于莱茵衣藻来说，在低浓度碳酸氢钠（0～2.00mmol/L）条件下，添加 10.0mmol/L AZ，编码 CAH1 基因表达上调；然而，在添加高浓度碳酸氢钠（4.00～16.00mmol/L）条件下，添加 10.0mmol/L AZ，编码 CAH1 基因表达量下调（图 2-9）。因为 AZ 是碳酸酐酶胞外酶特异性抑制剂（Moroney et al.，1985），在低浓度碳酸氢钠（0～2.00mmol/L）条件下，藻体本需要胞外碳酸酐酶的 CCM 机制来利用低浓度的无机碳，添加 10.0mmol/L AZ，却使碳酸酐酶胞外酶活性抑制，微藻始终处于低浓度的无机碳的微环境中，微藻为了应对这种环境，不得不上调编码 CAH1 基因表达量。高浓度碳酸氢钠可以完全抑制莱茵衣藻碳酸酐酶胞外酶活性，添加 10.0mmol/L AZ 对莱茵衣藻碳酸酐酶胞外酶活性没有叠加的抑制效应，所以只呈现出高浓度无机碳对莱茵衣藻编码碳酸酐酶胞外酶基因表达的抑制现象。

不同于编码 CAH1 基因的表达趋势，在低浓度碳酸氢钠（0 和 0.50mmol/L）条件下，添加 10.0mmol/L AZ，编码 CAH3 基因表达上调；而在较高浓度碳酸氢钠（2.0～16.0mmol/L）条件下，添加 10.0mmol/L AZ，编码 CAH3 基因表达呈下降趋势（图 2-9）。一般认为，添加 AZ，抑制了碳酸酐酶胞外酶的活性（Moroney et al.，1985），造成微藻吸收利用无机碳的能力下降，相应地，微藻的光合生产能力也下降，最终将造成编码 CAH3 基因表达量下降。

添加 10.0mmol/L AZ，对编码 CAH5 基因表达的影响明显，造成编码 CAH5 基因表达量急剧下调（图 2-9）。因为添加 AZ，抑制了碳酸酐酶胞外酶的活性（Moroney et al.，1985），造成微藻吸收利用的无机碳减少，相应地，微藻的线粒体代谢也将变缓，造成编码 CAH5 基因表达下调。

添加 10.0mmol/L AZ，对蛋白核小球藻编码碳酸酐酶基因表达所带来的影响与莱茵衣藻的基本相近，但也略有差异。添加 10.0mmol/L AZ 造成蛋白核小

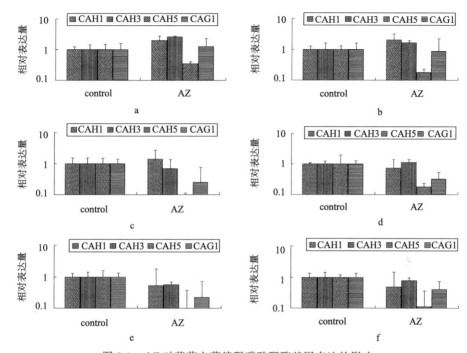

图 2-9 AZ 对莱茵衣藻编码碳酸酐酶基因表达的影响

Fig. 2-9 The effect of AZ on gene expression of encoded for carbon anhydrase in *Chlamydomonas reinhardtii*

a. 0 $NaHCO_3$; b. 0.50mmol/L $NaHCO_3$; c. 2.00mmol/L $NaHCO_3$;
d. 4.00mmol/L $NaHCO_3$; e. 8.00mmol/L $NaHCO_3$; f. 16.00mmol/L $NaHCO_3$

球藻的编码 CAH1 基因表达量全部上调（图 2-10）。这可能与高浓度碳酸氢钠不能完全抑制蛋白核小球藻碳酸酐酶胞外酶活性有关。AZ 的添加才能使碳酸酐酶胞外酶完全被抑制，与对照相比，添加 10.0mol/L AZ，编码 CAH1 基因表达被上调。

同理，添加 10.0mmol/L AZ，蛋白核小球藻的编码 CAH3 和 CAH5 基因表达量都呈下降趋势。在低浓度碳酸氢钠（0～2.00mmol/L）条件下，添加 10.0mmol/L AZ，编码 CAG1 基因表达略上调；然而，在高浓度碳酸氢钠条件下（4.00～16.00mmol/L），添加 10.0mmol/L AZ，编码 CAG1 基因表达量却略有下降（图 2-10）。

三、乙酰唑胺与碳酸氢钠共同作用下的碳酸酐酶基因表达

在添加 10.0mmol/L AZ 的条件下，不论莱茵衣藻还是蛋白核小球藻，它们的编码 CAH1 和 CAH5 基因表达量都随添加碳酸氢钠浓度的增加而呈下降趋势

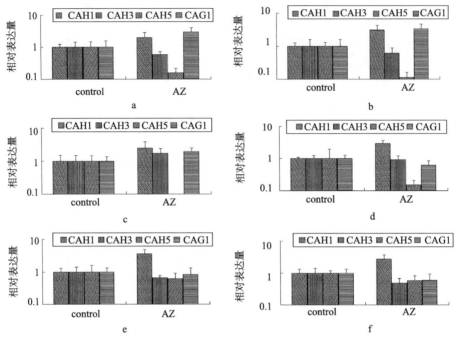

图 2-10　AZ 对蛋白核小球藻编码碳酸酐酶基因表达的影响

Fig. 2-10　The effect of AZ on gene expression of encoded for carbon anhydrase in *Chlorella pyrenoidosa*

a. 0 $NaHCO_3$; b. 0.50mmol/L $NaHCO_3$; c. 2.00mmol/L $NaHCO_3$;
d. 4.00mmol/L $NaHCO_3$; e. 8.00mmol/L $NaHCO_3$; f. 16.00mmol/L $NaHCO_3$

(图 2-11a、图 2-11c、图 2-12a 及图 2-12c),这跟未添加 AZ 的条件下,编码 CAH1 和 CAH5 基因表达随碳酸氢钠浓度梯度的变化趋势相似。值得说明的是:在添加 10.0mmol/L AZ 的条件下,编码 CAH1 和 CAH5 基因表达的规律性更强。因为 AZ 是碳酸酐酶胞外酶特异性抑制剂(Moroney et al.,1985),添加高浓度(10.0mmol/L)AZ,微藻的胞外碳酸酐酶受到了较大抑制,造成微藻吸收利用无机碳的能力下降,相应地,微藻的光合生产能力也下降,最终将造成编码叶绿体碳酸酐酶(CAH3)基因表达量下降;且添加 AZ 所带来的抑制作用大小与它本身的碳酸酐酶胞外酶活性大小呈正相关关系。

但是,在添加 10.0mmol/L AZ 的条件下,编码 CAH3 和 CAG1 基因对碳酸氢钠浓度变化不敏感(图 2-11b,图 2-11d,图 2-12b 及图 2-12d),莱茵衣藻和蛋白核小球藻的编码 CAH3 和 CAG1 基因表达都随碳酸氢钠浓度的变化呈微小波动(图 2-11b、图 2-11d、图 2-12b 及图 2-12d)。因为添加 10.0mmol/L AZ,抑制了碳酸酐酶胞外酶的活性(Moroney et al.,1985),造成微藻吸收利用的无机碳

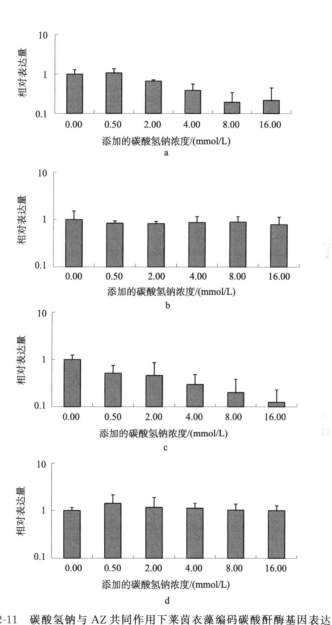

图 2-11 碳酸氢钠与 AZ 共同作用下莱茵衣藻编码碳酸酐酶基因表达

Fig. 2-11 The gene expression of encoded for carbon anhydrase in *Chlamydomonas reinhardtii* treated with AZ and different concentrations of NaHCO$_3$

a. CAH1; b. CAH3; c. CAH5; d. CAG1

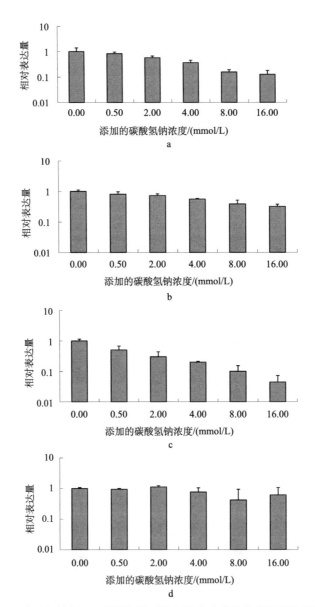

图 2-12　碳酸氢钠与 AZ 共同作用下蛋白核小球藻编码碳酸酐酶基因表达

Fig. 2-12　The gene expression of endoded for carbon anhydrase in Chlorella pyrenoidosa treated with AZ and different concentrations of $NaHCO_3$

a. CAH1; b. CAH3; c. CAH5; d. CAG1

减少，并对微藻产生了系列影响，最终也造成微藻的编码 CAH3 和 CAG1 基因表达量下降。

第四节　微藻碳酸酐酶基因表达对 pH 的响应

一、pH 对碳酸酐酶基因表达的影响

将培养基的 pH 分别用盐酸和 NaOH 调为 5.3、6.0、6.5、7.5、8.0 和 9.0，对莱茵衣藻和蛋白核小球藻进行培养。3 天后，用实时荧光定量 PCR 技术测定它们的基因表达。

本研究显示：在莱茵衣藻中，编码胞外碳酸酐酶 CAH1 基因在酸性（低 pH）条件下表达较低；而在碱性（高 pH）条件下，编码 CAH1 基因的表达有所上升（图 2-13a）。这可能跟微藻生活的水体中的 CO_2 含量变化有关。

因为水体中不同形态的无机碳之间可以相互转化，且处于动态平衡之中。一般情况下，水体中各种形态的无机碳比例变化主要受 pH 影响（图 2-14）。在水体由酸性到碱性的过渡过程中，CO_2 占总无机碳的比例存在先下降后上升的变化过程（Larkum et al.，1989；Skirrow and Whitfield，1975）。

CAH1 位于细胞周质。已有研究显示，在缺乏无机碳源的情况下，编码碳酸酐酶胞外酶基因将受到诱导，促使编码 CAH1 基因表达上调（Fujiwara et al.，1990；Fukuzawa et al.，1990；Moroney et al.，2011；Yoshioka et al.，2004）。

除了 pH5.3 外，随着水体 pH 的升高，莱茵衣藻的编码 CAH3 基因表达呈先上升后下降的趋势，整体呈抛物线变化（图 2-13b）。CAH3 位于叶绿体的类囊体中，参与光合作用，其主要功能是催化 HCO_3^- 快速脱水形成 CO_2，供 Rubisco 进行同化作用（Duanmu et al.，2009）。在酸性条件下，微藻生长受到了较大抑制，而随着环境 pH 的升高，在中性甚至碱性条件下，微藻生长都比较旺盛。相应地，编码叶绿体碳酸酐酶 CAH3 的基因表达也具有与其生长相近的变化趋势。

与编码 CAH3 的基因表达类似，编码线粒体碳酸酐酶 CAH5 的基因表达也是随 pH 的升高呈抛物线变化（图 2-13c）。与 CAH3 不同的是，CAH5 参与线粒体电子传递链，它可能在微藻的呼吸作用方面具有重要意义。

γ 类碳酸酐酶（CAG1）也在线粒体电子传递链中起重要作用（Hewett-Emmett and Tashian，1996；Klodmann et al.，2010；Sunderhaus et al.，2006）。CAG1 的基因表达随 pH 的升高变化较小（图 2-13d），这主要是因为它的表达量小，存在着较大的测定误差，较小的变化难以识别。

而在同样的 pH 处理下，碳酸酐酶在蛋白核小球藻中的基因表达具有与莱茵衣藻中的相似的趋势，但也存在一些差异。编码 CAH1 基因在强酸性（pH=5.3）条件下表达较低；而在弱酸性、中性及碱性条件下，CAH1 基因的表达上升（图 2-15a）。编码 CAH1 基因之所以在蛋白核小球藻中与莱茵衣藻中存在较大差异，这可能跟微藻的种类有关。

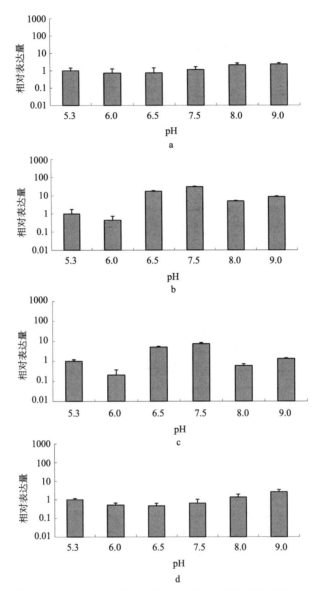

图 2-13 不同 pH 处理下莱茵衣藻的编码碳酸酐酶基因表达

Fig. 2-13 The gene expression of endoded for carbon anhydrase in *Chlamydomonas reinhardtii* under different pH

a. CAH1; b. CAH3; c. CAH5; d. CAG1

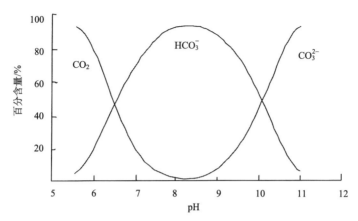

图 2-14　不同 pH 条件下水体中各无机碳分量的百分含量

注：图片来源 Larkum et al.，1989

Fig. 2-14　Percentage of DIC in water under different pH

From Larkum et al.，1989

蛋白核小球藻的编码 CAH3 基因表达随着水体 pH 的升高而上升（图 2-15b），这跟微藻生物量的变化趋势相似。因为 CAH3 位于叶绿体的类囊体中，参与光合作用，其主要功能是催化 HCO_3^- 快速脱水形成 CO_2，供 Rubisco 进行同化作用（Duanmu et al.，2009）。

与编码 CAH3 的基因表达类似，蛋白核小球藻的编码 CAH5 基因表达也是随 pH 的升高呈缓慢上升的趋势（图 2-15c）。CAH5 参与线粒体电子传递链，它与 CAH3 呈对应的互补关系。与莱茵衣藻一样，编码 CAG1 的基因表达随 pH 的升高变化较小（图 2-15d）。

二、乙酰唑胺与 pH 共同作用下的碳酸酐酶基因表达

在 pH 处理的培养基中再添加 AZ 对莱茵衣藻和蛋白核小球藻进行培养。3 天后，用实时荧光定量 PCR 技术测定它们的基因表达。

在添加 10.0mmol/L AZ 的条件下，编码碳酸酐酶基因表达随 pH 升高却具有另一番变化趋势。因为 AZ 是碳酸酐酶胞外酶特异性抑制剂（Moroney et al.，1985），添加高浓度（10.0mmol/L）AZ，微藻的胞外碳酸酐酶受到了抑制，微藻为适应添加 AZ 所带来的抑制作用，各种编码碳酸酐酶基因的表达必将做出相应调整。整体来看，莱茵衣藻的四个编码碳酸酐酶基因（CAH1、CAH3、CAH5 和 CAG1）表达量都随 pH 的升高呈先上升后下降的抛物线变化趋势（图 2-16）。

而蛋白核小球藻在添加 10.0mmol/L AZ 的条件下，它的四种编码碳酸酐酶基因（CAH1、CAH3、CAH5 和 CAG1）表达量都随 pH 的升高呈先下降后上升

的变化趋势(图 2-17)。可以看出，除了 pH5.3 外，碱性环境有利于编码碳酸酐酶基因的表达。

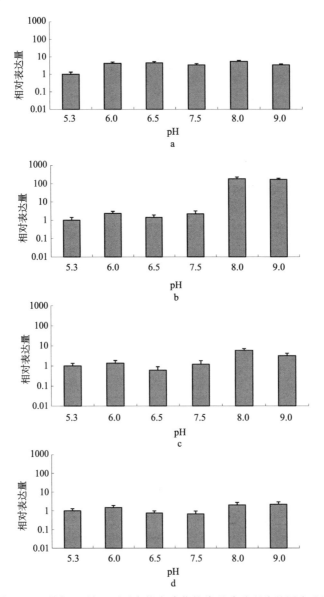

图 2-15　不同 pH 处理下蛋白核小球藻的编码碳酸酐酶基因表达图
Fig. 2-15　The gene expression of encoded for carbon
anhydrase under different pH treatment in *Chlorella pyrenoidosa*
a. CAH1; b. CAH3; c. CAH5; d. CAG1

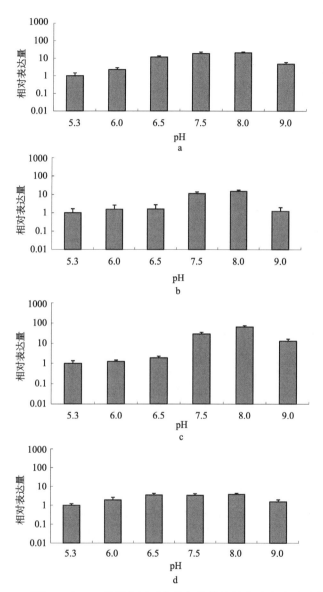

图 2-16 不同 pH 与 AZ 共同作用下莱茵衣藻的编码碳酸酐酶基因表达图

Fig. 2-16 The gene expression of encoded for carbon anhydrase in *Chlamydomonas reinhardtii* treated with AZ and different pH

a. CAH1; b. CAH3; c. CAH5; d. CAG1

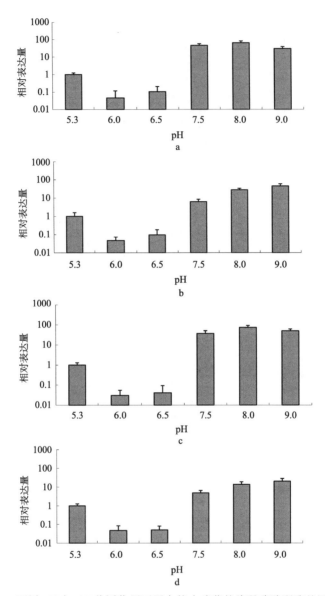

图 2-17 不同 pH 与 AZ 共同作用下蛋白核小球藻的编码碳酸酐酶基因表达图
Fig. 2-17 The gene expression of encoded for carbon anhydrase in
Chlorella pyrenoidosa treated with AZ and different concentrations of pH
a. CAH1; b. CAH3; c. CAH5; d. CAG1

第五节 微藻碳酸酐酶基因表达对氟的响应

设置 0mmol/L、0.1mmol/L、1mmol/L、10mmol/L、50mmol/L、100mmol/L、200mmol/L 七个 F^- 浓度梯度处理培养莱茵衣藻,10 天后,用实时荧光定量 PCR 技术测定它的碳酸酐酶基因表达。荧光定量 PCR 的结果(图 2-18)显示,在 F^- 为浓度 0~1mmol/L 范围内 4 种编码碳酸酐酶基因相对表达情况没有显著差异($n=9$,$P>0.05$)。编码胞外碳酸酐酶基因 CAH1 在 F^- 浓度为 10~100mmol/L 范围内相对表达量随 F^- 浓度升高而显著上升($n=9$,$P<0.05$),在 F^- 浓度为 50mmol/L 时其相对表达量为对照组的 17.2 倍,在 F^- 浓度为 100mmol/L 时为对照组的 33.9 倍。编码叶绿体碳酸酐酶基因 CAH3 和 γ 家族碳酸酐酶 CAG1 变化情况类似,均为在 F^- 浓度为 10mmol/L 时显著高于其他处理组($n=18$,$P<0.05$),编码 CAH3 相对表达量为对照组的 12.1 倍,编码 CAG1 相对表达量为对照组的 10.6 倍,在 F^- 为浓度 50mmol/L 和 100mmol/L 时,编码 CAH3 相对表达量为对照组的 4.3 倍和 4.1 倍,编码 CAG1 相对表达量则为对照组的 3.8 倍和 3.6 倍。编码线粒体碳酸酐酶基因 CAH5 的相对表达量在 F^- 浓度为 0.1mmol/L 和 1mmol/L 时分别为对照组的 27% 和 37%,

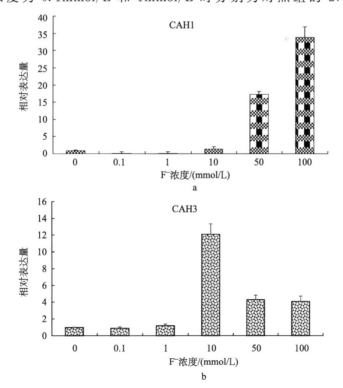

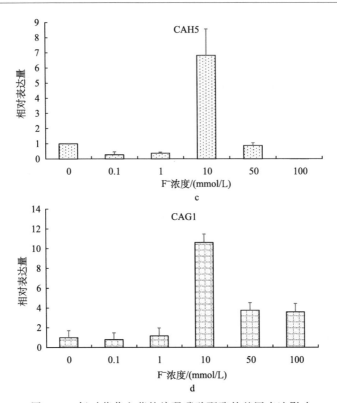

图 2-18　氟对莱茵衣藻的编码碳酸酐酶的基因表达影响

Fig. 2-18　The effect of fluoride on gene expression of encoded for carbon anhydrase in *Chlamydomonas reinhardtii*

a. CAH1；b. CAH3；c. CAH5；d. CAG1

50mmol/L 时差异显著（$n=18$，$P<0.05$），为对照组的 6.8 倍，在 F^- 浓度高于 50mmol/L 范围内，其相对表达量呈下降趋势。

Fujiwara 等的研究发现编码 CAH1 基因表达量有随光周期变化而显著变化的现象，编码 CAH2 基因表达量则比较稳定，基本不受光周期变化过程影响（Fujiwara et al.，1996）。在本试验中，CAH1 亦受 F^- 浓度影响相对表达量有显著变化，表现出诱导酶的特征。编码 CAH1 基因相对表达量的变化趋势与胞外酶活性的变化趋势在 F^- 浓度为 10～100mmol/L 范围内相一致（见第六章），且变化量均较大，表明胞外酶受到高浓度 F^- 胁迫时，活性和表达量都会增加。

CAH3 位于莱茵衣藻叶绿体类囊体膜中，其主要功能是快速脱水转化 HCO_3^- 成 CO_2 转运到类囊体腔，供给 Rubisco 进行同化作用（Duanmu et al.，2009）。CAG1 是线粒体 NADH 的组成亚基，是呼吸作用电子传递链的组成部分，直接参与呼吸作用的过程。在 F^- 浓度为 10mmol/L 处理组中，这两种基因相对表达量均显著增加，在 F^- 浓度为 50mmol/L 和 100mmol/L 处理组中，两种基因相对表达量也较对照组高出 4 倍左右，表明高氟胁迫能够促进这两种基因的表达。

CAH5 位于线粒体膜上，由同源性分析可见 CAH5 与其他基因组碳酸酐酶明显不同，如起始密码子为异亮氨酸，长度仅由 61 个氨基酸残基组成。线粒体具有独立的结构和相对独立的遗传物质，可以自主复制一部分 DNA 片段（董荣等，2008）。推测线粒体膜上的碳酸酐酶可能为线粒体自主合成。在 F^- 浓度为 0.1mmol/L 和 1mmol/L 培养的莱茵衣藻编码 CAH5 基因相对表达量低于对照组，可见较低的 F^- 浓度下，莱茵衣藻线粒体的编码 CAH5 基因转录速度减缓，而在 10mmol/L F^- 的作用下，编码 CAH5 基因相对表达量增加，说明在此浓度 F^- 作用下，莱茵衣藻线粒体活动能力增强，进行的分解有机物产生能量的过程加快，由此产生的过多 CO_2 溶解于线粒体基质中，导致线粒体内 HCO_3^- 浓度增高，为加快将 HCO_3^- 转运到线粒体外，线粒体中的编码 CAH5 基因表达出高于平时的量，合成更多的线粒体碳酸酐酶用于转运线粒体中过量 HCO_3^- 至细胞质中。在 100mmol/L F^- 胁迫下，其表达量受到明显抑制，可能原因为线粒体受高氟的影响，编码 CAH5 基因的表达受到抑制或转录成的 mRNA 稳定性降低更容易降解，导致编码 CAH5 基因的表观表达量较低。

在本研究中，编码胞外碳酸酐酶 CAH1 基因相对表达量较三种编码胞内碳酸酐酶基因的相对表达量变化大，可见胞外碳酸酐酶表达机制受 F^- 浓度变化影响更大，胞内碳酸酐酶则由于所处的细胞内环境相对稳定，变化范围较小。

参 考 文 献

董荣，白艳玲，周腊梅，徐海津，张秀明，乔明强. 2008. 线粒体遗传工程及其意义. 生物学通报，43(3)：13-16.

范国昌，钱凯先. 1998. 叶绿体基因工程研究的理想材料——衣藻. 生物学通报，33(2)：13-15.

赵焕英，包金凤. 2007. 实时荧光定量 PCR 技术的原理及其应用研究进展. 中国组织化学与细胞化学杂志，16(4)：492-497.

周莉娟，郑伟文. 2004. 香蕉花叶心腐病检测的两步法及一步法 RT—PCR 比较研究. 江西农业大学学报，25：776-779.

Baymann F, Zito F, Kuras R, Minai L, Nitschke W, Wollman F A. 1999. Functional characterization of *Chlamydomonas* mutants defective in cytochrome *f* maturation. The Journal of Biological Chemistry，274：22957-22967.

Boynton J E, Gillham N W, Harris E H, et al. 1988. Chloroplast transformation in *Chlamydomonas* with high velocity microprojectiles. Science，240(4858)：1534-1538.

Duanmu D Q, Wang Y J, Spalding M H. 2009. Thylakoid lumen carbonic anhydrase (CAH3) mutation suppresses air-dier phenotype of LCIB mutant in *Chlamydomonas reinhardtii*.

Plant Physiology, 149: 929-937.

Fujiwara S, Fukuzawa H, Tachiki A, Miyachi S. 1990. Structure and differential expression of 2 genes encoding carbonic anhydrase in *Chlamydomonas reinhardtii*. Proceedings of the National Academy of Sciences of the United States of America, 87: 9779-9783.

Fujiwara S, Ishida N, Tsuzuki M. 1996. Circadian expression of the carbonic anhydrase gene, Cah1, in *Chlamydomonas reinhardtii*. Plant Molecular Biology, 32: 745-749.

Fukuzawa H, Fujiwara S, Tachiki A, Miyachi S. 1990. Nucleotide sequences of 2 genes CAH1 and CAH2 which encode carbonic anhydrase polypeptides in *Chlamydomonas reinhardtii*. Nucleic Acids Research, 18: 6441-6442.

Grant D, Chiang K S. 1980. Physical mapping and characterization of *Chlamydomonas* mitochondrial DNA molecules: their unique ends, sequence homogeneity, and conservation. Plasmid, 4: 82-96.

Harris E H, Stern D B, Witman G. 2009. The *Chlamydomonas* Sourcebook. Academic Press.

Hewett-Emmett D, Tashian R E. 1996. Functional diversity, conservation, and convergence in the evolution of the α-, β-, and γ-carbonic anhydrase gene families. Molecular Phylogenetics and Evolution, 5(1): 50-77.

Hippler M, Redding K, Rochaix J D. 1998. *Chlamydomonas* genetics, a tool for the study of bioenergetic pathways. Biochimica et Biophysica Acta, 1367(1-3): 1-62.

Karlsson J, Clarke A K, Chen Z Y, et al. 1998. A novel alpha-type carbonic anhydrase associated with the thylakoid membrane in *Chlamydomonas reinhardtii* is required for growth at ambient CO_2. Embo Journal, 17: 1208-1216.

Klodmann J, Sunderhaus S, Nimtz M, Jänsch L, Braun H P. 2010. Internal architecture of mitochondrial complex I from *Arabidopsis thaliana*. The Plant Cell Online, 22: 797-810.

Larkum A, Roberts G, Kuo J, Strother S. 1989. Biology of seagrasses: A treatise on the biology of seagrasses with special reference to the Australian region. Amsterdam: Elsevier Science Publishers BV, 686-722.

Lewin R A. 1957. Four new species of *Chlamydomonas*. Canadian Journal of Botany, 35(3): 321-326.

Lumbreras V, Purton S. 1998. Recent advances in *Chlamydomonas transgenics*. Protist, 149: 23-27.

Markelova A G, Sinetova M P, Kupriyanova E V, Pronina N A. 2009. Distribution and functional role of carbonic anhydrase Cah3 associated with thylakoid membranes in the chloroplast and pyrenoid of *Chlamydomonas reinhardtii*. Russian Journal of Plant Physiology, 56: 761-768.

Meldrum N U, Roughton F J. 1933. Carbonic anhydrase. Its preparation and properties. The Journal of physiology, 80: 113-142.

Merchant S S, Prochnik S E, Vallon O, et al. 2007. The *Chlamydomonas* genome reveals the evolution of key animal and plant functions. Science, 318: 245-251.

Michaelis G, Vahrenholz C, Pratje E. 1990. Mitochondrial DNA of *Chlamydomonas reinhardtii*: the gene for apocytochrome *b* and the complete functional map of the 15. 8 kb DNA. Molecular and General Genetics, 223: 221-216.

Moroney J V, Ma Y, Frey W D, et al. 2011. The carbonic anhydrase isoforms of *Chlamydomonas reinhardtii*: intracellular location, expression, and physiological roles. Photosynthesis Research, 109: 133-149.

Moroney J V, Husic H D, Tolbert N. 1985. Effect of carbonic anhydrase inhibitors on inorganic carbon accumulation by *Chlamydomonas reinhardtii*. Plant Physiology, 79: 177-183.

Quarmby L. 2000. Meeting Report: 9th International Conference on the Cell and Molecular Biology of *Chlamydomodas*. Protist, 151: 193-199.

Rochaix J D. 1978. Restriction endonuclease map of the chloroplast DNA of *Chlamydomonas reinhardtii*. Journal of Molecular Biology, 126: 597-617.

Sager R, Granick S. 1953. Nutritional studies with *Chlamydomonas reinhardtii*. Annals of the New York Academy of Sciences, 56: 831-838.

Skirrow G, Whitfield M. 1975. The effect of increases in the atmospheric carbon dioxide content on the carbonate ion concentration of surface ocean water at 25℃. Limnology and Oceanography, 20: 103-108.

Smith K S, Ferry J G. 2000. Prokaryotic carbonic anhydrases. Fems Microbiology Reviews, 24: 335-366.

Sunderhaus S, Dudkina N V, Jansch L, et al. 2006. Carbonic anhydrase subunits form a matrix-exposed domain attached to the membrane arm of mitochondrial complex I in plants. Journal of Biological Chemistry, 281: 6482-6488.

Suzuki T, Higgins P J, Crawford D R. 2000. Control selection for RNA quantitation. Biotechniques, 29: 332-337.

Williams T G, Turpin D H. 1987. The role of external carbonic anhydrase in inorganic carbon acquisition by *Chlamydomonas reinhardtii* at alkaline pH. Plant Physiology, 83: 92-96.

Yoshioka S, Taniguchi F, Miura K, Inoue T, Yamano T, Fukuzawa H. 2004. The novel Myb transcription factor LCR1 regulates the CO_2-responsive gene Cah1, encoding a periplasmic carbonic anhydrase in *Chlamydomonas reinhardtii*. The Plant Cell, 16: 1466-1477.

第三章 胞外碳酸酐酶对稳定碳同位素分馏的影响

摘　要

微藻利用不同形式的无机碳可产生不同的稳定碳同位素分馏。微藻碳酸酐酶胞外酶(CAex)是CO_2浓缩机制(CCM)的重要组成部分，它可能是影响微藻稳定碳同位素分馏的一个重要因素。3-磷酸甘油酸(PGA)是Calvin循环的第一个稳定的中间产物，直接测定它的稳定碳同位素值$\delta^{13}C_{PGA}$，可以减少后续代谢的分馏作用产生的误差，能更准确地反映微藻利用溶解性的无机碳(DIC)的$\delta^{13}C$值产生分馏的影响。

在偏酸性环境(pH为5.3~6.0)和偏碱性环境(pH为7.5~9.0)下微藻CAex活性较低，在pH为6.5时微藻CAex活性最高。在pH为7.5~8.0时，微藻的$\delta^{13}C_{PGA}$和$\delta^{13}C$都趋于稳定。pH对稳定碳同位素分馏的影响大于CAex。pH对微藻^{13}C的影响与对碳酸酐酶胞外酶的影响存在"相差"。在pH为6~8时，微藻的$\delta^{13}CPGA$值能准确地反映水体可溶性无机碳与稳定碳同位素组成关系。

在低浓度HCO_3^-(0~2mmol/L)中，添加AZ处理的藻体$\delta^{13}C$值要比未添加AZ的藻体$\delta^{13}C$值偏负约10‰。考虑两者生长速率的巨大差异带来的同位素分馏差异，这个值接近9‰的理论值。在高浓度HCO_3^-(8~20mmol/L)下，微藻$\delta^{13}C$值与添加的$NaHCO_3$浓度呈较好的线性关系；无论添加AZ与否，微藻细胞主要利用的是添加的HCO_3^-。添加8mmol/L HCO_3^-足以完全抑制碳酸酐酶胞外酶活力。

微藻CAex活性随AZ浓度的增加而减少。0.1mmol/L AZ可以完全抑制CAex活性。AZ浓度小于0.1mmol/L时，藻体有CAex活性；当AZ浓度大于0.1mmol/L时，CAex完全无活性，$\delta^{13}C$值则明显偏负。CAex影响微藻稳定碳同位素分馏具有明显的剂量效应。

本研究可为微藻碳酸酐酶生物地球化学作用的后续研究提供理论依据和体系保障。

Chapter 3 The effect of extracellular carbonic anhydrase on stable carbon isotope fraction

Abstract

Microalgae can produce stable carbon isotope fractionation between the utilization different forms of inorganic carbon. Microalgae extracellular carbonic anhydrase (CAex) is an important part of CO_2 concentrating mechanism (CCM) and it is also a possible vital factor of affecting microalgae stable carbon isotope fractionation. Phosphoglyceric acid (PGA) is the first stable intermediates in Calvin cycle. Direct determination of $\delta^{13}C_{PGA}$ can reduce the error from fractionation of the stable carbon isotope in the following metabolism. The $\delta^{13}C_{PGA}$ can more accurately reflect the $\delta^{13}C$ value produced by the stable carbon fractionation in the process of microalgae utilization of DIC.

Microalgal CAex activity was low, both in the acidic environment (pH5.3~6.0) and the alkaline environment (pH7.5~9.0). The microalgal CAex activity was the highest at pH 6.5. The microalgal $\delta^{13}C$ and $\delta^{13}C_{PGA}$ value were stable during pH 7.5~8.0. The influence of pH on stable carbon isotope fractionation was greater than that of CAex. The influence of pH on $\delta^{13}C$ had a " phase difference" from that on extracellular carbonic anhydrase. Microalgal $\delta^{13}C_{PGA}$ value can be more accurately reflects the relationship between dissolved inorganic carbon in aquatic medium and the stable carbon isotope composition during pH 6.0~8.0.

The algal $\delta^{13}C$ value treated with acetazolamide (AZ) is more negative about 10‰ than that without AZ under the low concentration of HCO_3^- (0~2mmol/L). Considering the isotope fractionation resulted from the great difference of the growth rate between the treatment with AZ or without AZ, this difference of algal $\delta^{13}C$ was close to the theoretical value of 9‰. Under high concentration of bicarbonate (8~20mmol/L), microalgal $\delta^{13}C$ had a good linear relationship with the concentration of bicarbonate addition; and whether or not added AZ, the algae mainly use the bicarbonate added. The addition of 8 mmol/L bicarbonate can completely inhibit the activity of CAex.

Microalgal CAex activity decreased with increasing AZ concentration.

0.1mmol/L AZ can completely inhibit the activity of CAex. Microalgae had considerable activity of CAex under less than 0.1mmol/L AZ. Microalgae had no activity of CAex and their $\delta^{13}C$ values are obviously negative biased under more than 0.1mmol/L AZ. Microalgal CAex influencing the stable carbon isotopic fractionation has obvious dose effects.

This chapter will provide the theoretical basis and system guarantee for thesequential research on microalgal carbonic anhydrase biogeochemical action.

藻类稳定碳同位素变化已成为水生态环境研究的重要工具之一。海洋有机质（主要来源于浮游藻类）中的 $\delta^{13}C$ 常用来指示古环境、古气候的变化。通过对海洋、湖泊中的藻类同位素信息的提取，还可以有效地指示藻类利用碳源策略、水体食物链的时空变化，甚至还可以表征水体初级生产力，而这些指示作用的精确度是建立在藻类稳定碳同位素分馏机制的深刻了解的基础上的。

现有的藻类稳定碳同位素分馏模型对水环境 CO_2 浓度的估算存在较大偏差。目前众多的藻类稳定碳同位素分馏模型是建立在水环境中 CO_2 浓度与藻 $\delta^{13}C$ 值呈线性关系的基础上的。虽然既考虑到了水体 CO_2 浓度及环境因素，也考虑到了生长速率、细胞形态等因素（Francois et al.，1993；Laws et al.，1995；Rau et al.，1997；Popp et al.，1998；Burkhardt et al.，1999；Cassar et al.，2006），甚至还考虑到了藻体的无机碳的主动获取（未考虑到碳酸酐酶胞外酶的作用）（Keller and Morel，1999；Laws et al.，2002），但依据这些模型在一些水体中用来预测藻类碳同位素分馏时，依然存在较大的偏差，甚至存在预测值与实测值不相关的情况。这意味着还有一些重要因素不被认识，较强的生化动力学分馏作用的碳酸酐酶胞外酶（extracelluar CA，CAex）在其中可能起着重要作用。

第一节　碳酸酐酶胞外酶对稳定碳同位素分馏作用及其识别基础

碳酸酐酶（EC4.2.1.1）在许多生命体中被发现，包括动物、植物、真细菌、原核细菌和病毒（Hewett-Emmett and Tashian，1996）。碳酸酐酶（CAs）存在的证据遍布所有的单细胞微生物。从淡水到海洋环境，包括绿藻纲（Amoroso et al.，1996；Fisher et al.，1996；Ramazanov et al.，1995；Sültemeyer et al.，1995）、硅藻纲（Nimer，1997；Rotatore et al.，1995；Sültemeyer et al.，1993）、红藻门（Aizawa and Miyachi，1986）和定鞭藻纲（Nimer and Merret，1996）。此外，CA 活性也在苔藓和它们的共生光合生物中检测到（Badger et al.，1993；Palmqvist，1993；Palmqvist et al.，1994；Palmqvist and Badger，1996；Wu et al.,2006)。

很多微藻具有胞内酶和胞外酶。细胞质以及叶绿体中的胞内酶一般来说是过量的,且与无机碳的浓缩机制无关,因此,它不能调控光合作用,也不能影响碳的同位素分馏(Edwards and Mohamed,1973;Randall and Bouma,1973)。但是,碳酸酐酶胞外酶(extracelluar CA,CAex)可以引起无机碳的主动转运和浓缩(CCM),这是藻类最重要、最复杂的无机碳主动获取方式之一(Colman et al.,2002)。

很多研究证明微藻 CAex 存在于 *Chlamydomonas*、*Dunaliella*、*Scenedesmus*、*Chlorella chlorococcum littora*、*Nitzschia ruttneri* Hust 和某些海藻中(Aizawa and Miyachi,1986;Badger and Price,1994;Pesheva et al.,1994,Sültemeyer et al.,1993;Suzuki and Ikawa,1993;Williams and Colman,1994,1995)。在这些研究中,大部分都检测出 CAex 有加快无机碳(Ci)的相互转化的功能。当细胞要适应低浓度 Ci 时,CAex 活性在几小时内会突然增加。这就是对 CCM 高活性的诱导(Badger and Price,1992,1994;Sültemeyer et al.,1993)。与此相反,最近报道了蓝藻的一个被 α 基因型编码的 CAex 是在高浓度 CO_2 生长状态下被诱导的(Soltes-Rak et al.,1997)。一般而言,胞外酶和胞内酶的比率在真核藻类物种中变化很大。例如在莱茵衣藻中该比率多达 99%,而在杜氏盐藻中该比例为 25%~40%(Sültemeyer et al.,1995;Amoroso et al.,1996)。

已经在莱茵衣藻(*Chlamydomonas reinhardtii*)和盐生杜氏藻(*Dunaliella salina*)中检测到了 CAex 的确切位置(Coleman et al.,1991;Fisher et al.,1994,1996)。莱茵衣藻的胞外酶是可溶的,并且主要为位于细胞周质间隙空间内和细胞壁的内侧。相反,盐生杜氏藻的 CAex 是一类膜蛋白并且除非有清洁剂否则不溶于水。在高盐分处理的介质中,该酶能被释放。这说明它不是跨膜蛋白。从其疏水基团中可以预测该酶可通过离子交换而依附在膜的外围(Fisher et al.,1994,1996)。

CAex 存在的绝大部分证据是基于使用各类 CA 抑制剂的生理实验。乙酰唑胺(AZ)是含 1,3,4-噻二唑环的杂环磺酰胺类碳酸酐酶抑制剂,对 CA 有较好的抑制作用(肖忠海等,2008)。在这些研究中,AZ 在光合作用中的效应是基于 AZ 能缓慢穿透细胞的假设(Geib et al.,1996)。因此被抑制的 CA 仅在细胞表面。如果光合作用能被该处理抑制,那么 CAex 被认为在被研究的大型海藻的 Ci 摄取中起一定的作用。在高 pH 和低浓度无机碳的环境中,AZ 能明显抑制莱茵衣藻的光合作用(Moroney et al.,1985)。

众所周知,光能自养生物只有通过光合作用把无机碳固定下来才能维持生存。但 Emerson 等(1934)发现一种海洋杉藻(*Gigartina harvesyana*)在 CO_2 游离浓度达到 $360\mu mol/L$ 时才启动光合作用,这一浓度相当于自然海水中 CO_2 游离浓度的 36 倍。Tseng 等(1946)指出红藻软骨石花菜(*Gelidium cartilagineum*)在 CO_2 浓度达到 110μmol/L 时才使光饱和，这一 CO_2 浓度高出海水的 10 倍。其他水生高等植物的实验结果(Arens，1933，1936；Steeman-Nielsen，1946)也证实，水生植物的生存，绝不仅仅依赖于 CO_2 的吸收，必定存在其他获取无机碳的途径。藻类碳酸酐酶胞外酶的作用被认为是这些其他途径中最重要途径。碳酸酐酶在这个途径中行使了 CO_2 浓缩机制(CCM；Badger and Price，1992)。这种机制是在周围基质中的无机碳浓度(C_i)较有限时，CA 高度活跃，增加 CO_2 固定酶(Rubisco)(核酮糖-1，5 二磷酸羧化/加氧酶)周围的 CO_2 浓度，有利于羧化反应，从而能在较低的 CO_2 浓度下有效地进行光合作用。

碳酸酐酶的生化动力学分馏作用不容忽视。虽然，碳酸酐酶催化的 $CO_2 + H_2O \leftrightarrow H^+ + HCO_3^-$ 反应，所产生的碳同位素平衡分馏作用很小。但由于细胞的区室化和生化代谢的偶联反应，造成碳酸酐酶的生化动力学分馏作用很大，这一点在体外实验和转基因 C_4 植物的实验中也得到了初步证实(Marlier and O'Leary，1984；Paneth and Leary，1985；von Caemmerer et al.，2004；Cousins et al.，2006；McNevin et al.，2007)。

微藻质膜上分布的碳酸酐酶胞外酶(extracelluar CA，CAex)，能直接引起无机碳的浓缩，它的区室化特征使其具有影响碳同位素分馏的客观条件，它的活性在不同的藻类中差异很大，相差可达几十倍、甚至几百倍(Colman et al.，2002)，这使得碳酸酐酶胞外酶对碳的同位素分馏影响具有可识别性。藻类通过 CA 作用，将水中溶解的 HCO_3^- 催化转化为可直接利用的 CO_2。此时的 HCO_3^- 既有来源于水中的无机盐，也有大气的 CO_2 溶解于水中转化来的，使 HCO_3^- 的 $\delta^{13}C$ 值不同，反映在藻类光合作用上，就是其有机碳的 $\delta^{13}C$ 值。

微藻的稳定碳同位素组成主要取决于无机碳转运和羧化作用。羧化作用对碳同位素分馏作用的研究已非常充分，因此本研究就微藻 CA 胞外酶和无机碳转运的生理生化特点来设置研究内容，严格控制培养条件，力图准确识别无机碳(DIC)对微藻 CA 胞外酶和稳定碳同位素分馏的影响。通过比较在微藻培养液中加入碳酸酐酶胞外酶抑制剂 AZ 与否，来验证微藻碳酸酐酶胞外酶对稳定碳同位素分馏的影响以及影响的程度。以微藻的 CAex 作研究对象来辨识碳酸酐酶在植物同位素分馏中的作用，研究成果将对现有的微藻同位素分馏模型进行更新和校正，为用 $\delta^{13}C$ 解析水环境变化增添新的内容。研究碳酸酐酶作用下的稳定碳同位素分馏之间的关系，探索生物大分子与环境的协同作用，这样既可以加深对酶的地球化学作用的认识，也可以丰富对酶的生理生化机制的认识。

第二节 pH对微藻稳定碳同位素分馏的影响

一、研究材料和测定方法

以莱茵衣藻(*Chlamydomonas reinhardtii*)和小球藻(*Chlorella vulgaris*)作为实验材料,培养基为SE培养基。在无菌操作条件下,将藻种接种于事先高温高压灭菌(120℃,30min)的50mL培养液于200mL的锥形瓶中。将接种好微藻的锥形瓶置于25.0℃±1.0℃,安装了双排日光灯的架子上培养12h/12h的光照、黑暗交替,97μmol/(m^2·s)的光照强度,继代培养,手摇锥形瓶,每天3~5次。

pH梯度设置为5.3、6.0、6.5、7.5、8.0和9.0(用盐酸和氢氧化钠调节),每一种藻的每一pH梯度平行培养5瓶。培养条件是培养液采用SE(液态)培养液;白昼时间为12h/12h;室温保持24℃±1.0℃;光照强度为97μmol/(m^2·s)。每种微藻按上述条件分别扩大培养6天,换处理液。为加强处理效果,每6天换一次处理液,如此处理两次,活化培养两天,即可开始测定各项指标。选择其中3瓶作为平行样分别测定其CAex活性、叶绿素含量、蛋白质含量。余下两瓶测定藻体的$\delta^{13}C$值。提取PGA的微藻平行培养。

叶绿素a测定方法如下(丛海兵等,2007):取5~10mL混合均匀的藻液在1600g离心力条件下离心10min,弃去上清液,加入5~10mL的95%乙醇(乙醇量依微藻生长情况而定)。将离心管放置于2~4℃的冰箱中避光浸提24小时。24小时之后,将该离心管在1600g离心力下离心10min。将离心后的上清液倒入比色皿中,在分光光度计上测定萃取液在630nm、647nm、664nm和750nm波长处的吸光度。

采用考马斯亮蓝比色法测定微藻的蛋白质含量(郭颖娜和孙卫,2008)。

碳酸酐酶胞外酶活力测定按Wilbur和Adesron(1948)的方法进行。将实验需要量的微藻储备液在1600g下离心10min,去掉上清液,将收集的微藻悬浮在3mL的0℃的巴比妥缓冲液中(pH约为8.30)作为碳酸酐酶胞外酶液待测。之后将0.4mL酶液加入到4.5mL的0℃的巴比妥缓冲液中,再加入3mL冰浴的饱和CO_2蒸馏水,记下这时巴比妥缓冲液pH从8.20下降到7.20所需要的时间T,没有加入微藻的情况下所需时间为T_0,作为空白。碳酸酐酶胞外酶活力按下面公式计算:CAex=($T_0/T-1$)/Chl-a。活力单位为Wilbur-Adesron units·(mgChl-a)$^{-1}$(Wilbur and Adesron,1948)。

微藻PGA(3-磷酸甘油酸)的提取。将培养并处理至对数生长期的微藻光照1h后在1600g离心力下离心10min,弃去上清液。用20mL蒸馏水洗涤藻体振荡离心两次。将藻体用 20mL 蒸馏水溶解并煮沸 5min。冷却后将藻液在 13 000r/min 转速下离心 5min，保留清液。将该清液流过 200~400 目强碱性氯离子树脂，待清液与树脂交换完毕后用 15mL 0.06mol/L 盐酸洗脱。使用 70℃ 水浴在通风橱下将洗脱液中的水分挥发干，即可得到微藻 PGA 晶体。

微藻稳定碳同位素比值($\delta^{13}C$)的测定。将测定所需的微藻储备液在 1600g 下离心 10min，去掉上清液，用 3mL 1mol/L 盐酸浸泡微藻 24h 以去除无机碳。之后离心弃去酸液，用蒸馏水洗涤微藻两次离心收集微藻，将微藻放入超低温冰箱。用冷冻干燥机冻干微藻，取约 5mg 微藻样品，放入事先在马弗炉中烧制过的石英管，在石英管加入 3g 线状氧化铜后封管。将封好的装有微藻样品的石英管放入马弗炉中在 850℃ 放置 5h。冷却后将石英管中的二氧化碳气体提纯，封入样品管中，使用质谱仪 MAT252 测定 $\delta^{13}C$ 值(稳定碳同位素标准使用的是 Pee Dee Belemnite,‰ PDB)。其结果表示为

$$\delta^{13}C(‰)=(^{13}C/^{12}C_{样品}/^{13}C/^{12}C_{标准}-1)\cdot 1000$$

二、pH 对微藻生长的影响

莱茵衣藻和小球藻在 pH 梯度(5.3、6.0、6.5、7.5、8.0 和 9.0)处理下叶绿素 a 含量见图 3-1。其中，未处理过的 SE 培养液的 pH 为 6.5，可作为空白对照。

随着 pH 的升高，莱茵衣藻和小球藻的叶绿素 a 含量都在上升(图 3-1)。尤其是莱茵衣藻，在 pH 高于 6.5 后其叶绿素 a 含量急剧上升。这说明高 pH 环境有利于微藻的光合作用。莱茵衣藻和小球藻在 pH 梯度处理下蛋白质含量见图 3-2。微藻蛋白质含量随 pH 变化的趋势与叶绿素 a 含量类似，都是随着 pH 的升高含量增加。莱茵衣藻的蛋白质含量在 pH 高于 6.5 后急剧上升。并且莱茵衣

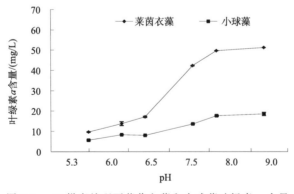

图 3-1　pH 梯度处理下莱茵衣藻和小球藻叶绿素 a 含量

Fig. 3-1　The content of chlorophyll a in *Chlamydomonas reinhardtii* and *Chlorella vulgaris* under pH gradient treatment

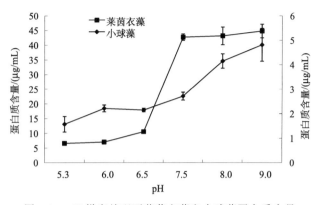

图 3-2 pH 梯度处理下莱茵衣藻和小球藻蛋白质含量

Fig. 3-2　The content of protein in *Chlamydomonas reinhardtii* and *Chlorella vulgaris* under pH gradient treatment

藻的叶绿素 a 含量和蛋白质含量都明显高于小球藻的。pH 为 8~9 的环境适于微藻的生长。可能是由于 pH 为 8.0~9.0 时有利于 CO_2 转化为 HCO_3^-，培养基中有着较高浓度的无机碳，可作为微藻生长的无机碳源。随着 pH 的升高，莱茵衣藻和小球藻的叶绿素 a 含量和蛋白质含量都在上升，说明偏碱环境适宜微藻的生长。

三、pH 对微藻碳酸酐酶胞外酶活力的影响

莱茵衣藻和小球藻在 pH 处理下 CAex 活性见图 3-3。总体而言，莱茵衣藻和小球藻在各 pH 中的生长趋势类似，都在 pH 为 6.5 时 CAex 活性最大。此时

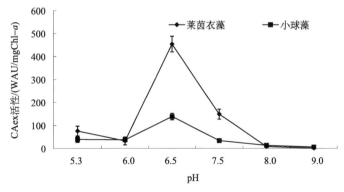

图 3-3　pH 梯度处理下莱茵衣藻和小球藻碳酸酐酶胞外酶活性

Fig. 3-3　The activity of CAex in *Chlamydomonas reinhardtii* and *Chlorella vulgaris* under pH gradient treatment

的 pH 为正常条件下的 SE 培养液 pH。可以看出，虽然在 pH 为 7.5、8.0 和 9.0 时，莱茵衣藻和小球藻都生长得较好，但此时的 CAex 活性却比较低。在酸性环境下，pH 影响无机碳存在形式，培养液中无机碳主要以 CO_2 形式存在，有足够的量进入细胞内，因此胞外碳酸酐酶活性低；在碱性 pH 时无机碳主要以 HCO_3^- 形式存在，且碱性环境下培养液中的 CO_2 可以源源不断地转化为 HCO_3^- 作为微藻的碳源，因此胞外碳酸酐酶活性也低。

pH 梯度处理下莱茵衣藻和小球藻培养液中各无机碳浓度见表 3-1。在 pH 为 5.3~6.0 时，微藻主要吸收的是 CO_2。在 pH 为 7.5~9.0 时微藻的无机碳源主要是 HCO_3^-。而在 pH 为 6.5 时微藻同时利用 CO_2 和 HCO_3^-。随着 pH 的升高，培养液中的总无机碳浓度在降低。

表 3-1　pH 梯度处理下莱茵衣藻和小球藻培养液中各无机碳浓度

Tab. 3-1　The concentration of different inorganic carbon species in medium of *Chlamydomonas reinhardtii* and *Chlorella vulgaris* under pH gradient treatment

	pH	$[HCO_3^-]$* /(mmol/L)	$[CO_2]$** /(mmol/L)	Total DIC /(mmol/L)
莱茵衣藻	5.30	0.08	2.85	2.93
	6.00	0.32	2.40	2.72
	6.50	0.80	0.61	1.41
	7.50	1.36	0.18	1.54
	8.00	1.52	0.06	1.58
	9.00	1.61	0.01	1.62
小球藻	5.30	0.08	2.28	2.36
	6.00	0.32	1.92	2.24
	6.50	0.80	0.61	1.41
	7.50	1.36	0.18	1.54
	8.00	1.52	0.06	1.58
	9.00	1.61	0.01	1.62

* 为培养液中 HCO_3^- 浓度；** 为培养液中 CO_2 浓度。

* is the HCO_3^- concentration in medium; ** is the CO_2 concentration in medium.

四、pH 对微藻稳定碳同位素分馏的影响

不同的 pH 对莱茵衣藻和小球藻 $\delta^{13}C$ 的影响如图 3-4。小球藻 $\delta^{13}C$ 值在 pH 为 5.3 时最偏正，而莱茵衣藻 $\delta^{13}C$ 值在 pH 为 7.5 时最偏正。在低 pH(5.3~6.5)环境培养下的微藻，微藻利用的无机碳源主要是游离的 CO_2，因此其 $\delta^{13}C$

值较负；而在高 pH（7.5～9.0）培养下的微藻，微藻利用的无机碳源主要是 HCO_3^-，因此其 $\delta^{13}C$ 值较偏正。

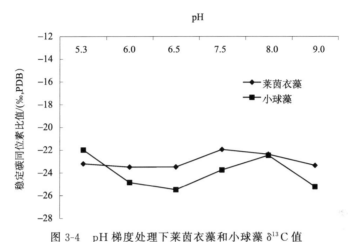

图 3-4　pH 梯度处理下莱茵衣藻和小球藻 $\delta^{13}C$ 值

Fig. 3-4　The $\delta^{13}C$ of *Chlamydomonas reinhardtii* and *Chlorella vulgaris* under pH gradient treatment

藻体的 $\delta^{13}C$ 值与 PGA 的 $\delta^{13}C_{PGA}$ 值相比，偏负 4‰～8‰。PGA 是光合作用中 Calvin 循环的第一个稳定中间产物，而藻体是后续光合代谢的产物，因此二者的 $\delta^{13}C$ 值存在转化和代谢的分馏差值。温度和 HCO_3^- 对由 Rubisco 引起的碳同位素分馏没有影响，PGA 分离纯化对 $\delta^{13}C$ 值也没有影响（Christeller et al.，1976）。PGA 是光合作用中 Calvin 循环的第一个稳定中间产物，而藻体是后续光合代谢的产物，因此二者的 $\delta^{13}C$ 值存在转化和代谢的分馏差值。

不同的 pH 对莱茵衣藻和小球藻 $\delta^{13}C_{PGA}$ 的影响如图 3-5。从图 3-5 中可以看出，在 pH 为 6.0～8.0 时，莱茵衣藻和小球藻 $\delta^{13}C_{PGA}$ 值均在 pH 为 7.5 时最偏正，在 pH 6.0 时最负，这表明微藻利用空气二氧化碳比例较高时，$\delta^{13}C$ 值偏负，而利用碳酸氢根离子比例较高时则呈现出偏正的现象。另外，除了 pH 为 5.3 之外，莱茵衣藻和小球藻 $\delta^{13}C_{PGA}$ 随 pH 变化的趋势类似，尤其是在 pH 为 6.0～8.0 时，莱茵衣藻与小球藻 $\delta^{13}C_{PGA}$ 值非常接近，平均相差仅为 0.14 ‰ PDB，与测定误差相同，表明莱茵衣藻与小球藻 $\delta^{13}C_{PGA}$ 在 pH 为 6.0～8.0 时相同，说明能够反映稳定碳同位素分馏的即时情况的 $\delta^{13}C_{PGA}$ 值能更准确反映 DIC 与稳定碳同位素比值的关系。

综合 pH 对莱茵衣藻与小球藻碳酸酐酶胞外酶活力以及 $\delta^{13}C_{PGA}$ 和 $\delta^{13}C$ 的影响研究结果可以看出，pH 对微藻稳定碳同位素分馏的影响大于碳酸酐酶胞外酶；同时还可以看出，pH 对微藻 ^{13}C 的影响与对碳酸酐酶胞外酶的影响存在"相

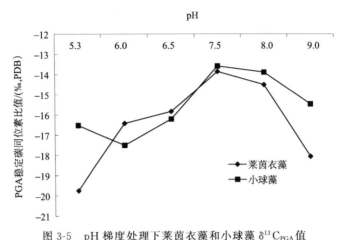

图 3-5　pH 梯度处理下莱茵衣藻和小球藻 $\delta^{13}C_{PGA}$ 值

Fig. 3-5　The $\delta^{13}C_{PGA}$ of *Chlamydomonas reinhardtii* and *Chlorella vulgaris* under pH gradient treatment

差",也即影响碳酸酐酶胞外酶活性的 pH 为影响微藻 ^{13}C 值的前一个处理,这可能因为碳酸酐酶胞外酶活性是即时值,而 $\delta^{13}C$ 值是累积值。而在 pH 为 7.5~8.0 时,两种微藻的 $\delta^{13}C_{PGA}$ 和 $\delta^{13}C$ 都趋于稳定。这为我们后期研究碳酸酐酶对微藻稳定碳同位素分馏研究提供依据。也即在研究碳酸酐酶对微藻稳定碳同位素分馏时,可选择 pH 为 8.0(接近岩溶湖泊和海洋环境)的培养条件。测定稳定碳同位素组成时,也只需简单测定藻体的 $\delta^{13}C$,而不需要繁琐地测定出微藻的 $\delta^{13}C_{PGA}$。

第三节　乙酰唑胺作用下重碳酸盐对微藻稳定碳同位素分馏的影响

一、培养条件和处理方法

将 $NaHCO_3$ 浓度梯度设置为 0mmol/L、0.5mmol/L、2mmol/L、8mmol/L、16mmol/L 和 20mmol/L,并给每一浓度梯度平行添加饱和乙酰唑胺溶液(AZ)处理。添加的 $NaHCO_3$ 的 $\delta^{13}C$ 均值为 -17.43 ‰。实验材料为莱茵衣藻和小球藻。培养条件是培养液采用 SE(液态)培养液;光周期为 12h/12h;室温保持 $24℃\pm1.0℃$;光照强度为 $97\mu mol/(m^2 \cdot s)$。每种微藻按上述条件分别扩大培养 10 天,换处理液。为加强处理效果,每 4 天换一次处理液,如此处理两次,活化培养两天,即可开始测定各项指标。分别测定其 CAex 活性、叶绿素含量、蛋白质含量和藻体的 $\delta^{13}C$ 值。

二、乙酰唑胺作用下重碳酸盐对微藻生长的影响

(一) 对微藻叶绿素 a 含量的影响

在无 AZ 的作用下，重碳酸盐对莱茵衣藻和小球藻叶绿素 a 含量的影响见图 3-6。莱茵衣藻的叶绿素 a 含量在 0.5mmol/L $NaHCO_3$ 处理下有一个峰值，这也是一个转折点，随着 $NaHCO_3$ 浓度的升高，叶绿素 a 含量平缓下降。这说明莱茵衣藻在 0.5mmol/L $NaHCO_3$ 环境下适宜光合作用。小球藻的叶绿素 a 含量与莱茵衣藻的表现明显不同，0~8mmol/L $NaHCO_3$，小球藻的叶绿素 a 含量随 $NaHCO_3$ 浓度的增加而迅速上升，随后保持在一个稳定水平，说明小球藻在高浓度 $NaHCO_3$（8~20mmol/L）环境下适宜光合作用，小球藻对高浓度重碳酸盐的耐受性高于莱茵衣藻。

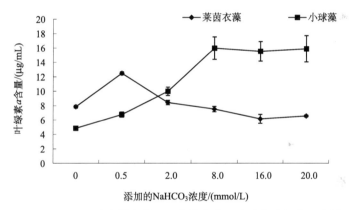

图 3-6　$NaHCO_3$ 浓度梯度处理下莱茵衣藻和小球藻叶绿素 a 含量

Fig. 3-6　The content of chlorophyll *a* in *Chlamydomonas reinhardtii* and *Chlorella vulgaris* under $NaHCO_3$ gradient treatment

在 AZ 的作用下，重碳酸盐对莱茵衣藻和小球藻叶绿素 a 含量的影响见图 3-7。莱茵衣藻叶绿素 a 含量随着 $NaHCO_3$ 浓度的升高而缓慢升高，当 $NaHCO_3$ 浓度超过 8mmol/L 时，叶绿素 a 含量不再增加而趋于稳定。小球藻的情况与莱茵衣藻的类似。在 8mmol/L 时莱茵衣藻和小球藻的叶绿素 a 含量均达到最高，说明在饱和 AZ 处理下 8mmol/L $NaHCO_3$ 浓度是微藻光合作用的饱和点。与未添加 AZ 的对应处理相比，莱茵衣藻对碳酸氢钠更具耐受性，高浓度 $NaHCO_3$ 不影响莱茵衣藻的光合作用(Ghoshal et al., 2002)。而对小球藻来说，添加 AZ 与不添加 AZ，叶绿素 a 含量都具有类似的变化趋势，这表明莱茵衣藻中碳酸酐酶在藻类生长中起的作用大于小球藻中的碳酸酐酶的作用。

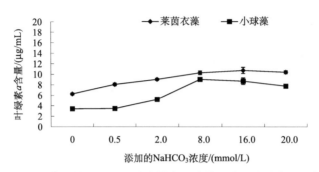

图 3-7　饱和 AZ 作用的 $NaHCO_3$ 浓度梯度下莱茵衣藻和小球藻叶绿素 a 含量

Fig. 3-7　The content of chlorophyll a in *Chlamydomonas reinhardtii* and *Chlorella vulgaris* under $NaHCO_3$ gradient treatment with saturated AZ

（二）对微藻蛋白质含量的影响

图 3-8 表示的是在没有 AZ 的情况下，添加不同浓度的 $NaHCO_3$ 对莱茵衣藻和小球藻蛋白质含量的影响。从图 3-8 可以看出，在添加 0～8mmol/L $NaHCO_3$ 的情况下，莱茵衣藻和小球藻蛋白质含量都随着 $NaHCO_3$ 浓度的增加而增加，8mmol/L $NaHCO_3$ 是一个转折点。当 $NaHCO_3$ 浓度大于 8mmol/L 时，微藻的蛋白质含量开始下降，说明 8mmol/L $NaHCO_3$ 是微藻生长的一个饱和点。

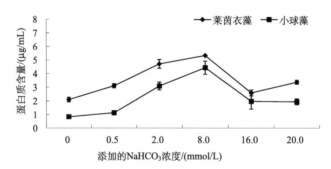

图 3-8　$NaHCO_3$ 浓度梯度处理下莱茵衣藻和小球藻蛋白质含量

Fig. 3-8　The content of protein in *Chlamydomonas reinhardtii* and *Chlorella vulgaris* under $NaHCO_3$ gradient treatment

图 3-9 表示的是在 AZ 的作用下，添加不同浓度的 $NaHCO_3$ 对莱茵衣藻和小球藻蛋白质含量的影响。低浓度下，莱茵衣藻和小球藻的蛋白质含量均随着 $NaHCO_3$ 浓度的升高而升高，在高浓度下趋于稳定。莱茵衣藻蛋白质含量在 16～20mmol/L $NaHCO_3$ 浓度时保持着较高值，小球藻在 8～20mmol/L $NaHCO_3$ 浓

度下保持着较高值,小球藻对碳酸氢钠更具耐受性。无论是莱茵衣藻还是小球藻,与未添加 AZ 的对应处理相比,添加 AZ 的处理对碳酸氢钠都具有更高的耐受性。

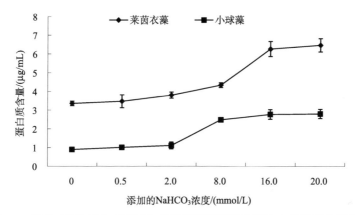

图 3-9　饱和 AZ 作用的 $NaHCO_3$ 浓度梯度下莱茵衣藻和小球藻蛋白质含量

Fig. 3-9　The content of protein in *Chlamydomonas reinhardtii* and *Chlorella vulgaris* under $NaHCO_3$ gradient treatment with saturated AZ

三、乙酰唑胺作用下重碳酸盐对微藻碳酸酐酶胞外酶活力的影响

图 3-10 表示的是在没有 AZ 的情况下,添加不同浓度的 $NaHCO_3$ 对莱茵衣藻和小球藻 CAex 的影响。低浓度 $NaHCO_3$(0~2mmol/L)下,莱茵衣藻的活性

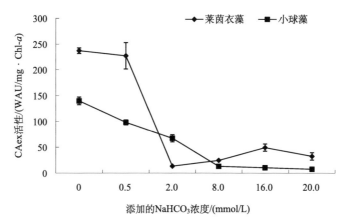

图 3-10　$NaHCO_3$ 浓度梯度处理下莱茵衣藻和小球藻 CAex 活性

Fig. 3-10　The activity of CAex in *Chlamydomonas reinhardtii* and *Chlorella vulgaris* under $NaHCO_3$ gradient treatment

随 NaHCO₃ 浓度的升高而降低，2mmol/L 是酶活性的转折点。在 2～20mmol/L NaHCO₃ 浓度中，莱茵衣藻的 CAex 活性几乎为零（考虑测定误差）。而小球藻 CAex 活性变化的趋势和莱茵衣藻类似，8mmol/L NaHCO₃ 是酶活性的转折点，从 8mmol/L 到 20mmol/L 浓度的 NaHCO₃ 处理中，小球藻 CAex 活性也几乎为零（考虑测定误差）。这说明高浓度的 HCO_3^- 会抑制微藻的 CAex 活性，而低浓度的 HCO_3^- 能诱导微藻的 CAex 活性。小球藻 CAex 要比莱茵衣藻的 CAex 更耐受高浓度的 HCO_3^-，这与前面叶绿素 a 研究结果一致。莱茵衣藻的叶绿素 a 含量变化的趋势和其 CAex 活性变化的趋势类似，这表明低浓度重碳酸盐环境下莱茵衣藻比小球藻更依赖 CAex 来摄取无机碳。

图 3-11 表示的是在 AZ 的作用下，添加不同浓度的 NaHCO₃ 对莱茵衣藻和小球藻胞外碳酸酐酶活性的影响。在误差范围内，两个微藻的 CAex 活性也几乎为零，这表明 AZ 可以完全抑制胞外碳酸酐酶活性。

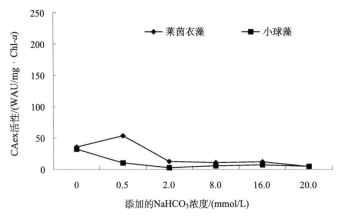

图 3-11　饱和 AZ 作用下 NaHCO₃ 浓度梯度下莱茵衣藻和小球藻 CAex 活性

Fig. 3-11　The activity of CAex in *Chlamydomonas reinhardtii* and *Chlorella vulgaris* under NaHCO₃ gradient treatment with saturated AZ

四、乙酰唑胺作用下重碳酸盐对微藻稳定碳同位素分馏的影响

图 3-12 表示的是在没有 AZ 的情况下，添加不同浓度的 NaHCO₃ 对莱茵衣藻和小球藻藻体 $\delta^{13}C$ 值的影响。从图 3-12 中可以看出，未添加 AZ 处理的藻体 $\delta^{13}C$ 值与添加 AZ 处理的藻体 $\delta^{13}C$ 值明显不同。在低浓度 HCO_3^-（0～2mmol/L）处理中，藻体 $\delta^{13}C$ 值各处理之间不存在显著差异，同一处理两种藻体之间也不存在显著差异；未添加 AZ 的藻体 $\delta^{13}C$ 值要比添加 AZ 的藻体 $\delta^{13}C$ 值偏正，两者差异约为 10‰（PDB）。另外，无论添加 AZ 与否，在 8mmol/L

HCO₃⁻ 处理中两种微藻的 $\delta^{13}C$ 值很接近（－25.4‰）。而在 16mmol/L 和 20mmol/L HCO₃⁻ 处理中，未添加 AZ 的藻体 $\delta^{13}C$ 值要比添加 AZ 的藻体 $\delta^{13}C$ 值偏负。由微藻的 $\delta^{13}C$ 值可以看出，8mmol/L HCO₃⁻ 处理能够完全抑制微藻的碳酸酐酶胞外酶，但这与实测的碳酸酐酶活力存在一个"相差"，也即 2mmol/L HCO₃⁻ 处理的微藻应该有活力。这与碳酸酐酶胞外酶活性是即时值，$\delta^{13}C$ 值是累积值有关。因为，培养4天时，我们能够测定出 2mmol/L HCO₃⁻ 处理的微藻碳酸酐酶胞外酶有较高的活力。

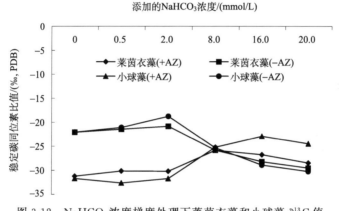

图 3-12　NaHCO₃ 浓度梯度处理下莱茵衣藻和小球藻 $\delta^{13}C$ 值

Fig. 3-12　The $\delta^{13}C$ of *Chlamydomonas reinhardtii* and *Chlorella vulgaris* under NaHCO₃ gradient treatment

表 3-2 表示的是 NaHCO₃ 浓度梯度处理下培养液中 HCO₃⁻ 和 CO₂ 转化情况。从表 3-2 中可以看出无论是莱茵衣藻还是小球藻的培养液中，总溶解性无机碳（total DIC）浓度随着添加的 NaHCO₃ 浓度的增大而增大。在低浓度 NaHCO₃（0～2mmol/L）处理中，由培养液中的 CO_2 转化而来的 HCO₃⁻ 所占比例较大，而在高浓度 NaHCO₃（8～20mmol/L）处理中，由培养液中的 CO_2 转化而来的 HCO₃⁻ 所占比例急剧减小，甚至出现了由小部分所添加的 HCO₃⁻，转化为 CO_2 而释放出培养体系的现象。

微藻的无机碳源有两个：一个来源于空气的 CO_2；另一个来源于添加的 NaHCO₃。而处于低浓度 NaHCO₃（0～2mmol/L）处理中的微藻主要是通过碳酸酐酶的 CCM 机制来获取无机碳；由于在没有 CAex 催化的情况下 CO_2 与 HCO₃⁻ 之间的转化会产生大约 10‰ 的稳定碳同位素分馏（Mook et al.，1974）；因此在低浓度 HCO₃⁻（0～2mmol/L）处理中添加 AZ 的藻体 $\delta^{13}C$ 值要比未添加 AZ 的藻体 $\delta^{13}C$ 值偏负 10‰。而处于高浓度 NaHCO₃（8～20mmol/L）处理中的微藻，无机碳源主要来源于 HCO₃⁻，这种 HCO₃⁻ 包含两部分，一部分来源于添加的

NaHCO$_3$，另一部分来源于空气中 CO$_2$ 转化的，这两部分的 HCO$_3^-$ 的 δ^{13}C 值有着较大的差异，添加的 NaHCO$_3$ 有着更负的 δ^{13}C 值，藻体利用两者的比例不同，藻体 δ^{13}C 值就呈现不同。由于藻体利用两者的比例与添加的 NaHCO$_3$ 浓度呈线性关系，因此，在高浓度 NaHCO$_3$（8～20mmol/L）处理中，能观察到微藻 δ^{13}C 值与添加的 NaHCO$_3$ 浓度成较好的线性关系。

表 3-2　NaHCO$_3$ 浓度梯度处理下培养液中 HCO$_3^-$ 和 CO$_2$ 转化情况
Tab. 3-2　The conservation situation between HCO$_3^-$ and CO$_2$ under NaHCO$_3$ gradient treatment

	pH	[HCO$_3^-$]* /(mmol/L)	[HCO$_3^-$]** /(mmol/L)	[HCO$_3^-$]*** /(mmol/L)	[CO$_2$]**** /(mmol/L)	Total DIC /(mmol/L)
莱茵衣藻	6.50	0.00	0.80	0.80	0.61	1.41
	6.87	0.50	1.60	1.10	0.52	2.12
	7.75	2.00	2.90	0.90	0.12	3.02
	8.54	8.00	8.20	0.20	0.06	8.26
	8.82	16.00	15.70	−0.30	0.06	15.76
	8.88	20.00	19.40	−0.60	0.06	19.46
小球藻	6.50	0.00	0.80	0.80	0.61	1.41
	6.87	0.50	1.45	0.95	0.47	1.92
	7.75	2.00	2.70	0.70	0.11	2.81
	8.54	8.00	8.20	0.20	0.06	8.26
	8.82	16.00	15.45	−0.55	0.06	15.51
	8.88	20.00	19.5	−0.50	0.06	19.56

＊培养液中添加的 HCO$_3^-$ 浓度；＊＊培养液中 HCO$_3^-$ 的实测浓度；＊＊＊由培养液中的 CO$_2$ 转化而来的 HCO$_3^-$ 浓度；＊＊＊＊培养液中游离的 CO$_2$ 浓度。H$_2$CO$_3$ 的一级解离常数是 6.38(25℃)。

＊ the added HCO$_3^-$ concentration in medium；＊＊ the determined HCO$_3^-$ concentration in medium；＊＊＊ the HCO$_3^-$ concentration converted by CO$_2$ in medium；＊＊＊＊ the dissociative CO$_2$ concentration in medium. The primary dissociation constant of H$_2$CO$_3$ is 6.38(25℃).

表 3-3 表示的是添加饱和 AZ 时 NaHCO$_3$ 浓度梯度处理下培养液中 HCO$_3^-$ 和 CO$_2$ 转化情况。从表 3-3 可以看出，添加饱和 AZ 时，培养液的酸度增加。总溶解性无机碳(total DIC)浓度也同样随着添加的 NaHCO$_3$ 浓度的增大而增大，总体无机碳转化趋势与未添加 AZ 的一致。由于酸度增加，在高浓度 HCO$_3^-$（8～20mmol/L）处理下的培养液更合适藻类的生长，因此，在添加 AZ 时，高浓度 HCO$_3^-$（8～20mmol/L）并没有抑制两个藻类的生长。至于在添加 AZ 时，高浓度 HCO$_3^-$（16～20mmol/L）处理下小球藻较莱茵衣藻的 δ^{13}C 值偏正，目前还不清楚它的机制。

添加 AZ 与不添加 AZ，在 8mmol/L HCO_3^- 处理下，两种微藻的 $\delta^{13}C$ 值相同，这表明两种藻类在添加 AZ 与不添加 AZ 的情况下都没有胞外碳酸酐酶的活力，这与前面的胞外碳酸酐酶活力测得的结果是一致的，也是碳酸酐酶胞外酶影响微藻同位素分馏的重要依据。

表 3-3　添加饱和 AZ 时 $NaHCO_3$ 浓度梯度处理下培养液中 HCO_3^- 和 CO_2 转化情况

Tab. 3-3　The conservation situation between HCO_3^- and CO_2 under $NaHCO_3$ gradient treatment with saturated AZ

	pH	$[HCO_3^-]^*$ /(mmol/L)	$[HCO_3^-]^{**}$ /(mmol/L)	$[HCO_3^-]^{***}$ /(mmol/L)	$[CO_2]^{****}$ /(mmol/L)	Total DIC /(mmol/L)
莱茵衣藻 +AZ	6.00	0.00	1.00	1.00	2.40	3.40
	6.40	0.50	1.80	1.30	1.72	3.52
	7.00	2.00	2.90	0.90	0.70	3.60
	7.62	8.00	8.40	0.40	0.48	8.88
	7.90	16.00	15.90	-0.10	0.48	16.38
	8.00	20.00	19.50	-0.50	0.47	19.97
小球藻 +AZ	6.00	0.00	0.80	0.80	1.92	2.72
	6.40	0.50	1.60	1.10	1.53	3.13
	7.00	2.00	2.80	0.80	0.67	3.47
	7.62	8.00	8.25	0.25	0.47	8.72
	7.90	16.00	15.70	-0.30	0.47	16.17
	8.00	20.00	19.50	-0.50	0.47	19.97

* 培养液中添加的 HCO_3^- 浓度；** 培养液中 HCO_3^- 的实测浓度；*** 由培养液中的 CO_2 转化而来的 HCO_3^- 浓度；**** 培养液中游离的 CO_2 浓度。H_2CO_3 的一级解离常数是 6.38(25℃)。

* The added HCO_3^- concentration in medium；** The determined HCO_3^- concentration in medium；
*** The HCO_3^- concentration converted by CO_2 in medium；**** The dissociative CO_2 concentration in medium. The primary dissociation constant of H_2CO_3 is 6.38(25℃).

第四节　乙酰唑胺对微藻稳定碳同位素分馏的影响

一、培养条件和处理方法

将乙酰唑胺（AZ）浓度梯度设置为 0、0.01mmol/L、0.1mmol/L、0.5mmol/L 和 1mmol/L，且在此基础上每一浓度梯度都添加 0.5mmol/L $NaHCO_3$。添加的 $NaHCO_3$ 的 $\delta^{13}C$ 均值为 -17.43‰。实验材料为莱茵衣藻和小球藻。培养条件是培养液采用 SE（液态）培养液；光周期为 12h/12h；室温保持 24℃±1.0℃；光照强度为 97μmol/($m^2 \cdot s$)。每种微藻按上述条件分别扩大培养 10 天,换处理液。为加强处理效果,每 4 天换一次处理液,如此处理两次,活化培养两天,即可开始测定各项指标。分别测定其 CAex 活性、叶绿素含量、蛋白质含量和藻体的 $\delta^{13}C$ 值。

二、乙酰唑胺对微藻生长的影响

图 3-13 表示的是不同浓度的 AZ 对莱茵衣藻和小球藻叶绿素 a 含量的影响。从图 3-13 中可以看出,在 0.1mmol/L AZ 作用下莱茵衣藻叶绿素 a 含量是最高的,浓度大于 0.1mmol/L 时 AZ 对莱茵衣藻叶绿素 a 含量的抑制变得较明显。小球藻的叶绿素 a 含量随着 AZ 浓度的升高而略有降低。不同浓度的 AZ 对莱茵衣藻和小球藻蛋白质含量的影响与对叶绿素 a 含量的影响相似(图 3-14)。

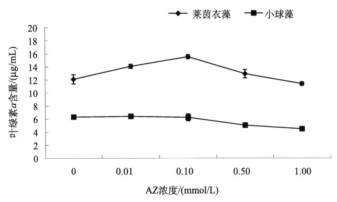

图 3-13　AZ 对莱茵衣藻和小球藻叶绿素 a 含量的影响
Fig. 3-13　The effect of AZ on the content of chlorophyll a in Chlamydomonas reinhardtii and Chlorella vulgaris

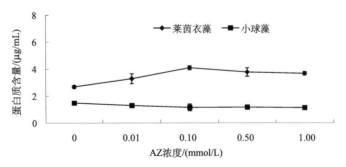

图 3-14　AZ 对莱茵衣藻和小球藻蛋白质含量的影响
Fig. 3-14　The effect of AZ on the content of protein in Chlamydomonas reinhardtii and Chlorella vulgaris

三、乙酰唑胺对微藻碳酸酐酶胞外酶活力的影响

图 3-15 表示的是 AZ 对莱茵衣藻和小球藻 CAex 活性的影响。从图 3-15 可知，莱茵衣藻在 AZ 浓度梯度处理下其 CAex 活性是随 AZ 浓度的增加而减少的。0.1mmol/L 是转折点，0.1~1mmol/L AZ 中，莱茵衣藻的 CAex 几乎没有活性。小球藻的情况类似于莱茵衣藻，当 AZ 浓度大于等于 0.1mmol/L 时，微藻的 CAex 活性就会完全受到抑制。

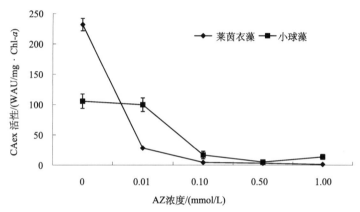

图 3-15 AZ 对莱茵衣藻和小球藻 CAex 活性的影响
Fig. 3-15 The effect of AZ on the activity of CAex in
Chlamydomonas reinhardtii and *Chlorella vulgaris*

四、乙酰唑胺对微藻稳定碳同位素分馏的影响

图 3-16 表示的是 AZ 对莱茵衣藻和小球藻 $\delta^{13}C$ 值的影响。从图 3-16 中可以看出，总体上看，除了 0.01mmol/L AZ 处理的莱茵衣藻之外，随着 AZ 浓度的增大，藻体 $\delta^{13}C$ 值减小。当 AZ 浓度为 0.5mmol/L 时，微藻的 $\delta^{13}C$ 值相比 0.1mmol/L 浓度的 AZ 处理时有大幅下降；而 AZ 浓度为 0.1mmol/L 时微藻的 CAex 就有着大幅度下降。这说明 0.5mmol/L AZ 这个点实现了碳酸酐酶胞外酶由部分抑制到完全抑制的转折。由于 CAex 被抑制，$\delta^{13}C$ 值偏负。由于碳酸酐酶胞外酶为即时值，而藻体 $\delta^{13}C$ 值为累积值，因此也存在着"相差"，也即微藻的 CAex 的转折点为 0.1mmol/L AZ 处理，而微藻的 $\delta^{13}C$ 值的转折点为 0.5mmol/L AZ 处理。

AZ 对莱茵衣藻和小球藻 $\delta^{13}C$ 值的影响与它们的无机碳存在形式、含量和比例无关。表 3-4 表示的是不同 AZ 浓度处理下培养液中 HCO_3^- 和 CO_2 转化情况。从表 3-4 中可以看出，AZ 的添加量的多少并没有影响各个处理培养液中无机碳存在形式、含量和比例。因此，AZ 处理对藻体稳定碳同位素分馏的效应应该与胞外

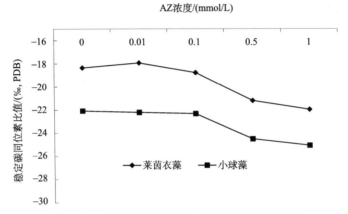

图 3-16　AZ 对莱茵衣藻和小球藻 $\delta^{13}C$ 值的影响

Fig. 3-16　The effect of AZ on the $\delta^{13}C$ of *Chlamydomonas reinhardtii* and *Chlorella vulgaris*

碳酸酐酶被抑制程度有关，进一步证明碳酸酐酶胞外酶对藻体稳定同位素的影响。

表 3-4　不同 AZ 浓度处理下培养液中 HCO_3^- 和 CO_2 转化情况

Tab. 3-4　The conservation situation between HCO_3^- and CO_2 under different AZ concentration

	pH	[AZ]*/(mmol/L)	[HCO_3^-]**/(mmol/L)	[HCO_3^-]***/(mmol/L)	[CO_2]****/(mmol/L)	Total DIC/(mmol/L)
莱茵衣藻	6.53	0.00	1.60	1.10	1.13	2.73
	6.47	0.01	1.50	1.00	1.22	2.72
	6.49	0.10	1.50	1.00	1.16	2.66
	6.43	0.50	1.40	0.90	1.25	2.65
	6.32	1.00	1.35	0.85	1.55	2.90
小球藻	6.53	0.00	1.45	0.95	1.03	2.48
	6.47	0.01	1.35	0.85	1.10	2.45
	6.49	0.10	1.35	0.85	1.05	2.40
	6.43	0.50	1.35	0.85	1.20	2.55
	6.32	1.00	1.30	0.80	1.49	2.79

* 培养液中的 AZ 浓度；** 培养液中 HCO_3^- 的实测浓度；*** 由培养液中的 CO_2 转化而来的 HCO_3^- 浓度；**** 培养液中游离的 CO_2 浓度。H_2CO_3 的一级解离常数是 6.38(25℃)。

* The added HCO_3^- concentration in medium； ** The determined HCO_3^- concentration in medium； *** The HCO_3^- concentration converted by CO_2 in medium ； **** The dissociative CO_2 concentration in medium. The primary dissociation constant of H_2CO_3 is 6.38(25℃).

参 考 文 献

丛海兵, 黄廷林, 周真明, 何文杰, 韩宏大. 2007. 藻类叶绿素测定新方法. 给水排水, 33: 28-32.

郭颖娜, 孙卫. 2008. 蛋白质含量测定方法的比较. 河北化工, 31(4): 36-37.

肖忠海, 王林, 汪海. 2008. 碳酸酐酶抑制剂乙酰唑胺的临床应用进展. 中国新药杂志, 17(16): 1390-1394.

Aizawa K, Miyachi S. 1986. Carbonic anhydrase aizawa and CO_2 concentrating mechanisms in microalgae and cyanobacteria. FEMS Microbiology Letters, 39: 215-233.

Amoroso G, Weber C, Sültemeyer D, Fock H. 1996. Intracellular carbonic anhydrase activities in *Dunaliella tertiolecta* (Butcher) and *Chlamydomonas reinhardtii* (Dangeard) in relation to inorganic carbon concentration during growth: further evidence for the existence of two distinct carbonic anhydrases associated with the chloroplasts. Planta, 199: 177-184.

Arens K. 1933. Physiologisch polarisierter massenaustausch und photosynthese bei submersen wasserpflanzen. I. Planta, 20: 621-658.

Arens K. 1936. Physiologisch polarisierter massenaustausch und photosynthese bei submersen wasserpflanzen. II. Die Ca(HCO_3^-)$_2$-assimilation. Jahrbucher fur Wissenschaftliche Botanik, 83: 513-560.

Badger D L, Price G D. 1994. The role of carbonic anhydrase in photosynthesis. Annual Review Plant Physiology, 45: 369-392.

Badger M R, Pfanz H, Büdel B, Heber U, Lange O L. 1993. Evidence for the functioning of photosynthetic CO_2-concentrating mechanisms in lichens containing green algal and cyanobacterial photobionts. Planta, 191: 57-70.

Badger M R, Price G D. 1992. The CO_2 concentrating mechanism in cyanobactiria and microalgae. Physiologia Plantarum, 84(4): 606-615.

Burkhardt S, Riebesell U, Zondervan I. 1999, Effects of growth rate, CO_2 concentration, and cell size on the stable carbon isotope fractionation in marine phytoplankton. Geochimica et Cosmochimica Acta, 63(22): 3729-3741.

Cassar N, Laws E, Popp B N. 2006. Carbon isotopic fractionation by the marine diatom phaeodactylum tricornutum under nutrient and light-limited growth conditions. Geochimica et Cosmochimica Acta, 70(21): 5323-5335.

Christeller J T, Laing W A, Troughton J H. 1976. Isotope discrimination by ribulose 1, 5-bisphosphate carboxylase-No effect of temperature or HCO_3^- concentration. Plant Physiology, 57: 580-582.

Coleman J R, Luinenburg I, Majeau N, Provart N. 1991. Sequence analysis and regulation of expression of a gene coding for carbonic anhydrase in *Chlamydomonas reinhardtii*. Canadian Journal of Botany-revue Canadienne de Botanique, 69: 1097-1102.

Colman B, Huertas I E, Bhatti S, Dason J S. 2002. The diversity of inorganic carbon acquisition mechanisms in eukaryotic microalgae. Functional Plant Biology, 29: 261-270.

Cousins A B, Badger M R, Von Caemmerer S. 2006. Carbonic anhydrase and its influence on carbon isotope discrimination during C_4 photosynthesis. Insight from antisense RNA in *Flaveria bidentis*. Plant Physiology, 41: 232-242.

Edwards G E, Mohamed A K. 1973. Reduction in carbonic anhydrase activity in zinc deficient leaves of *Phaseolus vulgaris* L. Crop. Science, 13: 351-354.

Emerson R, Green I. 1934. Manometric measurements of photosynthesis in the marine alaga Gigatina. The Journal of General Physiology, 17: 817-843.

Fisher M, Gokhman I, Pick U, Zamir A. 1996. A salt-resistant plasma membrane carbonic anhydrase is induced by salt in *Dunaliella salina*. Journal of Biological Chemistry, 271: 17718-17723.

Fisher M, Pick U, Zamir A. 1994. A salt-induced 60-kilodalton plasma membrane protein plays a potential role in the extreme halotolerance of the alga *Dunaliella*. Plant Physiology, 106: 1359-1365.

Francois R, Altabet M A, Goericke R, McCorkle D C, Brunet C, Poisson A. 1993. Changes in the $\delta^{13}C$ of surface water particulate organic matter across the subtropical convergence in the SW Indian Ocean. Global Biogeochemical Cycles, 7(3): 627-644.

Geib K, Golldack D, Gimmler H. 1996. Is there a requirement for an external carbonic anhydrase in the extremely acid-resistant green alga *Dunaliella acidophila*? European Journal of Phycology, 31: 273-284.

Ghoshal D, David Husic H, Goyal A. 2002. Dissolved inorganic carbon concentration mechanism in *Chlamydomonas moewusii*. Plant Physiology and Biochemistry, 40(4): 299-305.

Hewett-Emmett D, Tashian R E. 1996. Functional diversity, conservation, and convergence in the evolution of the α-, β-, and γ-carbonic anhydrase gene families. Molecular Phylogenetics and Evolution, 5: 50-77.

Keller K, Morel F M M. 1999. A model of carbon isotopic fractionation and active carbon uptake in phytoplankton. Marine Ecology Progress Series, 182: 295-298.

Laws E A, Popp B N, Bidigare R R, Kennicutt M C, Macko S A. 1995. Dependence of phytoplankton carbon isotopic composition on growth rate and $[CO_2]$aq: Theoretical consideration and experimental results. Geochimica et Cosmochimica Acta, 59(6): 1131-1138.

Laws E A, Popp B N, Cassar N, Tanimoto J. 2002. ^{13}C discrimination patterns in oceanic phytoplankton: likely influence of CO_2 concentration mechanisms, and implications for palaeoreconstructions. Functional Plant Biology, 29(2/3): 323-333.

Marlier J F, O'Leary M H. 1984. Carbon kinetic isotope effects on the hydraton of carbon dioxide and the dehydration of bicarbonate ion. Joural of American Chemical Society, 106: 5054-5057.

McNevin D B, Badger M R, Whitney S M, von Caemmerer S, Tcherkez G G B, Farquhar G D. 2007. Differences in carbon isotope discrimination of three variants of D-Ribulose-1, 5-bisphosphate carboxylase/oxygenase reflect differences in their catalytic mechanisms. Journal of Biological Chemistry, 282: 36068-36076.

Mook W G, Bommerson J C, Staverman W H. 1974. Carbon isotope fractionation between dissolved bicarbonate and gaseous carbon dioxide. Earth and Planetary Science Letters, 22: 169-176.

Moroney J V, Husic H D, Tolbert N E. 1985. Effect of carbonic anhydrase inhibitors on inorganic carbon accumulation by *Chlamydomonas reinhardtii*. Plant Physiology, 79: 177-183.

Nimer N A, Iglesias-Rodriguez M D, Merrett M J. 1997. Bicarbonate utilization by marine phytoplankton species. Journal of Phycology, 33(4): 625-631.

Nimer N A, Merrett M J. 1996. The development of a CO_2-concentrating mechanism in *Emiliania huxleyi*. New Phytologist, 133: 383-389.

Palmqvist K. 1993. Photosynthetic CO_2-use efficiency in lichens and their isolated photobionts: the possible role of a CO_2-concentrating mechanism. Planta, 191: 48-56.

Palmqvist K, Badger M R. 1996. Carbonic anhydrase(s) associated with lichens: in vivo activities, possible locations and putative roles. New phytologist, 132: 627-639.

Palmqvist K, Yu J W, Badger M R. 1994. Carbonic anhydrase activity and inorganic carbon fluxes in low and high Cicells of *Chlamydomonas reinhardtii* and *Scenedesmus obliquus*. Physiologia Plantarum, 90: 537-547.

Paneth P O, Leary M H. 1985. Carbon isotope effect on dehydration of bicarbonate ion catalyzed by carbonic anhydrase. Biochemistry, 24: 5143-5147.

Pesheva I, Kodama M, Dionisio-Sese M L, Miyachi S. 1994, Changes in photosynthetic characteristics induced by transferring air-grown cells of *Chlorococcum littorale* to high-CO_2 conditions. Plant and Cell Physiology, 35: 379-387.

Popp B N, Law E A, Dore J E, Hanson K L, Wakeham S G. 1998. Effect of phytoplankton cell geometry on carbon isotope fractionation. Geochimica et Cosmochimica Acta, 62(1): 69-77.

Ramazanov Z, Shiraiwa Y, Jiménez del Rio M, Rubio J. 1995. Effect of external CO_2 concentrations on protein synthesis in the green algae *Scenedesmus obliquus* (Turp.) Kütz and *Chlorella vulgaris* (Kosikov). Planta, 197: 272-277.

Randall P J, Bouma D. 1973. Zinc deficiency, carbonic anhydrase, and photosynthetesis in leves of spinach. Plant Physiology, 57: 229-232.

Rau G H, Riebesell U, Wolf-Gladrow D. 1997. CO_2(aq)-dependent photosynthetic ^{13}C fractionation in the ocean: A model versus measurements. Global Biogeochemical Cycles, 11(2): 267-278.

Rotatore R, Colman B, Kuzma M. 1995. The active uptake of carbon dioxide by the marine diatoms *Phaeodactylum ticornutum* and *Cyclotella sp*. Plant Cell &.Environment, 18: 913-918.

Sültemeyer D, Amoroso G, Fock H. 1995. Induction of intracellular carbonic anhydrases during the adaptation to low inorganic carbon concentrations in wild-type and ca-1 mutantcells of

Chlamydomonas reinhardtii. Planta, 196: 217-224.

Sültemeyer D, Schmidt C, Fock H P. 1993. Carbonic anhydrases in higher plants and aquatic microorganisms. Physiologia Plantarum, 88: 179-190.

Soltes-Rak E, Mulligan M E, Coleman J R. 1997. Identification and characterization of a gene encoding a vertebrate-type carbonic anhydrase in cyanobacteria. Journal of bacteriology, 179: 769-774.

Steemann-Nielsen E. 1946. Carbon sources in the photosynthesis of aquatic plant. Nature, 158: 594-596.

Suzuki K, Ikawa T. 1993. Oxygen enhanement of photosynthetic $14CO_2$ fixation in a freshwater diatom Nitzschia ruttneri. The Japanese Journal of Phycology, 41: 19-28.

Tseng C K, Sweeney B M. 1946. Physiological studies of Gelidium cartilagineum. I. Photosynthesis, with special reference to the carbon dioxide factor. American Journal of Botany, 33(9): 706-715.

Von Caemmerer S, Quinn V, Hancock N C, Price G D, Furbank R T, Ludwig M. 2004. Carbonic anhydrase and C4 photosynthesis: a transgenic analysis. Plant Cell and Environment, 27: 697-703.

Wilbur K M, Anderson N G. 1948. Electrometric and colorimetric determination of carbonic anhydrase. Journal Biological Chemistry, 176: 147-154.

Williams T G, Colman B. 1994. Rapid separation of carbonic anhydrase isozymes using cellulose acetate membrane electrophoresis. Journal of Experimental Botany, 45: 153-158.

Williams T G, Colman B. 1995. Quantification of the contribution of CO_2, HCO_3^-, and external carbonic anhydrase to photosynthesis at low dissolved inorganic carbon in Chlorella saccharophila. Plant physiology, 107(1): 245-251.

Wu Y Y, Zhao X Z, Li P P, Wang B L, Liu C Q. 2006. A study on the activities of carbonic anhydrase of two species of bryophytes, *Tortula sinensis* (Mull. Hal.) Broth. and *Barbula convoluta* Hedw. Cryptogamie Bryologie, 27 (3): 349-355.

第四章 微藻碳酸酐酶对碳源、碳汇的调节作用

摘　　要

　　水生生态系统是全球最大的生态系统，它所带来的初级生产力占到了全球净初级生产力的 50%，相应地，水生生态系统是全球最重要的碳源和碳汇之一。不同于陆生生态系统，水体环境缺乏 CO_2，生物为适应环境，形成了一些适应水体环境的机制：一些微藻形成了无机碳浓缩机制（CCMs）来提高 CO_2 浓度；一些微藻可以直接或间接利用环境中的碳酸氢根离子。这将有助于提高光合作用对二氧化碳的固定，进而提高初级生产力。在这些过程中，碳酸酐酶都起着非常关键的作用。有碳酸酐酶催化和无酶催化的藻体之间存在约 9‰的稳定碳同位素分馏差异。

　　利用双同位素示踪技术，通过比较莱茵衣藻和蛋白核小球藻碳同位素组成的变化，定量出微藻对不同无机碳源利用份额。通过添加两种抑制剂（乙酰唑胺（AZ）和 4,4'-二异硫氰-2,2'-二黄酸芪（DIDS）），计算微藻不同无机碳代谢途径份额。依据微藻生长速率、不同无机碳源利用份额和不同无机碳代谢途径份额，计算微藻的直接碳汇和间接碳汇。

　　不论莱茵衣藻还是蛋白核小球藻，它们利用添加的重碳酸盐占其总无机碳源的份额都较小。随着添加碳酸氢钠和 AZ 浓度的增加，利用添加的重碳酸盐份额呈增加趋势。添加 DIDS，会造成利用添加的重碳酸盐份额减小，在高浓度 DIDS（2.0mmol/L）条件下，微藻完全无法利用添加的碳酸氢根离子。微藻主要碳酸氢根离子利用途径来利用环境中的无机碳，它们的碳酸氢根离子利用途径所占的份额超过 76%，但是，添加过高浓度的碳酸氢钠（16.0mmol/L），会造成碳酸氢根离子利用途径所占的份额急剧减小。添加 AZ 或 DIDS，也会造成碳酸氢根离子利用途径所占的份额下降，在高浓度 DIDS（2.0mmol/L）条件下，直接利用碳酸氢根离子途径完全被抑制。

　　碳酸酐酶胞外酶主要是增加了微藻利用大气无机碳源的能力，而只是轻微增加了利用自然水体中固有的可溶性无机碳的能力。在海水和喀斯特自然水体中，微藻所利用的碳源主要来自大气 CO_2，它占到了微藻利用总无机碳份额的 92%。微藻利用水体固有的碳酸氢根离子可能与水体"遗失的碳汇"有关。

Chapter 4 The regulation on carbon source and carbon sequestration by microalgal carbonic anhydrase

Abstract

The aquatic ecosystem is the largest ecosystem and contributes to approximately 50% of the net primary productivity on earth. Accordingly, aquatic ecosystem is one of the most important carbon sinks and sources. Unlike the environment on land, CO_2 in aquatic media for microalgae is limited. The algae have formed a serial of mechanisms to adapt to the aquatic environment. Several microalgae have formed the carbon-concentrating mechanisms (CCMs) to increase the CO_2 concentrations. Some microalgae can utilize HCO_3^- directly or indirectly. Thus, these mechanisms can improve the fixation of the CO_2 during photosynthesis, subsequently, they can accelerate the net primary productivity. The carbonic anhydrase may play the key role in these processes. Approximately 9‰ of carbon isotope discrimination in algae was found between the catalyzed by carbonic anhydrase and uncatalyzed processes.

The proportion of two inorganic carbon sources consumed by microalgal species *Chlamydomonas reinhardtii* and *Chlorella pyrenoidosa* was quantified by comparing their stable carbon isotope compositions using the bidirectional isotope labeling method. The proportion of inorganic carbon utilization pathways was determined using two inhibitors: acetazolamide(AZ) and 4,4'-diisothiocyanatostilbene-2,2'-disulfonic acid(DIDS). Direct and indirect carbon sequestrations were calculated according to the algal growth rate and proportion of inorganic carbon sources and utilization pathways.

In both *C. reinhardtii* and *C. pyrenoidosa*, the proportion of the utilization on dissolved inorganic carbon (DIC) from the added bicarbonate (f_B) were very small. The f_B increased with the increasing $NaHCO_3$ and AZ added. However, the f_B decreased with the increasing DIDS added. Under 2.0 mmol/L DIDS condition, the algae cannot utilize DIC from the added $NaHCO_3$, completely. The microalgae mainly utilized DIC in aquatic environment via the bicarbonate utilization pathway. The proportion of the bicarbonate utilization pathway (f_b) was above 76%. Moreover, the f_b decreased sharply under excess added NaHCO$_3$. In addition, the f_b decreased with increasing AZ and DIDS. The bicarbonate utilization pathway way completely inhibited 2.0 mmol/L DIDS.

Extracellular carbonic anhydrase mainly increased the utilization of inorganic carbon from the atmosphere and slightly increased the utilization of bicarbonate pre-existing in nature aquatic media. In natural seawater and karst aquatic media, the microalgae mainly utilize CO$_2$ from the atmosphere, it account for 92% of the total carbon sequestration. Microalgae depleted the bicarbonate pre-existing in aquatic media may be involved in "missing carbon sink".

由于自然环境的多样性,微藻为适应这种多样性的环境进化出多样性的无机碳的吸收利用机制。在自然水体中,由碳酸酐酶参与的无机碳吸收(主要包括：二氧化碳自由扩散、碳酸氢根离子间接利用等)被认为是最重要的无机碳吸收模式(Axelsson et al.,1995;Colman et al.,2002)。由于不同微藻的碳酸酐酶活性及其利用无机碳的能力差异较大(Martin and Tortell,2008;Mercado et al.,1998),再加上不同微藻吸收无机碳的机制差异,这都造成不同微藻捕获大气CO$_2$的能力差异悬殊,并最终造成不同种类微藻的碳汇能力差异较大。为深入研究微藻碳酸酐酶对碳汇的调节作用,本研究通过设置不同浓度的碳酸酐酶胞外酶抑制剂 AZ、阴离子通道抑制剂 DIDS 作用和交互作用于微藻,探讨微藻吸收利用环境中的无机碳源的途径,定量计算微藻的各种碳汇能力,为人类寻找"丢失的碳汇"提供科学依据,为应对全球气候变化提供科学依据。

第一节 碳酸酐酶在无机碳代谢中的作用

一、碳酸酐酶与无机碳的转运

水体中可溶性无机碳(dissolved inorganic carbon,DIC)存在以下四种形态,分别是：CO_2,HCO_3^-,H_2CO_3 和 CO_3^{2-},它们之间存在如下化学平衡过程：

$$CO_2 + H_2O \leftrightarrow H_2CO_3 \leftrightarrow HCO_3^- + H^+ \leftrightarrow CO_3^{2-} + 2H^+ \tag{4-1}$$

自然水体中不同形态的无机碳之间可以相互转化,且处于动态平衡之中。一般情况下,自然水体中各种形态无机碳的比例变化主要受 pH、温度等因素的影响。

在常温条件下,pH<6.4 的酸性环境下,CO_2 占 DIC 的主体;随着 pH 升高,化学平衡向右移动,游离 CO_2 浓度逐渐降低,HCO_3^- 和 CO_3^{2-} 的浓度逐渐升高,在 6.4<pH<10.3 条件下,HCO_3^- 占 DIC 的主体;在 pH>10.3 的强碱性条件下，CO_3^{2-}占DIC的主体（Larkum et al.，1989；Skirrow and Whitfield，1975）。

虽然水体中存在CO_2，HCO_3^-，H_2CO_3和CO_3^{2-}四种形态的无机碳，但只有CO_2和HCO_3^-这两种形态的无机碳可被微藻吸收利用，另两种形态无机碳只有转化为这两种可被微藻利用的形态，才能最终成为可被生物固定的碳汇。

在自然水体中，CO_2占总DIC的量不到1‰（Riebesell et al.，1993），再加上CO_2在水体中的扩散速度很慢，已有报道显示，CO_2在水体中的扩散速度只有其在大气中扩散速度的万分之一（Larkum et al.，1989），总之，自然水体是一种缺乏CO_2的环境（Talling，1976）。HCO_3^-是水体中DIC的主要存在形式，尤其是在喀斯特岩溶湖泊水体中，它不仅是水体中二氧化碳的储存库，经自发脱水可保证游离二氧化碳源源不断地供应，而且还可以直接作为藻类外源无机碳的供应形式。伴随着微藻对水体中CO_2和HCO_3^-的不断消耗，其他形态的无机碳可以不断转化，以便满足微藻生长对碳源的需要。

CO_2作为线性非极性分子，呈电中性，它可以自由扩散进入细胞双层脂膜为微藻的光合作用所利用（图4-1）。

HCO_3^-是带负电的极性分子，它不能通过自由扩散进入细胞双层脂膜。HCO_3^-进入细胞需要通过直接转运或者间接转运的过程，然后才能为微藻光合作用所利用。HCO_3^-的直接转运指的是经过细胞质膜表面载体蛋白或阴离子交换蛋

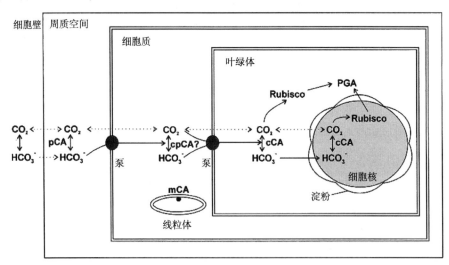

图4-1　微藻CO_2浓缩机制（CCM）模型

图片来源：Sültemeyer，1998

Fig. 4-1　The carbon dioxide centration mechanism (CCM) in algae

From Sültemeyer，1998

白，直接把 HCO_3^- 转运进入细胞，然后在细胞内经各种胞内碳酸酐酶转化为 CO_2 或由叶绿体膜蛋白直接把 HCO_3^- 主动转运到叶绿体内，再经叶绿体碳酸酐酶转化为 CO_2 供核酮糖-1,5 二磷酸羧化/加氧酶（Rubisco）固定（Sültemeyer，1998；Sharkia et al.，1994）。HCO_3^- 的间接转运是指 HCO_3^- 自发脱水而游离出 CO_2 的无机碳供应途径，HCO_3^- 在胞外碳酸酐酶的催化作用下迅速水解，它可以加快 CO_2 的形成和供应速度，然后再以 CO_2 的形式通过自由扩散进入细胞内，从而保证了细胞对 CO_2 的大量需求（Invers et al.，1999；Tsuzuki and Miyachi，1989）（图4-1）。

二、碳酸酐酶与光合作用

光合作用是地球上最重要的光能合成作用，是指陆生植物、藻类及一些细菌，在光照下，利用光合色素，把二氧化碳（或硫化氢）和水转化为有机物，并释放出氧气（或氢气）的过程（沈允钢，2006）。光合作用是一系列复杂的生物代谢反应的总和，是地球上最基本的物质代谢和能量代谢过程，也是生物界赖以生存的基础，总之，光合作用是碳的生物地球化学循环的最重要部分。

CA 在光合作用固定 CO_2 的过程中发挥着重要作用。自然条件下，CO_2 在大气中的含量很低，再加上 CO_2 自身的跨膜速度很慢等，这些因素都影响光合作用的速度。然而在有 CA 作用的情况下，它可以加快 CO_2 与 HCO_3^- 之间的相互转化，为光合作用的进行提供充足的 CO_2，进而促进生物体内的光合作用快速进行（Badger and Price，1994；Coleman，2004）。

在自然水体中，CO_2 占总 DIC 的量不到 1%（Riebesell et al.，1993），而且，已有相关研究结果报道，CO_2 在水体中的扩散速度只有其在大气中扩散速度的万分之一（Larkum et al.，1989），因此，自然水体是一种缺乏 CO_2 的环境（Talling，1976）。微藻为适应这种低 CO_2 的水体环境，它们形成了一种在细胞内提高 CO_2 浓度的机制——无机碳浓缩机制（Badger et al.，1998；Badger and Price，1992；Bozzo and Colman，2000；Colman et al.，2002；Giordano et al.，2005）。微藻 CCM 机制的关键作用就是浓缩 CO_2 及其把水体中的 HCO_3^- 转化为 CO_2，碳酸酐酶在其中发挥着关键作用。依靠 CCM 机制，微藻可以在水体低浓度 CO_2 的条件下达到较高的光合速率。

第二节 微藻无机碳利用途径的定量方法

一、微藻碳同位素测定方法及碳同位素修正模型

将各个处理后的微藻样品转入离心管（事先预处理过的）中，离心收集藻体

(1600g，10min)。加入 1.0mol/L 盐酸浸泡微藻 12 小时，离心弃去酸液，此过程再重复一次，以充分去除无机碳。接下来用蒸馏水洗涤微藻，再次离心收集微藻。最后，把处理后的样品放入冰箱，冷冻保存。同位素测试前，用冷冻干燥机充分干燥微藻样品。

取约 5mg 充分干燥的微藻样品，装入石英管(9×300mm)(850℃预灼烧 1h)中，然后加入 3g 线状氧化铜丝(850℃预灼烧 1h)，混匀(防微藻样品靠石英管壁造成灼烧过程中气体渗漏)，抽真空，封管。将封好的装有微藻样品的石英管放入马弗炉中灼烧(850℃，3h)，缓慢冷却后将石英管中获得的样品气体通过真空线装置提纯(经过酒精液氮、液氮等过程)。最终获得高纯二氧化碳气体样品，然后使用质谱仪 MAT252 测定其 $\delta^{13}C$ 值(稳定碳同位素标准使用的是 Pee Dee Belemnite，PDB)。其结果表示为

$$\delta^{13}C\ (‰) = [(R_{样品}/R_{标准}) - 1] \times 1000 \tag{4-2}$$

$R_{样品}$：样品中的 ^{13}C 与 ^{12}C 的比值；$R_{标准}$：标准品中的 ^{13}C 与 ^{12}C 的比值。

鉴于微藻生长的特殊性，我们很难分离新老藻体，造成最后测得的藻体碳同位素组成是一种包含处理前后的所有微藻藻体的混合物。为更加准确反映实验处理过程中新生成的藻体碳同位素组成，本研究建立了微藻碳同位素校正模型。

根据微藻培养前后的生物量增殖情况(一般采用蛋白质数据来表征生物量的变化)，培养前后的微藻碳同位素组成，我们建立了如下方程：

$$(N_0/N) \times \delta^{13}C_{N_0} + (1 - N_0/N) \times \delta^{13}C_{NA} = \delta^{13}C_N \tag{4-3}$$

N_0：初始接种时的微藻生物量；N：实验处理后的微藻生物量；$\delta^{13}C_{N_0}$：初始接种的微藻藻体碳同位素组成($\delta^{13}C$)；$\delta^{13}C_N$：实验处理后的微藻藻体碳同位素组成($\delta^{13}C$)；$\delta^{13}C_{NA}$：实验处理新生成的藻体碳同位素组成($\delta^{13}C$)。

本研究后续部分的藻体碳同位素组成数据都是经过此模型校正后的。

二、微藻利用不同碳源的份额计算方法

微藻的碳同位素组成可以反映其对环境中不同无机碳的利用(Chen et al.，2009)。一般情况下，微藻可利用的无机碳有两个来源，分别来自大气和水体。且对于每一种无机碳来源来说，都存在 CO_2 和 HCO_3^- 这两种形态的无机碳利用途径。具体到本实验中，莱茵衣藻和蛋白核小球藻都可以利用大气中的 CO_2 和水体中添加的 HCO_3^- 这两种无机碳源，再加上微藻对每种来源的无机碳都存在 CO_2 和 HCO_3^- 这两种无机碳利用途径。需要再补充一点的是，微藻在吸收利用 CO_2 和 HCO_3^- 这两种无机碳代谢途径之间存在约 9‰ 的稳定碳同位素分馏 (Emrich et al.，1970；Mook et al.，1974；Wu et al.，2012)。最终，微藻的稳定碳同位素组成可以表示为如下稳定碳同位素方程(详见公式 4-4 和公式 4-5)：

第四章 微藻碳酸酐酶对碳源、碳汇的调节作用

$$\delta_{TA} = (1 - f_{bai})\delta_a + f_{bai}(\delta_a + 9‰)(i = 1, 2) \tag{4-4}$$

$$\delta_{TB} = (1 - f_{bbi})\delta_{ai} + f_{bbi}(\delta_{ai} + 9‰)(i = 1, 2) \tag{4-5}$$

δ_{TA}是指微藻利用大气来源的无机碳，包括CO_2和HCO_3^-这两种利用途径；δ_{TB}是指微藻利用添加的碳酸氢钠来源的无机碳，同样包括CO_2和HCO_3^-这两种利用途径；f_{bai}是指微藻利用大气中的CO_2，溶于水体后，且进行HCO_3^-途径的份额；$(1 - f_{bai})$是指微藻利用大气中的CO_2，且进行CO_2途径的份额；f_{bbi}是指微藻利用添加的HCO_3^-，且进行HCO_3^-途径的份额；$(1 - f_{bbi})$是指微藻利用添加的HCO_3^-，且进行CO_2途径的份额。我们基于如下假设：在同一处理下，对于同一种微藻来说，它不论利用大气中的无机碳还是利用添加的无机碳，两种无机碳利用途径(CO_2或HCO_3^-)占利用总无机碳的份额都是相同的，因此，$f_{bai} = f_{bbi} = f_{bi}$。$\delta_a$为微藻完全利用大气中的$CO_2$，且完全进行$CO_2$途径时藻体的$\delta^{13}C$值；$(\delta_a + 9‰)$为微藻完全利用大气中的$CO_2$，且完全进行$HCO_3^-$途径时藻体的$\delta^{13}C$值；$\delta_{ai}$为微藻利用添加的某一无机碳源，且完全进行$CO_2$途径时藻体的$\delta^{13}C$值；$(\delta_{ai} + 9‰)$为微藻利用添加的某一无机碳源，且完全进行$HCO_3^-$途径时藻体的$\delta^{13}C$值。

对于自然水体中的微藻来说，它们都可以利用大气的无机碳源和水体中固有的无机碳源，且对于每一种来源的无机碳的利用，都存在CO_2和HCO_3^-两种无机碳利用途径，为此，我们建立如下两端元的同位素混合模型(详见公式4-6)：

$$\begin{aligned}\delta_{Ti} &= (1 - f_{Bi})\delta_{TA} + f_{Bi}\delta_{TB} \\ &= (1 - f_{Bi})[(1 - f_{bi})\delta_a + f_{bi}(\delta_a + 9‰)] \\ &\quad + f_{Bi}[(1 - f_{bi})\delta_{ai} + f_{bi}(\delta_{ai} + 9‰)](i = 1, 2)\end{aligned} \tag{4-6}$$

δ_{Ti}为添加已知$\delta^{13}C$的某一碳酸氢钠培养的微藻藻体的$\delta^{13}C$值，f_{Bi}为微藻利用添加的无机碳占总碳源的份额，$(1 - f_{Bi})$为微藻利用大气CO_2占总碳源的份额。

在同一条件下培养的同一种微藻，对于添加的两种标记的重碳酸盐来说，方程(4-6)可以分别表示如下：

$$\begin{aligned}\delta_{T1} &= (1 - f_{B1})[(1 - f_{b1})\delta_a + f_{b1}(\delta_a + 9‰)] \\ &\quad + f_{B1}[(1 - f_{b1})\delta_{a1} + f_{b1}(\delta_{a1} + 9‰)]\end{aligned} \tag{4-7}$$

$$\begin{aligned}\delta_{T2} &= (1 - f_{B2})[(1 - f_{b2})\delta_a + f_{b2}(\delta_a + 9‰)] \\ &\quad + f_{B2}[(1 - f_{b2})\delta_{a2} + f_{b2}(\delta_{a2} + 9‰)]\end{aligned} \tag{4-8}$$

在方程(4-7)和(4-8)中，δ_{T1}为添加第一种已知$\delta^{13}C$的碳酸氢钠培养的微藻藻体的$\delta^{13}C$值；δ_{T2}为添加第二种已知$\delta^{13}C$的碳酸氢钠培养的微藻藻体的$\delta^{13}C$值；f_{B1}为微藻利用添加的第一种碳酸氢钠占总碳源的份额；f_{B2}为微藻利用添加

的第二种碳酸氢钠占总碳源的份额；f_{b1} 为培养在添加第一种已知 $\delta^{13}C$ 的碳酸氢钠的培养液中的微藻利用碳酸氢根离子途径所占的份额；f_{b2} 为培养在添加第二种已知 $\delta^{13}C$ 的碳酸氢钠的培养液中的微藻利用碳酸氢根离子途径所占的份额；δ_{a1} 为培养在添加第一种已知 $\delta^{13}C$ 的碳酸氢钠的培养液中的微藻利用添加的重碳酸盐，且完全进行 CO_2 途径时藻体的 $\delta^{13}C$ 值；δ_{a2} 为培养在添加第二种已知 $\delta^{13}C$ 的碳酸氢钠的培养液中的微藻利用添加的重碳酸盐，且完全进行 CO_2 途径时藻体的 $\delta^{13}C$ 值；$(\delta_{a1}+9‰)$ 为培养在添加第一种已知 $\delta^{13}C$ 的碳酸氢钠的培养液中的微藻利用添加的重碳酸盐，且完全进行 CO_2 途径时藻体的 $\delta^{13}C$ 值；$(\delta_{a2}+9‰)$ 为培养在添加第二种已知 $\delta^{13}C$ 的碳酸氢钠的培养液中的微藻利用添加的重碳酸盐，且完全进行二氧化碳途径时藻体的 $\delta^{13}C$ 值。

基于以下两点认识：①不论添加哪种标记的碳酸氢钠，只要在同一培养条件下，同一种微藻利用添加的重碳酸盐所占总碳源的份额是相同的，由此可以得到：$f_{B1}=f_{B2}=f_B$。②不论添加哪种标记的碳酸氢钠，只要在同一培养条件下，同一种微藻利用碳酸氢根离子途径所占的份额是相同的，由此可以得到：$f_{b1}=f_{b2}=f_b$。在此基础上，由方程(4-7)与方程(4-8)做差、化简可得：

$$f_B = \frac{\delta_{T1}-\delta_{T2}}{\delta_{a1}-\delta_{a2}} \quad (4-9)$$

δ_{a1} 为培养在添加第一种已知 $\delta^{13}C$ 的碳酸氢钠的培养液中的微藻利用添加的重碳酸盐，且完全进行二氧化碳途径时藻体的 $\delta^{13}C$ 值；δ_{a2} 为培养在添加第二种已知 $\delta^{13}C$ 的碳酸氢钠的培养液中的微藻利用添加的重碳酸盐，且完全进行二氧化碳途径时藻体的 $\delta^{13}C$ 值。虽然 δ_{a1} 和 δ_{a2} 的数值很难获得，但是，它们之间的差别是由于添加的两种标记 $\delta^{13}C$ 的碳酸氢钠所造成的，因此 $(\delta_{a1}-\delta_{a2})$ 则可以换算成添加的同位素标记 1 的碳酸氢钠的 $\delta^{13}C$ 值(δ_{C1})与添加的同位素标记 2 的碳酸氢钠的 $\delta^{13}C$ 值(δ_{C2})的差，方程(4-9)可进一步表示为

$$f_B = \frac{\delta_{T1}-\delta_{T2}}{\delta_{C1}-\delta_{C2}} \quad (4-10)$$

δ_{C1} 为添加的同位素标记 1 的碳酸氢钠的 $\delta^{13}C$ 值；δ_{C2} 为添加的同位素标记 2 的碳酸氢钠的 $\delta^{13}C$ 值。

三、微藻利用不同形态无机碳的途径份额计算方法

上述方程(4-6)可以简化为

$$\delta_{Ti}=\delta_a+f_{Bi}(\delta_{ai}-\delta_a)+9‰f_{bi}(i=1,\ 2) \quad (4-11)$$

与方程(4-9)推导到方程(4-10)的情况类似，在方程(4-11)中，$\delta_{ai}-\delta_a$ 可以换算成添加的同位素标记的碳酸氢根离子的 $\delta^{13}C$ 值 δ_{Ci}(即 δ_{C1} 或 δ_{C2})与未添加碳酸

氢根离子、完全来自空气的培养液中无机碳 $\delta^{13}C$ 值 δ_{C0} 的差。在此基础上，我们建立如下方程：

$$\delta_{ai} - \delta_a = \delta_{Ci} - \delta_{C0} = D_i (i=1,2) \tag{4-12}$$

把方程(4-12)代入，方程(4-11)可以简化为

$$\delta_{Ti} = \delta_a + f_{Bi} D_i + 9‰ f_{bi} (i=1,2) \tag{4-13}$$

根据之前的研究，在添加高浓度碳酸氢钠(16.0mmol/L NaHCO$_3$)和充分抑制微藻的胞外碳酸酐酶活性(添加 10.0mmol/L AZ)条件下，微藻利用碳酸氢根离子的途径受到了完全抑制（Wu et al.，2012），得到此时的 $f_{bi}=0$，进而可以得到微藻完全利用大气中的二氧化碳，且完全进行二氧化碳途径时藻体的 $\delta^{13}C$ 值(δ_a)：

$$\delta_a = \delta_{Ti} - f_{Bi} D_i (i=1,2) \tag{4-14}$$

方程(4-14)中的 f_{Bi} 由方程(4-10)可以计算出，因此，最终可以计算出 δ_a 值。接下来，把 δ_a 值代入方程(4-13)，最后，可以计算出微藻利用碳酸氢根离子途径所占的份额(f_{bi})：

$$f_{bi} = 1000 (\delta_{Ti} - \delta_a - f_{Bi} D_i)/9 \ (i=1,2) \tag{4-15}$$

四、直接碳汇和间接碳汇的计算方法

综合以上关于藻类利用无机碳的两个来源（空气中的二氧化碳和溶液中固有的无机碳），再加上微藻利用无机碳的两种代谢途径（二氧化碳途径和碳酸氢根离子途径），我们可以做如下进一步论述：不论藻类采用二氧化碳利用途径还是采用碳酸氢根离子利用途径，只要是源于空气的无机碳被同化，我们就称之为藻类直接碳汇(CS_D)；而无论藻类采用二氧化碳利用途径还是采用碳酸氢根离子利用途径，只要是源于水体中固有的无机碳被同化，我们就称之为间接碳汇(CS_{ID})。直接碳汇直接移去大气中的二氧化碳，而间接碳汇通过移去水体中固有的无机碳，造成水体可溶性无机碳含量减少，根据开放体系中的气液动态平衡原理，大气二氧化碳将进入水体，最终是间接移去大气中的二氧化碳。直接碳汇和间接碳汇相加，就是微藻的总碳汇能力(CS_T)。为此，我们建立了如下模型用于计算微藻的碳汇能力。

$$CS_{A\text{-}ai} = (1 - f_{Bi}) \times (1 - f_{bi}) \times (P - 1) \tag{4-16}$$

$$CS_{A\text{-}bi} = (1 - f_{Bi}) \times f_{bi} \times (P - 1) \tag{4-17}$$

$$CS_{B\text{-}ai} = f_{Bi} \times (1 - f_{bi}) \times (P - 1) \tag{4-18}$$

$$CS_{B\text{-}bi} = f_{Bi} \times f_{bi} \times (P - 1) \tag{4-19}$$

$$CS_D = CS_{A\text{-}ai} + CS_{A\text{-}bi} \tag{4-20}$$

$$CS_{ID} = CS_{B\text{-}ai} + CS_{B\text{-}bi} \tag{4-21}$$

$CS_{A\text{-}ai}$：利用大气中的二氧化碳，且进行二氧化碳途径时的微藻碳汇能力；$CS_{A\text{-}bi}$：利用大气中的二氧化碳，且进行碳酸氢根离子途径时的微藻碳汇能力；$CS_{B\text{-}ai}$：利用添加的无机碳源，且进行二氧化碳途径时的微藻碳汇能力；$CS_{B\text{-}bi}$：利用添加的无机碳源，且进行碳酸氢根离子途径时的微藻碳汇能力；f_{Bi}：微藻利用添加的无机碳占总碳源的份额；f_{bi}：微藻利用碳酸氢根离子途径所占的份额；P：微藻处理前后，生物量增殖倍数（以接种时和处理后蛋白质含量变化来表征）；CS_D：直接碳汇；CS_{ID}：间接碳汇。

由此可知，利用本研究的双向同位素示踪技术，通过测定室内控制条件下纯培养的微藻碳同位素组成变化，可以成功地区分出了莱茵衣藻和蛋白核小球藻所吸收利用的不同无机碳来源及其份额。与pH漂移、无机碳平衡动力学、同位素分馏效应及光谱分析（Rost et al.，2007）等方法相比，这种双向同位素示踪技术可以定量计算微藻的各种碳汇。

第三节 无机碳利用途径对微藻碳汇的影响

一、重碳酸盐对微藻碳汇的影响

（一）实验设置

本研究以添加碳酸氢钠来模拟重碳酸盐对微藻碳汇的影响，碳酸氢钠浓度设置为0、0.5mmol/L、2.0mmol/L、4.0mmol/L、8.0mmol/L、16.0mmol/L。

（二）无机碳含量与组成变化

在碳酸氢钠浓度梯度处理下，培养液中的可溶性无机碳含量和碳同位素组成都随添加碳酸氢钠浓度的变化而变化（表4-1）。第一，水体中的碳酸氢根离子含

表4-1 培养液中不同形态无机碳含量

Tab. 4-1 The species of DIC in the original culture media

Treatment [NaHCO$_3$][①] /(mmol/L)	pH	[HCO$_3^-$][②] /(mmol/L)	[CO$_2$][③] /(mmol/L)	$\delta^{13}C_{DIC}$ /‰ (PDB)[④]	$\delta^{13}C_{DIC}$ /‰ (PDB)[⑤]
0.00	8.00	0.65	0.02	−11.8	−11.8
0.50	8.00	1.55	0.04	−13.7	−17.9
2.00	8.00	2.75	0.07	−14.8	−21.3

续表

Treatment [NaHCO₃]① /(mmol/L)	pH	[HCO₃⁻]② /(mmol/L)	[CO₂]③ /(mmol/L)	$\delta^{13}C_{DIC}$ /‰ (PDB)④	$\delta^{13}C_{DIC}$ /‰ (PDB)⑤
4.00	8.00	4.45	0.11	−15.2	−22.1
8.00	8.00	8.30	0.20	−15.9	−24.2
16.00	8.00	15.70	0.38	−16.5	−26.7

①向培养液中添加的 NaHCO₃ 浓度；②培养液中实测的 HCO₃⁻ 含量；③培养液中的 CO₂ 浓度；④向培养液中添加 δ^{13}C 为 −17.4‰ 的碳酸氢钠；⑤向培养液中添加 δ^{13}C 为 −28.4‰ 的碳酸氢钠。

①NaHCO₃ concentration added to the culture medium; ② HCO₃⁻ concentration in the culture medium; ③ CO₂ concentration in the culture medium; ④ NaHCO₃ added in the medium with an δ^{13}C value of −17.4‰; ⑤NaHCO₃ added in the medium with an δ^{13}C value of −28.4‰.

量随添加碳酸氢钠浓度的增加而相应地增加；第二，来源于大气中的二氧化碳占总无机碳的比例随添加碳酸氢钠浓度的增加而不断下降；第三，随着添加碳酸氢钠浓度的增加（尤其是在添加 16.0mmol/L 碳酸氢钠的条件下），水体中可溶性无机碳的 δ^{13}C 值越来越接近所添加的标记碳酸氢钠的 δ^{13}C 值。

(三) 微藻生物量变化

在碳酸氢钠浓度梯度处理下，微藻的生长速率呈规律性的变化。总体来看，不论莱茵衣藻还是蛋白核小球藻，它们的生物量都随添加碳酸氢钠浓度的增加而增加，由此可知，重碳酸盐能够提高微藻的光合生产能力（图 4-2）。具体来看，莱茵衣藻在添加 8.0mmol/L 碳酸氢钠条件下的生物量达到最大，但是，添加过

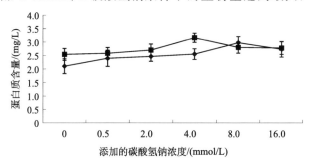

图 4-2 不同浓度重碳酸盐处理下的微藻蛋白质含量

(◆) 莱茵衣藻；(■) 蛋白核小球藻

Fig. 4-2 The content of protein in algae treated with different concentrations of bicarbonate

(◆) *Chlamydomonas reinhardtii*；(■) *Chlorella pyrenoidosa*

高浓度的碳酸氢钠，莱茵衣藻的生物量反而略有下降；而蛋白核小球藻的生物量在添加 4.0mmol/L 碳酸氢钠条件下达到最大，与莱茵衣藻相似，添加过高浓度的碳酸氢钠，蛋白核小球藻的生物量也是呈下降趋势。由此可以进一步说明，不论莱茵衣藻还是蛋白核小球藻，它们对重碳酸盐环境的耐受能力都是有限的。可能是因为过高浓度碳酸氢钠抑制了微藻的碳酸酐酶胞外酶活性（Moroney and Tolbert，1985），并最终抑制了微藻的生长。

（四）微藻碳同位素组成变化

从处理后的微藻碳同位素组成来看，藻体 $\delta^{13}C$ 值随添加碳酸氢钠浓度的变化而变化（表 4-2）。具体来看，在添加较低浓度碳酸氢钠（0～4.0mmol/L）条件下，微藻藻体的 $\delta^{13}C$ 值受所添加碳酸氢钠的影响较小；而在添加高浓度碳酸氢钠（8.0～16.0mmol/L）条件下，添加的碳酸氢钠对微藻藻体 $\delta^{13}C$ 值的影响较大，尤其是所添加的碳酸氢钠的 $\delta^{13}C$ 值越负，对藻体 $\delta^{13}C$ 值所带来的影响越明显。这可能是由于高浓度碳酸氢钠对微藻碳酸酐酶胞外酶的抑制作用，并进一步导致微藻稳定碳同位素分馏作用加大。

表 4-2　不同浓度重碳酸盐处理下的微藻碳同位素组成
Tab. 4-2　Stable carbon isotope composition of the algae treated with different concentrations of bicarbonate

[NaHCO$_3$]* /(mmol/L)	莱茵衣藻		蛋白核小球藻	
	δ_{T1}	δ_{T2}	δ_{T1}	δ_{T2}
0	−20.3±0.1	−20.3±0.1	−22.1±0.2	−22.1±0.2
0.50	−20.3±0.2	−20.5±0.2	−21.8±0.2	−22.1±0.1
2.00	−19.9±0.1	−20.8±0.1	−21.0±0.3	−21.8±0.2
4.00	−20.6±0.1	−22.4±0.2	−19.9±0.2	−21.6±0.3
8.00	−20.5±0.1	−23.5±0.2	−19.3±0.3	−23.4±0.2
16.00	−26.0±0.3	−28.5±0.3	−25.6±0.2	−28.4±0.1

* 培养液中添加的 NaHCO$_3$ 浓度。

注：δ_{T1}：添加 $\delta^{13}C$ 为 −17.4‰ 的 NaHCO$_3$ 的培养液所培养出的微藻 $\delta^{13}C$ 值；δ_{T2}：添加 $\delta^{13}C$ 为 −28.4‰ 的 NaHCO$_3$ 的培养液所培养出的微藻 $\delta^{13}C$ 值。

* Concentration of NaHCO$_3$ added in medium.

Note: δ_{T1}: The value of microalgae's $\delta^{13}C$ cultured in the medium added NaHCO$_3$ with an $\delta^{13}C$ value of −17.4‰; δ_{T2}: The value of microalgae's $\delta^{13}C$ cultured in the medium added NaHCO$_3$ with an $\delta^{13}C$ value of −28.4‰.

（五）微藻吸收利用不同无机碳的份额

水体中存在 CO_2，HCO_3^-，CO_3^{2-} 和 H_2CO_3 四种形态的可溶性无机碳，且它

们之间可以相互转化,最终只有 CO_2 和 HCO_3^- 可供微藻所利用。本研究利用双同位素示踪技术,通过添加两种标记 $\delta^{13}C$ 的 $NaH^{13}CO_3$,定量计算出了微藻利用添加的重碳酸盐所占总无机碳源的份额。

微藻利用添加的碳酸氢根离子占所利用的总无机碳份额(f_B)随添加碳酸氢钠浓度的变化而变化(表4-3)。总体来看,微藻对添加的重碳酸盐的利用份额(f_B)都随添加碳酸氢钠浓度的增加而变大。需要注意的是,在 16.0mmol/L 碳酸氢钠条件下,两种微藻的 f_B 都呈略下降的趋势,由此推断:过高浓度的碳酸氢钠对莱茵衣藻和蛋白核小球藻利用添加的碳酸氢根离子的能力都有所抑制,更深层次的原因可能是因为过高浓度碳酸氢钠条件下,微藻的碳酸酐酶胞外酶活性受到了抑制(Moroney and Tolbert,1985),并对微藻生长也带来了抑制作用所造成的,这从莱茵衣藻和蛋白核小球藻的蛋白质含量也可以看出(图4-2)。

表 4-3 不同浓度重碳酸盐处理下微藻增殖倍数和碳源利用份额

Tab. 4-3 The proliferating multiple of algae and the proportion of carbon sources treated with different concentrations of bicarbonate

[$NaHCO_3$]* /(mmol/L)	莱茵衣藻			蛋白核小球藻		
	P	f_B	f_b	P	f_B	f_b
0	3.51±0.27	0.00	0.96	4.24±0.22	0.00	0.77
0.50	4.00±0.30	0.02	0.97	4.31±0.22	0.03	0.81
2.00	4.11±0.18	0.09	1.00(1.05)③	4.51±0.23	0.08	0.92
4.00	4.26±0.20	0.18	1.00(1.02)③	5.28±0.17	0.17	1.00(1.10)
8.00	4.97±0.23	0.30	1.00(1.10)③	4.67±0.21	0.41	1.00(1.29)
16.00	4.56±0.29	0.24	0.45	4.64±0.23	0.28	0.52

* 培养液中添加的 $NaHCO_3$ 浓度。

注:P 为处理后的微藻生物量相对于接种时的增殖倍数;f_B 为微藻利用添加的无机碳占总碳源的份额;f_b 为微藻利用碳酸氢根离子途径所占的份额。

括号中的数据是计算得来的,而括号外面的数据是根据因微藻生长过程所带来的 $\delta^{13}C$ 误差矫正以后的(Burkhardt et al.,1999)。

* Concentration of $NaHCO_3$ added in medium.

Note:P:the proliferating multiple of the treated algae related to the beginning; f_B: the proportion of the utilization of DIC from the added HCO_3^- in the whole carbon sources used by the microalgae; f_b: the proportion of the HCO_3^- pathway in the whole carbon used by the microalgae.

The data in brackets were calculated values, whereas those out of brackets were correction values considering the error from the determination of $\delta^{13}C$ and the differences induced by growth status (Burkhardt et al., 1999).

微藻利用碳酸氢根离子途径所占的份额(f_b)也随添加的碳酸氢钠的浓度变化而变化(表 4-3)。一般情况下,f_b随添加碳酸氢钠浓度的增加而下降。两种微藻在较低浓度碳酸氢钠(0～8.0mmol/L)条件下,碳酸氢根离子的利用途径占主导($f_b>0.76$),但在过高浓度碳酸氢钠(16.0mmol/L)条件下,碳酸氢根离子的利用途径急剧下降,莱茵衣藻的f_b降为 0.45,蛋白核小球藻的f_b降为 0.52。因为在低浓度碳酸氢钠条件下,微藻的胞外碳酸酐酶活性较高,此时微藻利用碳酸氢根离子途径所占的份额也较高;随着添加碳酸氢钠浓度的增加,微藻的胞外碳酸酐酶活性也不断下降,由此带来微藻利用碳酸氢根离子途径所占的份额也呈不断下降的趋势。

(六) 微藻的碳汇能力

微藻的碳汇能力与其生物量呈正相关关系(表 4-4)。从不同来源的碳汇来看,一般情况下,微藻利用大气中的二氧化碳,且进行二氧化碳途径的碳汇能力($CS_{A\text{-}ai}$)较小;与之相反,利用大气中的二氧化碳,且进行碳酸氢根离子途径的碳汇能力($CS_{A\text{-}bi}$)较大。这可能跟碳酸酐酶胞外酶催化水体中不同形态无机碳之间的转化能力有关。一般情况下,微藻利用添加的碳酸氢钠,且进行二氧化碳途径的碳汇能力($CS_{B\text{-}ai}$)和微藻利用添加的碳酸氢钠,且进行碳酸氢根离子途径的碳汇能力($CS_{B\text{-}bi}$)都较小。

表 4-4 不同浓度重碳酸盐处理下莱茵衣藻和蛋白核小球藻的碳汇组成
Tab. 4-4 The composition of different carbon sequestrations in *Chlamydomonas reinhardtii* and *Chlorella pyrenoidosa* treated with different concentrations of bicarbonate

	Treatment [NaHCO$_3$]* /(mmol/L)	$CS_{A\text{-}ai}$	$CS_{A\text{-}bi}$	$CS_{B\text{-}ai}$	$CS_{B\text{-}bi}$	CS_D	CS_{ID}	$CS_T/\%$
	0.00	0.10	2.41	0.00	0.00	2.51	0.00	2.51 (80.71)
	0.50	0.08	2.85	0.00	0.07	2.93	0.07	3.00(96.46)
莱茵衣藻	2.00	0.00	2.83	0.00	0.28	2.83	0.28	3.11(100.00)
	4.00	0.00	2.68	0.00	0.58	2.68	0.58	3.26(104.82)
	8.00	0.00	2.78	0.00	1.19	2.78	1.19	3.97(127.65)
	16.00	1.49	1.22	0.47	0.39	2.70	0.86	3.56(114.47)

续表

Treatment	[NaHCO$_3$]*/(mmol/L)	CS$_{A-ai}$	CS$_{A-bi}$	CS$_{B-ai}$	CS$_{B-bi}$	CS$_D$	CS$_{ID}$	CS$_T$/%
蛋白核小球藻	0.00	0.76	2.48	0.00	0.00	3.24	0.00	3.24 (92.31)
	0.50	0.61	2.61	0.02	0.08	3.21	0.10	3.31 (94.30)
	2.00	0.24	2.99	0.02	0.26	3.23	0.28	3.51 (100.00)
	4.00	0.00	3.54	0.00	0.74	3.54	0.74	4.28 (121.94)
	8.00	0.00	2.18	0.00	1.49	2.18	1.49	3.67 (104.56)
	16.00	1.25	1.37	0.49	0.53	2.62	1.02	3.64 (103.70)

* 培养液中添加的 NaHCO$_3$ 浓度。

注：CS$_{A-ai}$：利用大气中的二氧化碳，且进行二氧化碳途径时的微藻碳汇能力；CS$_{A-bi}$：利用大气中的二氧化碳，且进行碳酸氢根离子途径时的微藻碳汇能力；CS$_{B-ai}$：利用添加的无机碳源，且进行二氧化碳途径时的微藻碳汇能力；CS$_{B-bi}$：利用添加的无机碳源，且进行碳酸氢根离子途径时的微藻碳汇能力；CS$_D$：直接碳汇，所有来源于大气中的无机碳被同化；CS$_{ID}$：间接碳汇，所有来源于水体中的无机碳被同化；CS$_T$：总碳汇，直接碳汇和间接碳汇的和。

括号中的数据是各个处理下的总碳汇能力相对于各自在自然条件下（2.0mmol/L NaHCO$_3$）的碳汇能力的百分比。

* Concentration of NaHCO$_3$ added in medium.

Note：CS$_{A-ai}$: the contributions of the CO$_2$ pathways using DIC from the atmosphere; CS$_{A-bi}$: the contributions of the HCO$_3^-$ pathways using DIC from the atmosphere; CS$_{B-ai}$: the contributions of the CO$_2$ pathways using DIC from the added NaHCO$_3$; CS$_{B-bi}$: the contributions of the HCO$_3^-$ pathways using DIC from the added NaHCO$_3$; CS$_D$: direct carbon sequestration; CS$_{ID}$: indirect carbon sequestration; CS$_T$: the total carbon sequestration.

The data in brackets were the percents of algal CS$_T$ to that of algae treated with 2.0 mmol/L NaHCO$_3$ addition.

不论莱茵衣藻还是蛋白核小球藻，两种微藻的间接碳汇（CS$_{ID}$）都随添加碳酸氢钠浓度的增加而上升（表4-4）。因为随着添加碳酸氢钠浓度的增加，微藻生长旺盛，它的总碳汇能力（CS$_T$）增加，相应地，CS$_{ID}$也增加。但是，在添加过高浓度碳酸氢钠（16.0mmol/L）的条件下，微藻的间接碳汇能力（CS$_{ID}$）反而有所下降。这与微藻在此时的生长速率下降趋势相同（图4-2）。

二、碳酸酐酶胞外酶对微藻碳汇的影响

（一）实验设置

AZ的浓度设置为：0.5mmol/L、10.0mmol/L，碳酸氢钠浓度设置为：0、0.5mmol/L、2.0mmol/L、4.0mmol/L、8.0mmol/L、16.0mmol/L。

(二) 碳酸酐酶胞外酶抑制剂

碳酸酐酶胞外酶对微藻的碳代谢具有非常重要的作用,乙酰唑胺(Acetazolamide,AZ)是碳酸酐酶胞外酶特异性抑制剂(Moroney et al., 1985),它能够抑制碳酸酐酶胞外酶对微藻吸收利用无机碳过程中的催化作用,最终影响微藻的生长。通过向培养液中添加不同浓度的 AZ,可以研究碳酸酐酶胞外酶对微藻生长所带来的影响。

(三) 微藻生物量变化

在添加 AZ 的情况下,整体来看,微藻的生物量随添加碳酸氢钠浓度的增加呈上升的趋势(图 4-3 和图 4-4)。本研究还发现:在高浓度碳酸氢钠(16.0mmol/L)条件下,添加 0.5mmol/L AZ 对莱茵衣藻生长的影响略大于添加 10.0mmol/L AZ 对莱茵衣藻生长的影响,这因为,在高浓度碳酸氢钠(16.0mmol/L)条件下,碳酸酐酶胞外酶的活力被完全抑制,碳酸酐酶胞外酶抑制剂对生长不发挥作用。

与不添加 AZ 的碳酸氢钠浓度梯度处理的微藻蛋白质含量(图 4-2)相比,蛋白核小球藻在添加高浓度碳酸氢钠(8.0～16.0mmol/L)条件下,添加少量 AZ(0.5mmol/L),却促进蛋白核小球藻的生长(图 4-3),说明此时高浓度碳酸氢钠对蛋白核小球藻的促进作用超过了低浓度(0.5mmol/L)AZ 所带来的抑制作用。

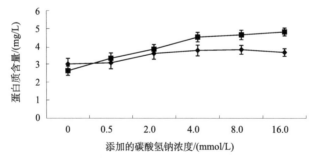

图 4-3 不同浓度重碳酸盐及 0.5mmol/L AZ 处理下的微藻蛋白质含量
(◆) 莱茵衣藻; (■) 蛋白核小球藻

Fig. 4-3 The content of protein in algae treated with 0.5mmol/L AZ and different concentrations of bicarbonate
(◆) *Chlamydomonas reinhardtii*; (■) *Chlorella pyrenoidosa*

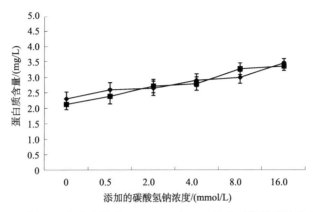

图 4-4　不同重浓度碳酸盐及 10.0mmol/L AZ 处理下的微藻蛋白质含量
（◆）莱茵衣藻；（■）蛋白核小球藻

Fig. 4-4　The content of protein in algae treated with 10.0mmol/L
AZ and different concentrations of bicarbonate

（◆）*Chlamydomonas reinhardtii*；（■）*Chlorella pyrenoidosa*

（四）微藻碳同位素组成变化

添加 AZ，碳酸酐酶胞外酶活性受到了抑制，对藻体 $\delta^{13}C$ 值所带来的影响很明显，造成藻体 $\delta^{13}C$ 值偏负（表 4-5，表 4-6）。具体来看，在同一碳酸氢钠浓度处理下，微藻的 $\delta^{13}C$ 值随添加 AZ 浓度的增加而偏负，尤其是在添加 10.0mmol/L AZ 的情况，造成藻体 $\delta^{13}C$ 值偏负较大。相比于未添加 AZ 的对照组（表 4-2），在低浓度碳酸氢钠（0～8.0mmol/L）条件下，添加 10.0mmol/L AZ，带来约 9‰ 的碳同位素分馏。因为 AZ 是碳酸酐酶胞外酶特异性抑制剂（Moroney et al.，1985），添加 AZ 所带来的稳定碳同位素分馏差异，正是微藻的碳酸酐酶胞外酶对稳定碳同位素分馏作用的贡献。而在高浓度碳酸氢钠（16.0mmol/L）条件下，添加 10.0mmol/L AZ，带来的碳同位素分馏较小。这可能与高浓度碳酸氢钠条件下的微藻碳酸酐酶胞外酶活性较低有关（Moroney and Tolbert，1985），因而造成添加 AZ 所带来的抑制效果不显著。整体来看，添加 10.0mmol/L AZ，抑制了碳酸酐酶胞外酶的活性，相应地，对微藻碳同位素分馏带来约 9‰ 的影响（Wu et al.，2012）。

从添加两种标记不同 $\delta^{13}C$ 值的碳酸氢钠来看，添加碳酸氢钠的 $\delta^{13}C$ 值越负，对应藻体的 $\delta^{13}C$ 值也越负。在相同条件下，由两种碳酸氢钠所造成的微藻碳同位素组成的差值也随添加碳酸氢钠浓度的增加而不断扩大。

表 4-5 不同浓度重碳酸盐和 0.5mmol/L AZ 共同处理下的微藻碳同位素组成

Tab. 4-5 Stable carbon isotope composition of the algae treated with 0.5 mmol/L AZ and different concentrations of bicarbonate

[NaHCO$_3$]* /(mmol/L)	莱茵衣藻		蛋白核小球藻	
	δ_{T1}②	δ_{T2}②	δ_{T1}②	δ_{T2}②
0	−23.0±0.2	−23.0±0.2	−22.7±0.1	−22.7±0.1
0.50	−23.1±0.2	−22.9±0.4	−23.6±0.2	−23.6±0.3
2.00	−23.1±0.2	−22.8±0.3	−22.8±0.2	−22.6±0.4
4.00	−25.5±0.3	−25.5±0.2	−25.4±0.2	−25.4±0.4
8.00	−27.6±0.2	−29.5±0.2	−28.0±0.2	−29.7±0.3
16.00	−31.5±0.2	−37.1±0.3	−30.4±0.4	−36.1±0.2

* 培养液中添加的 NaHCO$_3$ 浓度。

注：δ_{T1}：添加 $\delta^{13}C$ 为 −17.4‰ 的 NaHCO$_3$ 的培养液所培养出的微藻 $\delta^{13}C$ 值；δ_{T2}：添加 $\delta^{13}C$ 为 −28.4‰ 的 NaHCO$_3$ 的培养液所培养出的微藻 $\delta^{13}C$ 值。

* Concentration of NaHCO$_3$ added in medium.

Note: δ_{T1}: The value of microalgae's $\delta^{13}C$ cultured in the medium added NaHCO$_3$ with an $\delta^{13}C$ value of −17.4‰; δ_{T2}: The value of microalgae's $\delta^{13}C$ cultured in the medium added NaHCO$_3$ with an $\delta^{13}C$ value of −28.4‰.

表 4-6 不同浓度重碳酸盐和 10.0mmol/L AZ 共同处理下的微藻碳同位素组成

Tab. 4-6 Stable carbon isotope composition of the algae treated with 10.0 mmol/L AZ and different concentrations of bicarbonate

[NaHCO$_3$]* /(mmol/L)	莱茵衣藻		蛋白核小球藻	
	δ_{T1}②	δ_{T2}②	δ_{T1}②	δ_{T2}②
0	−31.5±0.2	−31.5±0.2	−32.7±0.3	−32.7±0.3
0.50	−29.4±0.3	−29.4±0.2	−31.3±0.2	−31.6±0.3
2.00	−28.7±0.2	−30.4±0.3	−29.6±0.5	−31.4±0.4
4.00	−27.9±0.3	−30.2±0.4	−29.7±0.2	−32.4±0.5
8.00	−29.0±0.4	−32.5±0.3	−29.9±0.3	−33.4±0.4
16.00	−31.5±0.4	−37.1±0.4	−30.9±0.5	−36.7±0.6

* 培养液中添加的 NaHCO$_3$ 浓度。

注：δ_{T1}：添加 $\delta^{13}C$ 为 −17.4‰ 的 NaHCO$_3$ 的培养液所培养出的微藻 $\delta^{13}C$ 值；δ_{T2}：添加 $\delta^{13}C$ 为 −28.4‰ 的 NaHCO$_3$ 的培养液所培养出的微藻 $\delta^{13}C$ 值。

* Concentration of NaHCO$_3$ added in medium.

Note: δ_{T1}: The value of microalgae's $\delta^{13}C$ cultured in the medium added NaHCO$_3$ with an $\delta^{13}C$ value of −17.4‰; δ_{T2}: The value of microalgae's $\delta^{13}C$ cultured in the medium added NaHCO$_3$ with an $\delta^{13}C$ value of −28.4‰.

(五) 微藻吸收利用不同无机碳源的份额

一般情况下,添加 AZ,造成微藻利用添加的碳酸氢根离子占所利用的总无机碳份额(f_B)增加(表 4-7),但也存在例外的情况,这应该是添加不同浓度 AZ 和碳酸氢钠复合作用的结果。原因如下:添加 AZ,抑制了碳酸酐酶胞外酶的活性(Moroney et al.,1985),造成微藻胞外碳酸酐酶催化碳酸氢根离子转化为二氧化碳的速率降低,甚至为 0,并最终造成微藻生长速率下降。此外,大气二氧化碳进入水体的过程中,碳酸酐酶胞外酶也发挥着重要作用,在碳酸酐酶胞外酶受到抑制的情况下,造成大气二氧化碳进入水体的过程也受到抑制,水体中来自大气二氧化碳的可溶性无机碳比例下降,造成添加的重碳酸盐在培养液中的总无机碳份额上升,由此带来 f_B 增大。而且随添加 AZ 浓度的增加,碳酸酐酶胞外酶受到的抑制作用增大,最终造成 f_B 增大。

表 4-7 不同浓度重碳酸盐和 0.5mmol/L AZ 共同处理下微藻增殖倍数和碳源利用份额
Tab. 4-7 The proliferating multiple of algae and the proportion of carbon sources treated with 0.5 mmol/L AZ and different concentrations of bicarbonate

[NaHCO₃]*/(mmol/L)	莱茵衣藻			蛋白核小球藻		
	P	f_B	f_b	P	f_B	f_b
0	3.36±0.21	0.00	0.66	2.95±0.17	0.00	0.69
0.50	3.43±0.22	0.00(−0.03)③	0.65	3.72±0.20	0.00	0.59
2.00	4.01±0.21	0.00(−0.03)③	0.66	4.28±0.17	0.00(−0.02)③	0.69
4.00	4.20±0.20	0.00	0.39	5.02±0.18	0.00	0.40
8.00	4.25±0.17	0.19	0.25	5.16±0.18	0.16	0.19
16.00	4.08±0.14	0.55	0.00	5.34±0.15	0.56	0.14

* 培养液中添加的 NaHCO₃ 浓度。

注:P 为处理后的微藻生物量相对于接种时的增殖倍数;f_B 为微藻利用添加的无机碳占总碳源的份额;f_b 为微藻利用碳酸氢根离子途径所占的份额。

括号中的数据是计算得来的,而括号外面的数据是根据因微藻生长过程所带来的 $\delta^{13}C$ 误差矫正以后的(Burkhardt et al.,1999)。

* Concentration of NaHCO₃ added in medium.

Note:P: the proliferating multiple of the treated algae related to the beginning;f_B: the proportion of the utilization of DIC from the added HCO_3^- in the whole carbon sources used by the microalgae;f_b: the proportion of the HCO_3^- pathway in the whole carbon used by the microalgae.

The data in brackets were calculated values, whereas those out of brackets were correction values considering the error from the determination of $\delta^{13}C$ and the differences induced by growth status (Burkhardt et al.,1999).

微藻利用碳酸氢根离子途径所占的份额(f_b)也随添加的碳酸氢钠和 AZ 的浓度变化而变化(表 4-8)。一般情况下,在同一碳酸氢钠浓度处理下,添加 AZ 也会造成 f_b 下降。因为添加 AZ,抑制了碳酸酐酶胞外酶的活性(Moroney et al.,1985),最终也造成微藻的碳酸氢根离子的利用途径(f_b)下降,尤其是在添加 10.0mmol/L AZ 的条件下,造成微藻的碳酸氢根离子的利用途径份额(f_b)急剧下降,此时 f_b 的最大值只有 0.24。需要特别指出的是,在添加较低浓度碳酸氢钠条件下,甚至出现了负数,这应该是由于实验误差造成的。

表 4-8 不同浓度重碳酸盐和 10.0mmol/L AZ 处理下微藻增殖倍数和碳源利用份额

Tab. 4-8 The proliferating multiple of algae and the proportion of carbon sources treated with 10.0 mmol/L AZ and different concentrations of bicarbonate

[NaHCO$_3$]*/(mmol/L)	莱茵衣藻			蛋白核小球藻		
	P	f_B	f_b	P	f_B	f_b
0	2.40±0.22	0.00	0.00 (−0.28)③	2.11±0.17	0.00	0.00 (−0.42)
0.50	2.88±0.24	0.00	0.00 (−0.05)③	2.53±0.25	0.03	0.00 (−0.24)
2.00	2.97±0.24	0.17	0.11	3.09±0.22	0.18	0.02
4.00	3.40±0.21	0.23	0.24	3.22±0.22	0.26	0.05
8.00	3.55±0.19	0.34	0.18	4.02±0.19	0.34	0.08
16.00	4.34±0.14	0.54	0.00	4.16±0.14	0.56	0.07

* 培养液中添加的 NaHCO$_3$ 浓度。

注:P 为处理后的微藻生物量相对于接种时的增殖倍数;f_B 为微藻利用添加的无机碳占总碳源的份额;f_b 为微藻利用碳酸氢根离子途径所占的份额。

括号中的数据是计算得来的,而括号外面的数据是根据因微藻生长过程所带来的 δ^{13}C 误差矫正以后的(Burkhardt et al.,1999)。

* Concentration of NaHCO$_3$ added in medium.

Note:P: the proliferating multiple of the treated algae related to the beginning;f_B: the proportion of the utilization of DIC from the added HCO$_3^-$ in the whole carbon sources used by the microalgae;f_b: the proportion of the HCO$_3^-$ pathway in the whole carbon used by the microalgae.

The data in brackets were calculated values, whereas those out of brackets were correction values considering the error from the determination of δ^{13}C and the differences induced by growth status (Burkhardt et al.,1999).

(六)微藻的碳汇能力和胞外碳酸酐酶对不同碳汇的影响

从不同来源的碳汇来看,一般情况下,微藻利用大气中的二氧化碳,且进行二氧化碳途径的碳汇能力(CS_{A-ai})较小;而利用大气中的二氧化碳,且进行碳酸氢根离子途径的碳汇能力(CS_{A-bi})较大。这同样可能与碳酸酐酶胞外酶催化水体中不同形态无机碳之间的转化能力有关。添加 AZ,造成 CS_{A-ai} 明显增加,而

CS_{A-bi} 下降。因为 AZ 是碳酸酐酶胞外酶抑制剂（Moroney et al., 1985），添加 AZ，造成微藻的碳酸酐酶胞外酶活性下降，相应地，水体中不同形态无机碳之间的转化能力下降。

一般情况下，微藻利用添加的碳酸氢钠，且进行二氧化碳途径的碳汇能力（CS_{B-ai}）和微藻利用添加的碳酸氢钠，且进行碳酸氢根离子途径的碳汇能力（CS_{B-bi}）较小，但随着添加 AZ 的增加，两者呈上升趋势（表 4-9，表 4-10）。这也与碳酸酐酶胞外酶催化水体中不同形态无机碳之间的转化能力有关。

表 4-9 不同浓度重碳酸盐和 0.5mmol/L AZ 处理下，莱茵衣藻和蛋白核小球藻的碳汇组成
Tab. 4-9 The composition of different carbon sequestrations in *Chlamydomonas reinhardtii* and *Chlorella pyrenoidosa* treated with 0.5mmol/L AZ and different concentrations of bicarbonate

Treatment	[NaHCO$_3$]* /(mmol/L)	CS_{A-ai}	CS_{A-bi}	CS_{B-ai}	CS_{B-bi}	CS_D	CS_{ID}	CS_T/%
莱茵衣藻	0.00	0.80	1.56	0.00	0.00	2.36	0.00	2.36(78.41)
	0.50	0.86	1.57	0.00	0.00	2.43	0.00	2.43(80.73)
	2.00	1.04	1.97	0.00	0.00	3.01	0.00	3.01(100.00)
	4.00	1.97	1.23	0.00	0.00	3.20	0.00	3.20(106.31)
	8.00	1.99	0.65	0.46	0.15	2.64	0.61	3.25(107.97)
	16.00	1.40	0.00	1.67	0.00	1.40	1.68	3.08(102.33)
蛋白核小球藻	0.00	0.60	1.35	0.00	0.00	1.95	0.00	1.95(59.45)
	0.50	1.11	1.61	0.00	0.00	2.72	0.00	2.72(82.93)
	2.00	1.03	2.25	0.00	0.00	3.28	0.00	3.28(100.00)
	4.00	2.43	1.59	0.00	0.00	4.02	0.00	4.02(122.56)
	8.00	2.81	0.66	0.56	0.13	3.47	0.69	4.16(126.83)
	16.00	1.63	0.26	2.10	0.34	1.90	2.44	4.34(132.32)

* 培养液中添加的 NaHCO$_3$ 浓度。

注：CS_{A-ai}：利用大气中的二氧化碳，且进行二氧化碳途径时的微藻碳汇能力；CS_{A-bi}：利用大气中的二氧化碳，且进行碳酸氢根离子途径时的微藻碳汇能力；CS_{B-ai}：利用添加的无机碳源，且进行二氧化碳途径时的微藻碳汇能力；CS_{B-bi}：利用添加的无机碳源，且进行碳酸氢根离子途径时的微藻碳汇能力；CS_D：直接碳汇，所有来源于大气中的无机碳被同化；CS_{ID}：间接碳汇，所有来源于水体中的无机碳被同化；CS_T：总碳汇，直接碳汇和间接碳汇的和。

括号中的数据是各个处理下的总碳汇能力相对各自在自然条件下（2.0mmol/L NaHCO$_3$）的碳汇能力的百分比。

* Concentration of NaHCO$_3$ added in medium.

Note: CS_{A-ai}: the contributions of the CO$_2$ pathways using DIC from the atmosphere; CS_{A-bi}: the contributions of the HCO$_3^-$ pathways using DIC from the atmosphere; CS_{B-ai}: the contributions of the CO$_2$ pathways using DIC from the added NaHCO$_3$; CS_{B-bi}: the contributions of the HCO$_3^-$ pathways using DIC from the added NaHCO$_3$; CS_D: direct carbon sequestration; CS_{ID}: indirect carbon sequestration; CS_T: the total carbon sequestration.

The data in brackets were the percents of algal CS_T to that of algae treated with 2.0 mmol/L NaHCO$_3$ addition.

表 4-10　不同浓度重碳酸盐和 10.0mmol/L AZ 处理下，莱茵衣藻和蛋白核小球藻的碳汇组成
Tab. 4-10　The composition of different carbon sequestrations in *Chlamydomonas reinhardtii* and *Chlorella pyrenoidosa* treated with 10.0mmol/L AZ and different concentrations of bicarbonate

Treatment	[NaHCO$_3$]* /(mmol/L)	CS$_{A\text{-}ai}$	CS$_{A\text{-}bi}$	CS$_{B\text{-}ai}$	CS$_{B\text{-}bi}$	CS$_D$	CS$_{ID}$	CS$_T$/%
莱茵衣藻	0.00	1.40	0.00	0.00	0.00	1.40	0.00	1.40 (71.07)
	0.50	1.88	0.00	0.00	0.00	1.88	0.00	1.88 (95.43)
	2.00	1.46	0.18	0.29	0.04	1.64	0.33	1.97 (100.00)
	4.00	1.40	0.45	0.41	0.13	1.85	0.55	2.40 (121.83)
	8.00	1.39	0.29	0.72	0.15	1.68	0.87	2.55 (129.44)
	16.00	1.52	0.00	1.82	0.00	1.52	1.82	3.34 (169.54)
蛋白核小球藻	0.00	1.11	0.00	0.00	0.00	1.11	0.00	1.11 (53.11)
	0.50	1.48	0.00	0.05	0.00	1.48	0.05	1.53 (73.21)
	2.00	1.69	0.03	0.36	0.01	1.72	0.37	2.09 (100.00)
	4.00	1.57	0.08	0.54	0.03	1.65	0.57	2.22 (106.22)
	8.00	1.83	0.15	0.96	0.08	1.98	1.04	3.02 (144.50)
	16.00	1.28	0.10	1.65	0.13	1.38	1.78	3.16 (151.20)

* 培养液中添加的 NaHCO$_3$ 浓度。

注：CS$_{A\text{-}ai}$：利用大气中的二氧化碳，且进行二氧化碳途径时的微藻碳汇能力；CS$_{A\text{-}bi}$：利用大气中的二氧化碳，且进行碳酸氢根离子途径时的微藻碳汇能力；CS$_{B\text{-}ai}$：利用添加的无机碳源，且进行二氧化碳途径时的微藻碳汇能力；CS$_{B\text{-}bi}$：利用添加的无机碳源，且进行碳酸氢根离子途径时的微藻碳汇能力；CS$_D$：直接碳汇，所有来源于大气中的无机碳被同化；CS$_{ID}$：间接碳汇，所有来源于水体中的无机碳被同化；CS$_T$：总碳汇，直接碳汇和间接碳汇的和。

括号中的数据是各个处理下的总碳汇能力相对于各自在自然条件下(2.0mmol/L NaHCO$_3$)的碳汇能力的百分比。

* Concentration of NaHCO$_3$ added in medium.

Note：CS$_{A\text{-}ai}$: the contributions of the CO$_2$ pathways using DIC from the atmosphere；CS$_{A\text{-}bi}$: the contributions of the HCO$_3^-$ pathways using DIC from the atmosphere；CS$_{B\text{-}ai}$: the contributions of the CO$_2$ pathways using DIC from the added NaHCO$_3$；CS$_{B\text{-}bi}$: the contributions of the HCO$_3^-$ pathways using DIC from the added NaHCO$_3$；CS$_D$: direct carbon sequestration；CS$_{ID}$: indirect carbon sequestration；CS$_T$: the total carbon sequestration.

The data in brackets were the percents of algal CS$_T$ to that of algae treated with 2.0 mmol/L NaHCO$_3$ addition as the natural condition.

一般情况下，间接碳汇(CS$_{ID}$)所占份额较小，而直接碳汇(CS$_D$)占主体。然而，在添加高浓度碳酸氢钠(16.0mmol/L)和高浓度 AZ(10.0mmol/L)的条件下，莱茵衣藻和蛋白核小球藻的 CS$_{ID}$ 都略大于 CS$_D$ (表 4-9，表 4-10)。

不论是否添加 AZ，微藻的间接碳汇 CS$_{ID}$ 都随添加碳酸氢钠浓度的增加而上

升(表 4-9,表 4-10)。因为随着添加碳酸氢钠浓度的增加,微藻生长旺盛,微藻的总碳汇能力(CS_T)增加,相应地,微藻的间接碳汇(CS_{ID})也增加。

综合上述研究发现,蛋白核小球藻与莱茵衣藻的无机碳代谢途径相同,它们都具有间接利用碳酸氢根离子和直接利用碳酸氢根离子的机制。不论是否添加 AZ,微藻利用添加的无机碳源,且进行碳酸氢根离子途径的微藻碳汇能力(CS_{B-bi})占总碳汇(CS_T)份额都不大。

从不同来源的碳汇来看,微藻所固定的碳汇主要来源于大气二氧化碳的直接碳汇。在模拟自然水体环境下(2.0mmol/L $NaHCO_3$),微藻的间接碳汇只占总碳汇的 8%。就算在添加较高浓度的碳酸氢钠条件下(8.0~16.0mmol/L),间接碳汇能力也不大于 40%。

在添加不同浓度碳酸氢钠的较大范围内(0~8.0mmol/L),微藻的碳汇能力变化并不大,具有无机碳浓缩机制的莱茵衣藻和蛋白核小球藻并没有过多利用水体中固有的 HCO_3^-,而是主要利用由胞外碳酸酐酶催化大气中二氧化碳转化来的碳酸氢根离子。也就是说,微藻利用大气中的二氧化碳,且进行碳酸氢根离子途径的碳汇(CS_{A-bi})是自然水体中碳汇的主体。这个结果与 Maberly(1992)关于石莼的光合作用不受碳酸氢根离子含量的影响的研究,结论是一致的。由此可知,全球变化虽然带来大气二氧化碳浓度和水体 DIC 含量的增加,但它对微藻碳汇能力的影响非常有限。

在海洋和喀斯特地区,自然水体具有高 HCO_3^- 含量、低 CO_2 含量的特征(Talling,1976)。这种水体环境极大限制了微藻对大气中二氧化碳的利用,进而影响微藻的光合速率。生物为适应低二氧化碳环境,满足光合作用对无机碳的需要,以莱茵衣藻和蛋白核小球藻为代表的微藻就形成了由碳酸酐酶参与的具有无机碳浓缩机制(CCM)(Miyachi et al.,1983;Moroney and Tolbert,1985)。微藻浓缩的无机碳浓度可为环境的 5~75 倍(Badger et al.,1998)。由于碳酸酐酶胞外酶的无机碳浓缩机制促进了微藻的生长,最终也提高了微藻的碳汇能力。又由于碳酸酐酶胞外酶能够催化大气中二氧化碳被微藻快速利用,因此,微藻的碳酸酐酶胞外酶又可以大大提高直接碳汇能力,而对间接碳汇能力只是轻微促进。

三、直接碳酸氢根离子利用途径对微藻碳汇的影响

4,4′-二异硫氰-2,2′-二黄酸芪(4,4-diisothiocyanatostilbene-2,2-disulfonate,DIDS)是阴离子通道特异性抑制剂,它能抑制阴离子进入细胞的过程(Cabantchik and Greger,1992;Huertas et al.,2000;Young et al.,2001)。通过向实验处理中添加不同浓度的 DIDS 来研究直接利用碳酸氢根离子的途径受到抑制的情况下,微藻的碳源利用策略及其碳汇组成变化,由此来反推直接碳酸氢根离子利用途径对微藻碳汇的影响。

(一) 实验设置

DIDS 的浓度设置：0、0.02mmol/L、0.1mmol/L、0.5mmol/L、2.0mmol/L。

(二) 微藻生物量变化

如图 4-5 所示，不同 DIDS 浓度处理下的微藻生长速率变化。随着添加 DIDS 浓度的增加，莱茵衣藻和蛋白核小球藻的生长速率都呈规律性的下降，尤其是在添加高浓度 DIDS(2.0mmol/L)条件下，两种微藻的生长都受到了极大抑制，生物量急剧下降。因为 DIDS 能够抑制阴离子进入细胞的过程（Cabantchik and Greger，1992），随着添加 DIDS 浓度的增加，碳酸氢根离子通过载体蛋白进入细胞的能力不断下降，直到添加足够浓度的 DIDS，最终抑制碳酸氢根离子无法直接进入微藻细胞。由本实验可以看出，阴离子抑制剂 DIDS 对微藻生长的剂量效应明显。

对比两种微藻的蛋白质含量，蛋白核小球藻的生长速率略大于莱茵衣藻的，这可能跟蛋白核小球藻的细胞较小，生长周期较短有关。

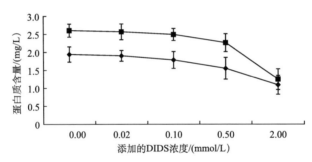

图 4-5　DIDS 处理下，微藻蛋白质含量变化
莱茵衣藻（◆），蛋白核小球藻（■）

Fig. 4-5　The content of protein in algae under different content of DIDS
Chlamydomonas reinhardtii（◆），*Chlorella pyrenoidosa*（■）

(三) 微藻碳同位素组成变化

从处理后的微藻碳同位素数据来看，藻体稳定碳同位素组成随添加 DIDS 浓度的增加呈偏负的趋势（表 4-11）。具体来看，在添加较低浓度 DIDS（0～0.1mmol/L）条件下，微藻藻体 $\delta^{13}C$ 值受所添加 DIDS 的影响较小；而在添加高浓度 DIDS（0.5～2.0mmol/L）条件下，DIDS 对微藻藻体 $\delta^{13}C$ 值的影响较大，这是由阴离子抑制剂 DIDS 的剂量效应造成的。

添加 DIDS，对两种微藻的具体影响略有差异，添加低浓度 DIDS（0～0.50mmol/L）对莱茵衣藻的稳定碳同位素组成影响很小（表 4-11），但高浓度

DIDS(2.0mmol/L)带来的影响较大,莱茵衣藻的碳同位素组成偏负约 8.2‰ (PDB)。2.0mmol/L DIDS 是莱茵衣藻碳同位素组成变化的转折点。

蛋白核小球藻的碳同位素组成受 DIDS 的影响跟莱茵衣藻的相近(表 4-11)。值得注意的是,0.5mmol/L DIDS 是蛋白核小球藻碳同位素组成变化的转折点。由此可说明蛋白核小球藻对 DIDS 更加敏感。

表 4-11 DIDS 处理下微藻碳同位素组成

Tab. 4-11 Stable carbon isotope composition of algae under different content of DIDS

处理	[DIDS]/(mmol/L)	δ_{T1}	δ_{T2}
莱茵衣藻	0	-20.3 ± 0.2	-22.6 ± 0.3
	0.02	-20.5 ± 0.2	-22.7 ± 0.2
	0.10	-20.3 ± 0.4	-22.8 ± 0.3
	0.50	-20.2 ± 0.3	-23.5 ± 0.3
	2.00	-29.5 ± 0.7	-29.7 ± 0.5
蛋白核小球藻	0	-21.8 ± 0.3	-23.3 ± 0.3
	0.02	-21.5 ± 0.4	-23.3 ± 0.5
	0.10	-21.3 ± 0.5	-23.5 ± 0.3
	0.50	-23.1 ± 0.4	-25.1 ± 0.3
	2.00	-31.2 ± 0.5	-31.5 ± 0.6

注:δ_{T1}:添加 $\delta^{13}C$ 为 $-17.4‰$ 的 $NaHCO_3$ 的培养液所培养出的微藻 $\delta^{13}C$ 值;δ_{T2}:添加 $\delta^{13}C$ 为 $-28.4‰$ 的 $NaHCO_3$ 的培养液所培养出的微藻 $\delta^{13}C$ 值。

Note:δ_{T1}: The value of microalgae's $\delta^{13}C$ cultured in the medium added $NaHCO_3$ with an $\delta^{13}C$ value of $-17.4‰$; δ_{T2}: The value of microalgae's $\delta^{13}C$ cultured in the medium added $NaHCO_3$ with an $\delta^{13}C$ value of $-28.4‰$.

从添加两种标记的不同 $\delta^{13}C$ 值的碳酸氢钠来看,添加碳酸氢钠的 $\delta^{13}C$ 值越负,对应藻体的 $\delta^{13}C$ 值也越负。但在添加过高浓度 DIDS 条件下,微藻藻体的碳同位素组成都不再随添加的标记碳酸氢钠的 $\delta^{13}C$ 值而变化,这也印证了阴离子通道特异性抑制剂 DIDS 在高浓度条件下能够充分抑制微藻的阴离子通道吸收途径(Cabantchik and Greger,1992)。

(四)微藻吸收利用不同无机碳碳源的份额及不同形态无机碳途径的份额

不论莱茵衣藻还是蛋白核小球藻,微藻利用添加的碳酸氢钠占总碳源的份额(f_B)都随添加 DIDS 的浓度而变化(表 4-12)。具体来看,在低浓度 DIDS($0\sim 0.5$mmol/L)条件下,两种微藻的 f_B 都在 $0.14\sim 0.30$,但是,在添加高浓度 DIDS(2.0mmol/L)条件下,两种微藻的 f_B 都接近于 0。这也再次印证了在高浓度 DIDS 条件下能够充分抑制微藻的阴离子吸收通道,最终造成微藻只能利用大气的中的碳源。

表 4-12　DIDS 处理下微藻的增殖倍数、碳源利用份额及不同形态无机碳的途径
Tab. 4-12　The proliferating multiple of algae and variation in the carbon utilization pathway and carbon sources under different content of DIDS

处理	[DIDS]/(mmol/L)	P	f_B	f_b
莱茵衣藻	0	3.24±0.21	0.21	1.00
	0.02	3.18±0.15	0.21	1.00
	0.10	2.98±0.23	0.23	1.00
	0.50	2.59±0.31	0.30	1.00
	2.00	1.82±0.26	0.01	0.00
蛋白核小球藻	0	4.35±0.18	0.14	0.81
	0.02	4.29±0.22	0.17	0.86
	0.10	4.16±0.17	0.20	0.90
	0.50	3.78±0.25	0.18	0.69
	2.00	2.09±0.29	0.02	0.00

注：P 为处理后的微藻生物量相对于接种时的增殖倍数；f_B 为微藻利用添加的无机碳占总碳源的份额；f_b 为微藻利用碳酸氢根离子途径所占的份额。

Note: P: the proliferating multiple of the treated algae related to the beginning; f_B: the proportion of the utilization of DIC from the added HCO_3^- in the whole carbon sources used by the microalgae; f_b: the proportion of the HCO_3^- pathway in the whole carbon used by the microalgae.

在添加低浓度 DIDS(0～0.5mmol/L)的条件下，莱茵衣藻几乎全部利用碳酸氢根离子途径($f_b=1$)；而蛋白核小球藻利用碳酸氢根离子途径所占的份额(f_b)略低，f_b 在 0.69～0.90。而在添加 2.0mmol/L DIDS 的条件下，微藻的阴离子吸收通道受到了充分抑制，造成两种微藻利用碳酸氢根离子途径所占的份额(f_b)都接近于 0。

(五) 微藻的碳汇能力

微藻的碳汇能力与其生长速率呈正相关关系(表 4-13)，具体到不同来源的碳汇来说，直接碳汇(CS_D)占主体，尤其是在添加较低浓度 DIDS(0～0.1mmol/L)条件下，莱茵衣藻和蛋白核小球藻的直接碳汇占总碳汇的比例都在 75% 以上。添加 DIDS，首先是造成微藻的直接碳汇能力下降；而对间接碳汇的影响可分为两种情况：低浓度 DIDS(0～0.5mmol/L)对间接碳汇只产生轻微影响，但在添加高浓度 DIDS(2.0mmol/L)条件下，造成间接碳汇能力直接降到 0 附近(表 4-13)。

表 4-13 DIDS 处理下微藻的碳汇组成

Tab. 4-13 The composition of different carbon sequestrations in algae under different content of DIDS

处理	[DIDS]/(mmol/L)	CS_{A-ai}	CS_{A-bi}	CS_{B-ai}	CS_{B-bi}	CS_D	CS_{ID}	$CS_T/\%$
莱茵衣藻	0.00	0.00	1.77	0.00	0.47	1.77	0.47	2.24 (100.00)
	0.02	0.00	1.72	0.00	0.46	1.72	0.46	2.18 (97.32)
	0.10	0.00	1.52	0.00	0.46	1.52	0.46	1.98(88.39)
	0.50	0.00	1.11	0.00	0.48	1.11	0.48	1.59(70.98)
	2.00	0.81	0.00	0.01	0.00	0.81	0.01	0.82(36.61)
蛋白核小球藻	0.00	0.55	2.33	0.09	0.38	2.88	0.47	3.35(100.00)
	0.02	0.38	2.35	0.08	0.48	2.73	0.56	3.29(98.21)
	0.10	0.25	2.28	0.06	0.57	2.53	0.63	3.16(94.33)
	0.50	0.71	1.57	0.16	0.35	2.28	0.50	2.78 (82.99)
	2.00	1.07	0.00	0.02	0.00	1.07	0.02	1.09(32.54)

注：CS_{A-ai}：利用大气中的二氧化碳，且进行二氧化碳途径时的微藻碳汇能力；CS_{A-bi}：利用大气中的二氧化碳，且进行碳酸氢根离子途径时的微藻碳汇能力；CS_{B-ai}：利用添加的无机碳源，且进行二氧化碳途径时的微藻碳汇能力；CS_{B-bi}：利用添加的无机碳源，且进行碳酸氢根离子途径时的微藻碳汇能力；CS_D：直接碳汇，所有来源于大气中的无机碳被同化；CS_{ID}：间接碳汇，所有来源于水体中的无机碳被同化；CS_T：总碳汇，直接碳汇和间接碳汇的和。括号中的数据是各个处理下的总碳汇能力相对于各自不添加 DIDS 条件下的碳汇能力的百分比。

Note：CS_{A-ai}: the contributions of the CO_2 pathways using DIC from the atmosphere；CS_{A-bi}: the contributions of the HCO_3^- pathways using DIC from the atmosphere；CS_{B-ai}: the contributions of the CO_2 pathways using DIC from the added $NaHCO_3$；CS_{B-bi}: the contributions of the HCO_3^- pathways using DIC from the added $NaHCO_3$；CS_D: direct carbon sequestration；CS_{ID}: indirect carbon sequestration；CS_T: the total carbon sequestration.

The data in brackets were the percents of algal CS_T to that of algae treated without DIDS.

一般情况下，微藻利用大气转化来的碳酸氢根离子途径的碳汇(CS_{A-bi})是微藻总碳汇(CS_T)的主体。随着添加 DIDS 浓度的增加，微藻的阴离子通道受到的抑制不断增强，造成 CS_{A-bi} 呈不断下降的趋势。在添加低浓度 DIDS(0～0.5mmol/L)条件下，两种微藻利用大气转化来的碳酸氢根离子占总碳汇的比例较大，CS_{A-bi} 都在 0.56 以上。但是，在添加高浓度 DIDS(2.0mmol/L)条件下，微藻几乎无法利用大气转化来的碳酸氢根离子（$CS_{A-bi}=0$），但此时的微藻利用大气二氧化碳的能力(CS_{A-ai})上升，这与高浓度 DIDS 能够完全抑制微藻的阴离子通道有关(Cabantchik and Greger, 1992)。

DIDS 是阴离子通道特异性抑制剂,它抑制碳酸氢根离子直接进入细胞的过程(Merrett et al.,1996)。莱茵衣藻和蛋白核小球藻的直接利用 HCO_3^- 途径主要来自于大气 CO_2 所转化来的 HCO_3^- 离子,添加低浓度 DIDS,只是轻微影响微藻利用碳酸氢根离子途径的微藻碳汇能力,只有在添加高浓度 DIDS 的条件下,才能对微藻利用碳酸氢根离子途径的碳汇能力带来明显影响。

四、碳酸酐酶胞外酶和直接碳酸氢根离子利用途径对微藻碳汇的复合影响

(一)实验设置

AZ 的浓度设置:0、0.1mmol/L、1.0mmol/L,DIDS 的浓度设置:0、0.5mmol/L,正交设计。

(二)微藻生物量变化

在不同浓度 AZ 与 DIDS 的交叉处理下,微藻的生长速率变化较大(图 4-6)。整体来看,随着添加 AZ 或 DIDS 浓度的增加,莱茵衣藻和蛋白核小球藻的生长速率都呈下降的趋势,尤其是在同时添加 AZ(1.0mmol/L)和 DIDS(0.5mmol/L)的共同作用下,两种微藻的生长都受到了较大抑制,生物量都明显下降。具体来看,莱茵衣藻对 AZ 更敏感,蛋白核小球藻对 DIDS 更敏感。具体原因分析如下:莱茵衣藻的碳酸酐酶胞外酶活性较强,而 AZ 是碳酸酐酶胞外酶的特异性抑制剂,因此,莱茵衣藻受 AZ 的影响较大;而蛋白核小球藻的碳酸酐酶胞外酶活性较弱,相应地,它的阴离子交换更活跃,而 DIDS 是阴离子通道抑制剂,最终造成蛋白核小球藻受 DIDS 的影响更大。

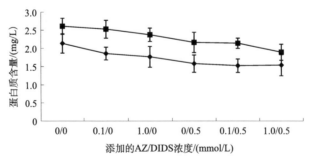

图 4-6 AZ 与 DIDS 处理下微藻蛋白质含量变化
莱茵衣藻(◆);蛋白核小球藻(■)
Fig. 4-6 The content of protein in algae under different content of AZ and DIDS
Chlamydomonas reinhardtii (◆); *Chlorella pyrenoidosa* (■)

(三) 微藻碳同位素组成变化

从处理后的微藻稳定碳同位素组成的数据来看，不论莱茵衣藻还是蛋白核小球藻，它们的藻体稳定碳同位素组成都随添加 AZ 或者 DIDS 浓度的增加呈偏负的趋势(表 4-14)。具体来看，不论是否添加 DIDS，AZ 都对莱茵衣藻的稳定碳同位素组成影响较大。这跟莱茵衣藻的碳酸酐酶胞外酶活性较强有关。在添加 0.5 mmol/L DIDS 的条件下，AZ 对蛋白核小球藻的稳定碳同位素组成的影响也变大。从添加两种标记的不同 $\delta^{13}C$ 值的碳酸氢钠来看，添加碳酸氢钠的 $\delta^{13}C$ 值越负，对应藻体的 $\delta^{13}C$ 值也越负。

表 4-14 AZ 与 DIDS 处理下微藻的稳定碳同位素组成
Tab. 4-14 Stable carbon isotope composition of algae under different content of AZ and DIDS

	处理 AZ/DIDS/(mmol/L)	δ_{T1}	δ_{T2}
莱茵衣藻	0/0	−19.1±0.2	−20.3±0.3
	0.10/0	−25.1±0.3	−25.4±0.4
	1.00/0	−25.4±0.4	−26.4±0.5
	0/0.50	−21.1±0.2	−24.0±0.3
	0.10/0.50	−27.2±0.5	−29.9±0.4
	1.00/0.50	−28.3±0.5	−31.8±0.5
蛋白核小球藻	0/0	−21.3±0.4	−22.3±0.3
	0.10/0	−22.0±0.4	−22.3±0.5
	1.00/0	−22.7±0.4	−23.6±0.4
	0/0.50	−22.9±0.4	−24.5±0.3
	0.10/0.50	−25.9±0.6	−27.8±0.5
	1.00/0.50	−28.0±0.5	−30.2±0.5

注：δ_{T1}：添加 $\delta^{13}C$ 为 −17.4‰的 NaHCO₃ 的培养液所培养出的微藻 $\delta^{13}C$ 值；δ_{T2}：添加 $\delta^{13}C$ 为 −28.4‰的 NaHCO₃ 的培养液所培养出的微藻 $\delta^{13}C$ 值。

Note：δ_{T1}: The value of microalgae's $\delta^{13}C$ cultured in the medium added NaHCO₃ with an $\delta^{13}C$ value of −17.4‰；δ_{T2}: The value of microalgae's $\delta^{13}C$ cultured in the medium added NaHCO₃ with an $\delta^{13}C$ value of −28.4‰.

(四) 微藻吸收利用不同无机碳碳源的份额及不同形态无机碳途径的份额

不论莱茵衣藻还是蛋白核小球藻，微藻利用添加的碳酸氢钠占总碳源的份额 (f_B) 都随添加 AZ 和 DIDS 的浓度的变化而变化(表 4-15)。具体来看，添加

0.5mmol/L DIDS 造成两种微藻的 f_B 都升高。

表 4-15　AZ 与 DIDS 处理下微藻的增殖倍数、碳源利用份额及不同无机碳利用途径
Tab. 4-15　The proliferating multiple of algae and variation in the carbon utilization pathway and carbon sources under different content of AZ and DIDS

	处理 AZ/DIDS/(mmol/L)	P	f_B	f_b
莱茵衣藻	0/0	3.50±0.26	0.11	0.97
	0.10/0	3.12±0.23	0.03	0.25
	1.00/0	2.89±0.29	0.09	0.26
	0/0.50	2.71±0.24	0.26	0.86
	0.10/0.50	2.58±0.27	0.25	0.17
	1.00/0.50	2.46±0.32	0.32	0.10
蛋白核小球藻	0/0	4.27±0.26	0.09	0.77
	0.10/0	4.19±0.25	0.03	0.65
	1.00/0	4.03±0.24	0.08	0.61
	0/0.50	3.64±0.28	0.15	0.63
	0.10/0.50	3.55±0.27	0.18	0.32
	1.00/0.50	3.21±0.30	0.20	0.10

注：P 为处理后的微藻生物量相对于接种时的增殖倍数；f_B 为微藻利用添加的无机碳占总碳源的份额；f_b 为微藻利用碳酸氢根离子途径所占的份额。

Note: P: the proliferating multiple of the treated algae related to the beginning; f_B: the proportion of the utilization of DIC from the added HCO_3^- in the whole carbon sources used by the microalgae; f_b: the proportion of the HCO_3^- pathway in the whole carbon used by the microalgae.

在不添加 AZ 和 DIDS 的条件下，莱茵衣藻几乎全部利用碳酸氢根离子途径（$f_b=0.97$）；而蛋白核小球藻的略低（$f_b=0.77$）。随着添加 AZ 浓度的增加，两种微藻的 f_b 下降明显，尤其是在添加 0.5mmol/L DIDS 的条件下，微藻的阴离子吸收通道也受到了抑制，随添加 AZ 浓度的增加，造成两种微藻的 f_b 下降更加明显。

（五）微藻的碳汇能力

微藻的碳汇能力与其生长速率呈正相关关系（表 4-15，表 4-16），在不添加 AZ 和 DIDS 的情况下，微藻利用大气转化来的碳酸氢根离子途径的碳汇（CS_{A-bi}）是微藻总碳汇（CS_T）的主体。具体到不同来源的碳汇来说，直接碳汇（CS_D）为微藻总碳汇（CS_T）的主体，在莱茵衣藻中，CS_D 占 CS_T 的比例不低于 0.68，而在蛋白核小球藻中，CS_D占CS_T的比例不低于0.79。但是，添加AZ或DIDS，都造成CS_D下降明显。在这两种微藻中，不添加AZ和DIDS的条件下，CS_{A-bi}占CS_T的比例都大于0.68，但是，在添加1.0mmol/L AZ和0.5mmol/L DIDS的条件下，CS_{A-bi}占CS_T的比例下降明显，都在0.16以下。

表4-16　AZ与DIDS处理下微藻的碳汇组成

Tab. 4-16　The composition of different carbon sequestrations in algae under different content of AZ and DIDS

	处理 AZ/DIDS/(mmol/L)	CS_{A-ai}	CS_{A-bi}	CS_{B-ai}	CS_{B-bi}	CS_D	CS_{ID}	CS_T/%
莱茵衣藻	0/0	0.06	2.17	0.01	0.27	2.23	0.27	2.50 (100.00)
	0.10/0	1.55	0.52	0.04	0.01	2.06	0.06	2.12 (84.80)
	1.00/0	1.27	0.45	0.13	0.05	1.72	0.17	1.89 (75.60)
	0/0.50	0.17	1.09	0.06	0.39	1.26	0.45	1.71 (68.40)
	0.10/0.50	0.99	0.20	0.32	0.07	1.19	0.39	1.58 (63.20)
	1.00/0.50	0.89	0.10	0.42	0.05	1.00	0.46	1.46 (58.40)
蛋白核小球藻	0/0	0.68	2.29	0.07	0.23	2.97	0.30	3.27 (100.00)
	0.10/0	1.09	2.01	0.03	0.06	3.10	0.09	3.19 (97.55)
	1.00/0	0.09	1.69	0.10	0.15	2.78	0.25	3.03 (92.66)
	0/0.50	0.83	1.42	0.14	0.24	2.26	0.38	2.64 (80.73)
	0.10/0.50	1.44	0.67	0.30	0.14	2.11	0.44	2.55 (77.98)
	1.00/0.50	1.58	0.18	0.40	0.05	1.77	0.44	2.21 (67.58)

注：CS_{A-ai}：利用大气中的二氧化碳，且进行二氧化碳途径时的微藻碳汇能力；CS_{A-bi}：利用大气中的二氧化碳，且进行碳酸氢根离子途径时的微藻碳汇能力；CS_{B-ai}：利用添加的无机碳源，且进行二氧化碳途径时的微藻碳汇能力；CS_{B-bi}：利用添加的无机碳源，且进行碳酸氢根离子途径时的微藻碳汇能力；CS_D：直接碳汇，所有来源于大气中的无机碳被同化；CS_{ID}：间接碳汇，所有来源于水体中的无机碳被同化；CS_T：总碳汇，直接碳汇和间接碳汇的和。括号中的数据是各个处理下的总碳汇能力相对于各自不添加DIDS条件下的碳汇能力的百分比。

Note：CS_{A-ai}: the contributions of the CO_2 pathways using DIC from the atmosphere; CS_{A-bi}: the contributions of the HCO_3^- pathways using DIC from the atmosphere; CS_{B-ai}: the contributions of the CO_2 pathways using DIC from the added $NaHCO_3$; CS_{B-bi}: the contributions of the HCO_3^- pathways using DIC from the added $NaHCO_3$; CS_D: direct carbon sequestration; CS_{ID}: indirect carbon sequestration; CS_T: the total carbon sequestration.

The data in brackets were the percents of algal CS_T to that of algae treated without AZ or DIDS.

第四节　微藻碳汇组成与"遗失的碳汇"

由于水环境和种群多样性，造成微藻吸收无机碳的机制也具有多样性，并形成了一系列利用无机碳的方式。莱茵衣藻及一些微藻已被证实具有碳酸酐酶胞外酶和直接利用碳酸氢根离子的机制(Colman et al., 2002; Sültemeyer et al., 1991),通过本研究,可以证实蛋白核小球藻也具有碳酸酐酶胞外酶和直接利用碳酸氢根离子的机制,并由此推测,其他的一些由碳酸酐酶胞外酶所催化的具有CCM机制的微藻也具有碳酸酐酶胞外酶和直接利用碳酸氢根离子的机制。

在海水和喀斯特自然水体中的微藻利用无机碳的过程中,由CA参与的对扩散的CO_2利用以及直接HCO_3^-利用途径等被认为是最重要的无机碳获取方式(Axelsson et al., 1995; Colman et al., 2002)。由于不同微藻的CA活性差异较大,再加上不同微藻的无机碳获取方式的不同,相应地,它们的无机碳获取能力差异也大(Mercado et al., 1998; Martin and Tortell, 2008)。不同的藻类因为不同的无机碳利用机制捕获大气中的二氧化碳能力也不同。表4-17表示的是莱茵衣藻和蛋白核小球藻在不同无机碳利用模式下的碳汇组成。

表4-17 莱茵衣藻和蛋白核小球藻在不同无机碳获取模式下的碳汇组成
Tab. 4-17 Carbon sequestration of *Chlamydomonas reinhardtii* and *Chlorella pyrenoidosa* with assuming multiple Ci uptake systems

AZ/DIDS	碳汇种类	只有直接碳酸氢根离子利用途径 /%	只有二氧化碳利用途径 /%	碳酸酐酶作用下的无机碳利用途径(胞外碳酸酐酶作用和直接无机碳利用途径) /%
莱茵衣藻	CS_T	68♀(100♂)	37♀(100♂)	100♀(100♂)
	CS_D	58♀(85♂)	36♀(98♂)	92♀(92♂)
	CS_{ID}	10♀(15♂)	1♀(2♂)	8♀(8♂)
蛋白核小球藻	CS_T	54♀(100♂)	31♀(100♂)	100♀(100♂)
	CS_D	45♀(83♂)	30♀(99♂)	93♀(93♂)
	CS_{ID}	9♀(17♂)	1♀(1♂)	7♀(7♂)

注:CS_D:直接碳汇,所有来源于大气中的无机碳被同化;CS_{ID}:间接碳汇,所有来源于水体中的无机碳被同化;CS_T:总碳汇,直接碳汇和间接碳汇和;♀:某一无机碳吸收模式的微藻相对于对照微藻的碳汇百分比;♂:括号中的数据是相对于总碳汇的百分比。

Note: CS_D: direct carbon sequestration; CS_{ID}: indirect carbon sequestration; CS_T: the total carbon sequestration; ♀ the data out of brackets were the percents of CS_T of algae with certain mode of Ci uptake system to that under the control; ♂ : The data in brackets were the percents of CS_T.

微藻行碳酸酐酶作用下的无机碳利用途径获得的碳汇比只行直接碳酸氢根离子利用途径或二氧化碳利用途径获得的碳汇多得多。只行二氧化碳利用途径的藻类获得的碳汇最少。碳汇的种类也因藻类无机碳的利用方式的不同有显著差异。

只行二氧化碳利用途径的藻类的间接碳汇只占总碳汇的1%～3%，而只行直接碳酸氢根离子利用途径的间接碳汇可达总碳汇的1/6。自然水体，不同的藻类行使各自的无机碳利用策略来应对无机碳的变化。

全球大气二氧化碳收支不平衡；目前报道出的全球每年"遗失的碳汇"约有0.4～4.0Gt标准碳（Gifford，1994）。众多学者对影响全球大气CO_2浓度的因素及其反馈机制的研究存在较大争议（Sabine et al.，2004；Schindler，1999）。碳汇估算的这一收支不平衡现状预示着还有一些重要因素未被发现。海洋是大气二氧化碳最重要的碳汇，大约吸收了全球化石燃料的燃烧和砍伐热带雨林所释放的二氧化碳的三分之一（Sabine et al.，2004；Siegenthaler and Sarmiento，1993）。之前估计的碳汇是基于大气二氧化碳通量，海洋里溶解的无机碳的通量是难以衡量和估算的（Tans et al.，1990），因此，对全球大气中的二氧化碳收支考察时，海洋中碳酸氢根离子通量变化对碳汇的贡献常被忽略。目前的研究表明，藻类利用已存在自然水体的HCO_3^-占总DIC的1%～17%。由于水体的无机碳的平衡作用，消耗HCO_3^-会使相应的二氧化碳补到水体，造成二氧化碳的封存。因此，水生植物的间接碳汇可能是自然水体的主要"遗失的碳汇"。此外，陆生植物的相关研究也表明，植物利用外源碳酸氢根离子占总碳汇份额在0～30%（Wu and Xing，2012），而植物对外源碳酸氢根离子吸收利用也被认为是一种"遗失的碳汇"。

参 考 文 献

沈允钢. 2006. 光合作用. 中国生物学文摘，20(2)：1.

Axelsson L，Ryberg H，Beer S. 1995. Two modes of bicarbonate utilization in the marine green macroalga *Ulva lactuca*. Plant，Cell & Environment，18(4)：439-445.

Badger M R，Andrews T J，Whitney S M，et al. 1998. The diversity and coevolution of Rubisco，plastids，pyrenoids，and chloroplast-based CO_2-concentrating mechanisms in algae. Canadian Journal of Botany，76(6)：1052-1071.

Badger M R，Price G D. 1992. The CO_2 concentrating mechanism in cyanobacteria and microalgae. Physiologia Plantarum，84(4)：606-615.

Badger M R，Price G D. 1994. The role of carbonic anhydrase in photosynthesis. Annual Review of Plant Biology，45：369-392.

Bozzo G G，Colman B. 2000. The induction of inorganic carbon transport and external carbonic anhydrase in *Chlamydomonas reinhardtii* is regulated by external CO_2 concentration. Plant，Cell & Environment，23(10)：1137-1144.

Burkhardt S，Riebesell U，Zondervan I. 1999. Effects of growth rate，CO_2 concentration，and cell size on the stable carbon isotope fractionation in marine phytoplankton. Geochimica et Cosmochimica Acta，63(22)：3729-3741.

Cabantchik Z I, Greger R. 1992. Chemical probes for anion transporters of mammalian cell membranes. American Journal of Physiology-Cell Physiology, 262(4): C803-C827.

Chen Z, Cheng H M, Chen X W. 2009. Effect of Cl$^-$ on photosynthetic bicarbonate uptake in two cyanobacteria *Microcystis aeruginosa* and *Synechocystis* PCC6803. Chinese Science Bulletin, 54(7): 1197-1203.

Coleman J R. 2000. Carbonic anhydrase and its role in photosynthesis. Photosynthesis: 353-367.

Colman B, Huertas I E, Bhatti S, Dason J S. 2002. The diversity of inorganic carbon acquisition mechanisms in eukaryotic microalgae. Functional Plant Biology, 29(3): 261-270.

Emrich K, Ehhalt D H, Vogel J C. 1970. Carbon isotope fractionation during the precipitation of calcium carbonate. Earth and Planetary Science Letters, 8(5): 363-371.

Gifford R M. 1994. The global carbon cycle: a viewpoint on the missing sink. Australian Journal of Plant Physiology, 21(1): 1-15.

Giordano M, Beardall J, Raven J A. 2005. CO_2 concentrating mechanisms in algae: mechanisms, environmental modulation, and evolution. Annual Review Plant Biology, 56: 99-131.

Huertas I E, Espie G S, Colman B, Lubian L M. 2000. Light-dependent bicarbonate uptake and CO_2 efflux in the marine microalga *Nannochloropsis gaditana*. Planta, 211(1): 43-49.

Invers O, Perez M, Romero J. 1999. Bicarbonate utilization in seagrass photosynthesis: role of carbonic anhydrase in *Posidonia oceanica* (L) Delile and *Cymodocea nodosa* (Ucria) Ascherson. Journal of Experimental Marine Biology and Ecology, 235(1): 125-133.

Larkum A W D, Roberts G, Kuo J, Strother S. 1989. Biology of seagrasses: A treatise on the biology of seagrasses with special reference to the Australian region. Amsterdam: Elsevier Science Publishers BV: 686-722.

Maberly S C. 1992. Carbonate ions appear to neither inhibit nor stimulate use of bicarbonate ions in photosynthesis by *Ulva lactuca*. Plant, Cell & Environment, 15: 255-260.

Martin C L, Tortell P D. 2008. Bicarbonate transport and extracellular carbonic anhydrase in marine diatoms. Physiologia Plantarum, 133(1): 106-116.

Mercado J M, Gordillo F J L, Figueroa F L, Niell F X. 1998. External carbonic anhydrase and affinity for inorganic carbon in intertidal macroalgae. Journal of Experimental Marine Biology and Ecology, 221(2): 209-220.

Merrett M J, Nimer N A, Dong L F. 1996. The utilization of bicarbonate ions by the marine microalga *Nannochloropsis oculata* (Droop) Hibberd. Plant, Cell & Environment, 19(4): 478-484.

Miyachi S, Tsuzuki M, Avramova S T. 1983. Utilization modes of inorganic carbon for photosynthesis in various species of *Chlorella*. Plant and cell physiology, 24(3): 441-451.

Mook W G, Bommerson J C, Staverman W H. 1974. Carbon isotope fractionation between dissolved bicarbonate and gaseous carbon dioxide. Earth and Planetary Science Letters, 22(2): 169-176.

Morant-Manceau A, Nguyen T L N, Pradier E, Tremblin G. 2007. Carbonic anhydrase activity and photosynthesis in marine diatoms. European Journal of Phycology, 42(3): 263-270.

Moroney J V, Husic H D, Tolbert N E. 1985. Effect of carbonic anhydrase inhibitors on inorganic carbon accumulation by *Chlamydomonas reinhardtii*. Plant Physiology, 79 (1): 177-183.

Moroney J V, Tolbert N E. 1985. Inorganic carbon uptake by *Chlamydomonas reinhardtii*. Plant Physiology, 77(2): 253-258.

Nimer N A, Iglesias-Rodriguez M D, Merrett M J. 1997. Bicarbonate utilization by marine phytoplankton species. Journal of Phycology, 33(4): 625-631.

Riebesell U, Wolf-Gladrow D A, Smetacek V. 1993. Carbon dioxide limitation of marine phytoplankton growth rates. Nature, 361: 249-251.

Rost B, Kranz S, Richter K U, Tortell P D. 2007. Isotope disequilibrium and mass spectrometric studies of inorganic carbon acquisition by phytoplankton. Limnology and Oceanography: Methods, 5: 328-337.

Sabine C L, Feely R A, Gruber N, et al. 2004. The oceanic sink for anthropogenic CO_2. Science, 305(5682): 367-371.

Schindler D W. 1999. Carbon cycling: The mysterious missing sink. Nature, 398: 105-107.

Sharkia R, Beer S, Cabantchik Z I. 1994. A membrane-located polypeptidd of *Ulva* sp. Which may be involved in HCO_3^- uptake is recognized by antibodies raised against the human red-blood-cell anion-exchange protein. Planta, 194(2): 247-249.

Siegenthaler U, Sarmiento J L. 1993. Atmospheric carbon dioxide and the ocean. Nature, 365: 119-125.

Skirrow G, Whitfield M. 1975. The effect of increases in the atmospheric carbon dioxide content on the carbonate ion concentration of surface ocean water at 25℃. Limnology and Oceanography, 20(1): 103-108.

Sültemeyer D. 1998. Carbonic anhydrase in eukaryotic algae: characterization, regulation, and possible function during photosynthesis. Canadian Journal of Botany, 76(6): 962-972.

Sültemeyer D F, Fock H P, Canvin D T. 1991. Active uptake of inorganic carbon by *Chlamydomonas reinhardtii*: evidence for simultaneous transport of HCO_3^- and CO_2 and characterization of active CO_2 transport. Canadian Journal Botany, 69(5): 995-1002.

Talling J F. 1976. The depletion of carbon dioxide from lake water by phytoplankton. The Journal of Ecology, 64(1): 79-121.

Tans P P, Fung I Y, Takahashi T. 1990. Observational contrains on the global atmospheric CO_2 budget. Science, 247(4949): 1431-1438.

Tsuzuki M, Miyachi S. 1989. The function of carbonic anhydrase in aquatic photosynthesis. Aquatic Botany, 34(1-3): 85-104.

Wu Y Y, Xing D K. 2012. Effect of bicarbonate treatment on photosynthetic assimilation of inorganic carbon in two plant species of Moraceae. Photosynthetica, 50(4): 587-594.

Wu Y Y, Xu Y, Li H T, Xing D K. 2012. Effect of acetazolamide on stable carbon isotope fractionation in *Chlamydomonas reinhardtii* and *Chlorella vulgaris*. Chinese Science Bulletin, 57(7): 786-789.

Young E, Beardall J, Giordano M. 2001. Inorganic carbon acquisition by *Dunaliella tertiolecta* (Chlorophyta) involves external carbonic anhydrase and direct HCO_3^- utilization insensitive to the anion exchange inhibitor DIDS. European Journal of Phycology, 36(01): 81-88.

第五章 微藻碳酸酐酶与生物岩溶作用

摘　　要

微藻参与下的岩溶动力系统，形成一个"水（H_2O）-岩（$CaMg(CO_3)_2$）-气（CO_2）-生（CH_2O）"相互作用的耦合体系。这个体系对全球碳循环有着重要的影响。微藻的胞外碳酸酐酶和阴离子通道系统使微藻二氧化碳浓缩机制得以实现，并对微藻在这个耦合体系中的作用起着重要的影响。同时，外界二氧化碳分压（P_{CO_2}）和 pH 强烈影响着微藻胞外碳酸酐酶活力，其在微藻与碳酸盐岩交互过程中起到重要的控制作用。

以莱茵衣藻、蛋白核小球藻和铜绿微囊藻为研究对象，采用室内模拟控制实验的方法，利用双向同位素示踪技术、生物溶蚀定量模型和反向识别技术，成功定量研究了微藻胞外碳酸酐酶、阴离子通道、P_{CO_2} 和水体 pH 对碳酸盐矿物生物溶蚀和沉积作用，及其胞外碳酸酐酶在不同条件下对岩溶体系各碳汇效应的影响。

微藻利用方解石碳源的份额随着乙酰唑胺（AZ）浓度的增加而增加，但是单位时间单位质量的莱茵衣藻对方解石 Mg^{2+} 的释放量却减小。因此，微藻的胞外碳酸酐酶对方解石生物溶蚀过程起到促进作用，但同时也能够通过催化利用水体的重碳酸盐而导致碳酸钙沉淀的产生。总体来看，微藻的胞外碳酸酐酶在方解石的生物溶蚀过程中起到负的碳汇效应，而在光合碳汇中起到正效应。

铜绿微囊藻利用方解石碳源的份额以及单位时间单位质量的藻体对方解石 Mg^{2+} 的释放量均随着 4,4′-二异硫氰-2,2′-二磺酸芪（DIDS）浓度增大而减小。微藻的阴离子通道一方面能够通过对水体重碳酸盐的利用来促进方解石的生物溶蚀；另一方面随着重碳酸盐的利用又会引起碳酸钙的沉淀。总体来看，阴离子通道对 $CaCO_3$ 沉淀的产生效应大于方解石的溶蚀效应，因此其对方解石生物岩溶过程中碳汇效应表现为消极的影响，而在光合碳汇中起到积极的影响。

外界环境的 P_{CO_2} 一方面影响胞外碳酸酐酶活力，另一方面控制着方解石的化学溶蚀作用。P_{CO_2} 越低，方解石的总溶蚀量越小，微藻的生物溶蚀作用越小，但却能促进微藻对方解石碳源的利用份额。

对于有较强胞外碳酸酐酶活力的莱茵衣藻来说，pH 越大，单位时间单位质量的微藻对方解石中 Mg^{2+} 的释放量越小；而只有较小或者不具有胞外碳酸酐酶活力的蛋白核小球藻和铜绿微囊藻反之。水体 pH 越高，微藻对方解石碳源的利用份额越小。总体来看，在生物岩溶过程中，水体 pH 越低，方解石溶蚀释放的无机碳被藻体利用和形成溶解性无机碳的量越大，其岩溶碳汇效应越大，光合碳汇越小。

Chaper 5 Microalgal carbonic anhydrase versus the Bio-Karst

The karst dynamic system with the involvement of mircroalgae had formed an interactive coupling system of "water (H_2O)-rock($CaMg(CO_3)_2$)-gas(CO_2)-biology (CH_2O)", which has a great influence on the global carbon cycle. Extracellular carbonic anhydrase (CAex) and the anion channel system had realized the carbon dioxide concentration mechanism (CCM) of microalgae, and play a major part in the coupling system. In addition, extracellular carbonic anhydraseis greatly affected by partial pressure of carbon dioxide (P_{CO_2}) and pH in the aquatic medium, and play an important role in control the interaction between the microalgae and carbonate rock.

Chlamydomonas reinhardtii (C.R), *Chlorella pyrenoedosa* (C.P), *Microcystis aeruginosa* (M.A) were used for experimental materials. The simulating control experiment, bidirectional isotope labelling tracer, the biological dissolution quantitative model cooperatecl reverse recognition technolog were employed to quantify the effect on the biological dissolution and precipitation of carbonate minerals, which resulted from CAex, the anion channel in alage, P_{CO_2} and pH in aquatic medium, and the effect of CAex on the carbon sink of karst.

The proportion of the utilization on carbon in magnesium calcite by microalgae increased, and the amounts of Mg^{2+} released from magnesium calcite per unit algal biomass and unit time by C.R decreased, accordingly, with the increasing concentrations of acetazolamide (AZ). Thus, the CAex of microalgae can promote the biological dissolution of calcite, while simultaneously, it can also catalyze and utilize the bicarbonate in aquatic medium, resulting the precipitation of calcium carbonate. Generally, the CAex of microalgae had a negative effect on the bio-karst carbon sink and a positive effect on the photosynthetic carbon sink.

Both the proportion of the utilization on carbon in magnesium calcite by microalgae and the amounts of Mg^{2+} released from magnesium calcite per unit algal biomass and unit time by $M.A$ decreased with the increasing 4, 4'-diisothiocyanatostilbene-2, 2'-disulfonic acid (DIDS). The anion channel, on the one hand, can promote the biological dissolution of calcite by ultilizing the bicarbonate; on the other hand, can lead to the precipitation of calcium carbonate with the ultilization on bicarbonate by microalgae. In summary, the anion channel in microalgae had greater effect on the calcium carbonate precipitation than on the dissolution of calcite, therefore, it had a negative effect on bio-karst carbon sink and a positive effect on photosynthetic carbon sink.

The P_{CO_2} in external environment, on the one hand, influence the activity of CAex, on the other hand, the chemical dissolution of calcite. Both the total dissolution amounts and the biological dissolution of microalgae decrease with the decreasing P_{CO_2}. However, low P_{CO_2} will promote the proportion of utilization on carbon in calcite by microalgae.

The amount of Mg^{2+} released from magnesium calcite per unit algal biomass and unit time by $C.R$ with great CAex activity decrease with the increasing pH, contrarily, that by the $C.P$ and $M.A$ with less or no CAex activity increase. In addition, the proportion of utilization on carbon in calcite decreases with the increasing pH in aquatic media. Overall, in the process of bio-karst, the amount of inorganic carbon utilized by microalgae and dissolved inorganic carbon in the aquatic medium form the dissolution of calcite, increase with the decrease in pH, thus, the effect of bio-karst carbon sink become more obvious and photosynthetic carbon sink become more negative under the low pH condition.

水生光合生物的参与下，碳酸盐岩溶蚀及其碳汇效应越来越受到广泛的关注。微藻一方面能够促进碳酸盐岩的溶蚀，增强"碳酸盐泵"的效应(Xie and Wu, 2014)；另一方面又能够利用水体的无机碳，将活性碳成分转变惰性有机碳成分，从而使碳长期保存在海洋里，形成"微型生物泵"(Chen, 2011)。

因此，研究微藻对碳酸盐岩的生物溶蚀作用对进一步认识全球碳循环模型有重要的指导意义，并有助于我们寻找迷失的碳汇。同时，对微藻的生物溶蚀机理和无机碳利用途径的研究，又能够帮助我们预测全球碳循环变化规律，并为碳收支情况和减排政策提供重要的依据。

第一节 碳酸酐酶对碳酸盐岩的溶蚀作用

碳酸盐岩是全球最大的碳库，其碳储量约 5×10^{21} mol(Bricker，1988)，占全球总碳量的 99.55%，分别是海洋和全世界植被的 1694 和 1 100 000 倍(Houghton and Woodwell，1989)。现代岩溶学研究结果表明，碳酸盐岩在积极的参与全球物质循环，且响应极其迅速(袁道先，1993，1995)。我国和世界碳酸盐岩地区因碳酸盐岩岩溶作用从大气中吸收的净 CO_2 总量，即碳酸盐岩岩溶作用对大气 CO_2 沉降的贡献，分别达 1800 万 t C/a，整个世界岩溶地区 1.1 亿 t C/a(刘再华，2000)。如果水生光合生物参与其中，碳酸盐岩溶蚀及其碳汇效应显得更加明显(Liu et al.，2010b)。微藻的碳酸酐酶(carbonic anhydrase，CA)能够快速催化 $CO_2+H_2O \longleftrightarrow HCO_3^-+H^+$ 可逆水合反应，加速无机碳向羧化酶活性部位扩散，提高 CO_2 固定效率，改变微藻细胞外微环境的 pH 和各种形式 DIC(CO_2、HCO_3^-、CO_3^{2-})的相对含量(Hewett-Emmett and Tashian，1996；Saarnio et al.，1998；郭敏亮和高煜珠，1989)，从而影响碳酸盐岩的溶蚀和 $CaCO_3$ 次生沉淀的产生，并进一步控制着与碳酸盐岩相关的生物成矿作用。

一、碳酸酐酶与碳酸盐岩的溶蚀

碳酸盐岩的溶蚀($CaCO_3+CO_2+H_2O \rightarrow Ca^{2+}+2HCO_3^-$)速率主要受三个过程的控制：①矿物表面的动力学溶蚀过程；②边界层中溶蚀出的 Ca^{2+}，HCO_3^- 和 CO_3^{2-} 的扩散速度和 CO_2 的富集含量；③CO_2 与 H^+ 和 HCO_3^- 之间的转换(Kaufmann and Dreybrodt，2007；Wallin and Bjerle，1989；Yadav et al.，2008)。$CO_2+H_2O \rightarrow H^++HCO_3^-$ 的动力学慢反应，在控制 $H_2O-CO_2-CaCO_3$ 系统中碳酸盐岩的溶蚀和沉淀过程中起着重要的限制作用(Dreybrodt et al.，1996；Liu and Dreybrod，1997)。而碳酸酐酶能够快速催化 CO_2 水合反应，在碳酸盐岩溶蚀过程中起到重要的作用。其一方面，提高 CO_2 与 H^+ 和 HCO_3^- 之间的转换速率；另一方面，通过促进微藻对 HCO_3^- 和 Ca^{2+} 的利用来提高相关离子的扩散速度，从而促进碳酸盐岩的溶蚀。

刘再华等利用旋转盘技术开展室内多条件变化研究发现，加入牛碳酸酐酶后，灰岩的溶解速率在高 CO_2 分压条件下可增加 10 倍，白云岩溶解速率增加主要在低 CO_2 分压条件下，可达 3 倍(Liu et al.，2005)。余龙江等利用微生物的胞外碳酸酐酶粗提酶与石灰岩相互反应发现，微生物 CA 粗提酶液体系对灰岩的溶蚀速率比酶液受抑制体系高 139%，比去离子水对照体系高 975%(余龙江等，2004)。李为等通过室内模拟石灰岩土壤系统，对接种莱茵衣藻的土柱进行不同淋滤条件的处理发现，以双蒸水进行淋滤的土柱，Ca^{2+} 总淋出量在 24 天时仅为以微藻培养液进行淋滤主柱的 11.9%,且淋出液碳酸酐酶的平均活性与 Ca^{2+} 总淋出量间有较好的相关性(李为等,2011)。此外,通过对微生物胞外碳酸酐酶进行类似的研究,也发现胞外碳酸酐酶明显促进了石灰土 Ca^{2+}、Mg^{2+} 的释放(Li et al.,2005)。在进行碳酸盐岩与莱茵衣藻(*Chlamydomonas reinhardtii*)和蛋白质核小球藻(*Chlorella pyrenoedosa*)的交互实验过程中,我们也发现在胞外碳酸酐酶未受到抑制的情况下,单位时间单位质量叶绿素 a 的 *C. reinhardtii* 和 *C. pyrenoedosa* 对灰岩 Mg^{2+} 的释放量分别为:3.37×10^{-4} mg/(μg·d)和 2.44×10^{-4} mg/(μg·d),而当胞外碳酸酐酶受到抑制时 *C. reinhardtii* 和 *C. pyrenoedosa* 对灰岩 Mg^{2+} 的释放量分别仅为:1.99×10^{-4} mg/(μg·d)和 2.19×10^{-4} mg/(μg·d)(Xie and Wu,2014)。

二、碳酸酐酶与碳酸钙的沉积

微藻对碳酸盐沉积的主要机制可归结为 2 种:①微藻在通过光合作用利用水体中的 CO_2 和 HCO_3^- 同时,改变着海洋或湖泊的局部水体环境,使水体中的 pH 升高、DIC 的平衡体系发生移动,从而导致碳酸钙沉淀的产生($Ca^{2+} + HCO_3^- \longleftrightarrow CaCO_3 + CH_2O + O_2$)(李为等,2009)。②微藻通过阴离子通道吸收水体中的 HCO_3^- 进入细胞内,再通过细胞膜上的 Ca^{2+}/H^+ 离子通道将细胞外的 H^+ 转运到细胞内将 HCO_3^- 转化为 CO_2($HCO_3^- + H^+ \longleftrightarrow CO_2 + H_2O$),从而改变细胞表面微环境的 pH 和 Ca^{2+} 离子浓度,促使 $CaCO_3$ 的产生($Ca^{2+} + HCO_3^- \rightarrow CaCO_3 + H^+$)(Hammes and Verstraete,2002)。碳酸酐酶在这两种 $CaCO_3$ 的微藻沉淀机制中都起到重要的作用。其中胞外碳酸酐酶广泛地分布在微藻细胞外,特别是绿藻,能够将细胞表面的 HCO_3^- 快速地转化为 CO_2,再通过渗透作用进入细胞内供微藻进行光合作用,而此过程中将减少水体中的一个 H^+,从而升高细胞表面微环境的 pH 促进 $CaCO_3$ 沉淀。胞内碳酸酐酶同样广泛分布在微藻细胞内,能够催化细胞内 HCO_3^- 的水合反应,降低细胞内 H^+ 的浓度,诱导 Ca^{2+}/H^+ 离子通道转运速度的提高,从而促进 $CaCO_3$ 的产生。

1996 年 Miyamoto 等首次报道在牡蛎(*Pinctada fucata*)珍珠层的 nacrein(一种可溶性有机质蛋白质)中发现了 2 个功能域:一个是 CA,一个是 Gly-Xaa-Asn (Xaa=Asp,Asn 或 Glu)重复域,不过,CA 域被 Gly-Xaa-Asn 重复域插入分成 2 个子域,由此认为,nacrein 的实际功能是作为基质蛋白质,其重复 Gly-Xaa-Asn 域可能连接钙,同时作为 CA 催化 HCO_3^- 的形成,因而参与了珍珠层 $CaCO_3$ 晶体的形成(Miyamoto et al.,1996)。随后,Li 等的研究发现,单位毫克的小球藻的胞内和胞外碳酸酐酶粗提酶对碳酸钙的沉淀量分别为 160mg 和 72mg,而单位毫克的 *Citrobacter freundii* 的纯碳酸酐酶将诱导 225mg 的 $CaCO_3$ 沉淀的产生(Li et al.,2012)。Sharma 等也发现,与没有碳酸酐酶催化的情况相比，碳酸酐酶明显促进了 $CaCO_3$ 的沉淀(Sharma and Bhattacharya，2010)。同时，有学者在密闭培养体系中研究喜钙念珠藻(*Nosioc calciola Breb*)对 $CaCO_3$ 的沉积作用时，发现较高胞外 CA 活性的喜钙念珠藻能使桂林岩溶试验场水样的碳酸钙沉积速率达到 22.4mg/(d·L)，并推测喜钙念珠藻在胞外 CA 的作用下使 HCO_3^- 脱水形成 CO_2 而被利用进行光合作用，并引起水体内 CO_2 平衡体系的移动从而促进 $CaCO_3$ 的沉积(李强等，2005)。Liu 等在喀斯特和非喀斯特水体中对 *Oocystis solitaria* 进行相应的实验，也得出相似的结果，认为在具有高 Ca^{2+}、高 pH 和低 CO_2 浓度特点的喀斯特水体中微藻的胞外碳酸酐酶能够催化利用 HCO_3^-，从而导致 $CaCO_3$ 沉淀的产生(Liu et al.，2010a)。在研究微藻对碳酸盐岩的溶蚀过程中，我们也发现胞外碳酸酐酶对水体中 $CaCO_3$ 沉淀的产生有潜在的促进作用(Xie and Wu，2014)。

三、碳酸酐酶与生物成矿作用

微生物成矿作用是指微生物及其代谢作用所产生的有机质参与成矿或分异、聚集元素形成矿床或矿化菌体自身直接堆积形成有用矿床的作用。微生物的活动及其代谢作用，一方面能改变成矿的物理和化学环境，促进金属元素的迁移和富集；另一方面，微生物机体及其生命活动可吸附和吸收成矿元素并矿化，而在有利成矿部位直接沉淀和聚集成矿(李冬玉和黄建华，2009)。近些年的研究表明，除了油、气等可燃矿床外，某些微细粒浸染状金矿、红层铜矿和细砂铀矿，以及磷块岩等非金属矿床，具有大量微生物参与成矿过程。

藻类作为微生物的典型代表，是水体有机物的主要制造者，即第一生产力。它既是水体氧化环境的唯一制造者，又是聚集矿物的最主要的生物群。藻菌生物成矿，主要通过以下几个途径：①细胞的吸附作用，它们是通过细胞表面的活性基团的化学作用，和藻体多糖的物理黏结作用；②藻细胞的吸收作用，主要是通过藻类生命活动在胞内形成的有机组分，金属络合物和螯合物；③通过藻类生命活动，特别是光合作用，放出 O_2，促进环境的氧化性增强，改变离子平衡系统，导致环境的 pH 上升，氧化还原电位 E_h 下降，有利于多种金属氧化物和盐类的形成和沉淀，这是聚矿的最强大的动力，是聚矿的最基本原因；④多种有机质的络合、螯合、溶解、运移、再聚矿沉积，腐殖质和二价金属形成的配位物的稳定性符合 Irving-Williams 次序：Hg＞Cu＞Ni＞Zn＞Co＞Mn＞Cd＞Ca＞Mg(刘志礼和刘雪娴，1999)。碳酸酐酶在微藻的生物成矿过程中起到重要的作用，特别是对于聚矿作用最强的第三种途径。碳酸酐酶能够快速催化 HCO_3^- 水合反应($HCO_3^- + H^+ \longleftrightarrow CO_2 + H_2O$)进行光合作用并放出 O_2，导致环境的 pH 上升、E_h 下降，从而导致金属氧化物和盐类的形成和沉淀。在进行微藻对碳酸盐岩溶蚀作用的实验过程中，我们发现，微藻胞外碳酸酐酶一方面能够促进碳酸盐岩的溶蚀，另一方面又能够促进 $CaCO_3$ 的沉淀，而这过程促进了碳酸盐岩中 Mg^{2+} 的释放，导致 $CaCO_3$ 纯度更高的方解石的生成(Xie and Wu，2014)。

第二节 微藻与碳酸盐岩交互作用的定量方法

当前，微藻与碳酸盐岩的交互作用主要集中在定性方面的研究，很少能够定量，然而定量研究对估算生物参与下的岩溶碳汇效应有着重要的作用。本节主要介绍通过生物溶蚀定量模型和双向同位素示踪技术来对微藻与碳酸盐岩的交互作用进行定量研究的方法。

一、碳酸盐岩的微藻生物溶蚀作用定量方法

首先，测定培养液和单位叶绿素 a 藻体对应的 Ca^{2+}、Mg^{2+} 含量。由于处理培养过程中的藻体会与方解石微粒混合，较难分离。因此，采取以下方法进行测定：取 10mL 混合均匀的藻液，在 $1600g$ 离心力条件下离心 10min，弃去上清液，加入 5~10mL 的 95%乙醇，测定其叶绿素 a 的含量。测定后，将所有萃取液转入特氟龙杯里。再向含有沉淀的离心管里加入 5mL 1mmol/L 的 HCl 使方解石微粒溶解，待溶解完全，于 $1600g$ 离心力条件下离心 10min，弃上清液，加入 5mL 超纯水(Merck Millipore，German)混匀，全部转入特氟龙杯。将特氟龙杯放置于电热板上加热(80℃)，使乙醇和水至近干，加入 3mL HNO_3 和 1mL H_2O_2，在水热反应釜里高温(180℃)高压消解 8h，再加酸定容，用 ICP-OES (Vista MPX，USA)进行测定(Huang and Schulte，1985；Topper and Kotuby-Amacher，1990)。

由于方解石释放的离子一部分被藻体吸附或吸收利用；一部分残留在培养液中。因此，从最初培养液到取样时碳酸盐矿物的离子总释放量可以通过式(5-1)求得：

$$T_i = (C_i - C_0) + (Ch_i - Ch_0) \cdot Cp \tag{5-1}$$

其中 T_i 是不同处理时间下单位体积培养液中碳酸盐矿物的各离子总释放量(mg/L)；C_i 是不同处理时间下培养液各离子浓度；C_0 是起始时各离子浓度(mg/L)；Ch_0 和 Ch_i 是起始和不同处理时间下培养液的叶绿素 a 浓度(μg/L)；Cp 是单位质量叶绿素 a 的藻体的 Mg^{2+} 含量(mg/μg)。

为了消除物理和化学作用对 Ca^{2+}、Mg^{2+} 浓度的影响，我们利用归一法校正不同处理时间下直接测定的培养液中的 Ca^{2+}、Mg^{2+} 浓度，见公式(5-2)：

$$CT_i = C_i \cdot C_0 / CO_i \tag{5-2}$$

其中 C_i 和 CT_i 分别是校正前和校正后不同处理时间下含有方解石和藻体的培养

液的 Ca^{2+} 和 Mg^{2+} 浓度（mg/L）；C_0 是培养液初始 Ca^{2+} 和 Mg^{2+} 离子浓度（mg/L）；CO_i 是不同处理时间下只添加矿粉而没有藻体的培养液的 Ca^{2+} 和 Mg^{2+} 离子浓度；i 表示不同取样时间。

然后，确定不同处理时间下方解石的 Mg^{2+} 释放量。由于方解石释放的 Mg^{2+} 一部分被藻体吸附或吸收利用；一部分残留在培养液中。因此，从最初培养液到取样时方解石的 Mg^{2+} 释放量可以通过公式(5-3)计算得到：

$$N_i = (CT_i - C_0) + (Ch_i - Ch_0) \cdot Cp \tag{5-3}$$

其中 N_i 是不同处理时间下单位体积培养液中方解石的 Mg^{2+} 释放量（mg/L）；C_0 是起始时 Mg^{2+} 浓度（mg/L）；Ch_0 和 Ch_i 是起始和不同处理时间下培养液的叶绿素 a 浓度（μg/L）；Cp 是单位质量叶绿素 a 的藻体的 Mg^{2+} 含量（mg/μg）。

接着，建立不同处理时间下培养液中叶绿素 a 浓度（Ch_i）随时间（T）变化的方程(SigmaPlot 10.0)，即 $Ch_i = f(T)$，并对此方程积分，获得生物累计作用时间（PT_i）方程，如公式(5-4)。

$$PT_i = \int_0^T f(T)d(T) \tag{5-4}$$

在此基础上，求出不同处理时间下的生物累计作用时间（PT_i）。将不同处理时间的 Mg^{2+} 释放量（N_i）与对应的生物累计作用时间（PT_i）建立方程(SigmaPlot 10.0)，即 $N_i = f(PT_i)$。最后，对建立的方程求导，获得单位叶绿素 a 的藻体在单位时间内对方解石 Mg^{2+} 的释放量 P_i(mg/(μg·h))，即：

$$P_i = \frac{d(N_i(PT_i))}{d(PT_i)} \tag{5-5}$$

二、微藻对碳酸钙碳源利用份额的定量方法

微藻碳同位素组成可以反映其对环境中无机碳的利用（Chen et al., 2009）。本实验条件下微藻可利用的无机碳有两个来源，分别是添加的方解石无机碳源和固有的无机碳源。相应地，我们利用双向同位素示踪技术（添加标记 $δ^{13}C$ 的方解石），可以定量两种碳源所占的比例，两端元的同位素混合模型可以表示为

$$δT_i = (1 - f_{Bi})δA + f_{Bi}δB_i \quad (i = 1, 2) \tag{5-6}$$

在方程(5-6)中，$δT_i$ 为微藻藻体的 $δ^{13}C$ 值，$δA$ 为假定微藻完全利用固有的无机碳源时藻体的 $δ^{13}C$ 值，$δB_i$ 为假定微藻完全利用添加的方解石碳源时藻体的 $δ^{13}C$ 值，f_{Bi} 为微藻利用添加的方解石碳源占总碳源的份额。

对于添加的两种标记的方解石来说，方程(5-6)可以分别表示如下：

$$δT_1 = (1 - f_{B_1})δA + f_{B_1}δB_1 \tag{5-7}$$

$$δT_2 = (1 - f_{B_2})δA + f_{B_2}δB_2 \tag{5-8}$$

在方程(5-7)和(5-8)中，δT_1 和 δT_2 分别为添加第一种或第二种已知 $\delta^{13}C$ 的方解石培养的微藻藻体的 $\delta^{13}C$ 值，δA 为假定微藻完全利用固有的无机碳源时藻体的 $\delta^{13}C$ 值，δB_1 和 δB_2 分别为假定微藻完全利用添加的第一种或第二种方解石时藻体的 $\delta^{13}C$ 值，f_{B_1} 和 f_{B_2} 分别为微藻利用添加的第一种或第二种方解石碳源占总碳源的份额。

在联立以上两个方程求解的过程中，我们需要注意：不论添加哪种标记的方解石，只要在同一培养条件下，同一种微藻利用添加的方解石碳源所占总碳源的份额是相同的，由此可以得到：$f_{B_1}=f_{B_2}=f_B$。因此，联立方程(5-7)和(5-8)，最终可以化简为

$$f_B=(\delta T_1-\delta T_2)/(\delta B_1-\delta B_2) \quad (5-9)$$

在方程(5-9)中，f_B 为微藻利用添加的方解石碳源占总碳源的份额，δT_1 和 δT_2 分别为添加第一种或第二种已知 $\delta^{13}C$ 值的方解石培养的微藻藻体的 $\delta^{13}C$ 值，δB_1 和 δB_2 分别为假定微藻完全利用添加的第一种或第二种方解石时藻体的 $\delta^{13}C$ 值。

方程(5-9)中 $(\delta B_1-\delta B_2)$ 则可以换算成添加的同位素标记1的方解石的 $\delta^{13}C$ 值(δC_1)与添加的同位素标记2的方解石的 $\delta^{13}C$ 值(δC_2)的差，方程(5-9)可进一步表示为

$$f_B=(\delta T_1-\delta T_2)/(\delta C_1-\delta C_2) \quad (5-10)$$

因此，根据添加的两种标记方解石的 $\delta^{13}C$(δC_1 和 δC_2)，及其对应培养的微藻藻体的 $\delta^{13}C$ 值(δT_1 和 δT_2)，由方程(5-10)可以求出微藻利用添加的方解石碳源所占总碳源的份额(f_B)。

三、方解石生物溶蚀所释放的无机碳在各介质中分配的定量方法

方解石溶蚀所释放的碳(M)一部分残留在水体或进入大气(M')，一部分被藻体吸收利用($M_{藻}$)，还有一部分形成 $CaCO_3$ 再次沉淀(M_{CaCO_3})，如公式(5-11)所示：

$$M=M'+M_{藻}+M_{CaCO_3} \quad (5-11)$$

其中 M 可以根据 Mg^{2+} 的释放量(mg/L)求得，即：$M=N_{Mg}\cdot(m_C/m_{Mg})$。$N_{Mg}$ 表示培养过程中方解石 Mg 的最大释放量；m_C 表示方解石碳的含量；m_{Mg} 表示方解石 Mg 的含量。

$M_{藻}$ 可以根据微藻对方解石碳源的利用份额 f_B 和培养过程中 NPOC 增量(Δ_{NPOC})求得，即 $M_{藻}=f_B\cdot\Delta_{NPOC}$。

M_{CaCO_3} 可以根据 Mg^{2+} 和 Ca^{2+} 释放量的关系求得，即 $M_{CaCO_3}=[N_{Mg}\cdot(m_{Ca}/m_{Mg})-N_{Ca}]\cdot(m_C/m_{Ca})$。$N_{Ca}$ 表示培养过程中方解石 Ca 的释放量；m_{Ca} 表示碳

酸钙中 Ca 的含量；m_C表示碳酸钙中 C 的含量。

四、岩溶碳汇和光合碳汇计算方法

微藻参与下的岩溶体系碳汇作用可以分为两个部分，分别为：岩溶碳汇和光合碳汇。其中，岩溶碳汇(S_C)主要指因碳酸盐岩矿物溶蚀所吸收的大气 CO_2 进入水体形式 DIC 或者被微藻所利用形成有机碳。其可以通过公式(5-12)计算求得：

$$S_C = (M + M_藻) = M - M_{CaCO_3} \quad (5-12)$$

其中 S_C 表示岩溶作用吸收的大气 CO_2；M 表示碳酸盐岩矿物溶蚀过程中吸收的 CO_2 以 DIC 形式存在的量(mg/L)；$M_藻$表示碳酸盐岩矿物溶蚀过程中吸收的 CO_2 被微藻利用的量(mg/L)；M 表示碳酸盐岩矿物溶蚀过程中 C 的总释放量(mg/L)；M_{CaCO_3} 表示 $CaCO_3$ 沉淀过程中 C 的吸收量(mg/L)。

光合碳汇(S_P)主要指因微藻利用水体中的 DIC 形成有机碳而导致大气中的 CO_2 进入水体。其可以通过公式(5-13)计算求得：

$$S_P = (1 - 2f_B) \cdot \Delta_{NPOC} \quad (5-13)$$

其中 S_P 表示微藻光合固碳吸收的大气 CO_2 (mg/L)；f_B 表示微藻对碳酸盐岩矿物利用份额(mg/L)；Δ_{NPOC} 表示培养过程中不可吹灭有机碳(non-purgeable organic carbon, NPOC)增量(mg/L)。

第三节 碳酸酐酶在微藻生物溶蚀中的作用

胞外碳酸酐酶在碳酸盐岩的生物溶蚀过程中起到了重要的作用(Xie and Wu, 2014；李强等, 2012)。同时，其也是微藻无机碳浓缩机制(CCM)的一个重要的组成部分，它能够快速将水体中的主要无机碳存在形式(HCO_3^-)转化为 CO_2 来被藻体利用，并伴随有 $CaCO_3$ 沉淀的产生(Sharma and Bhattacharya, 2010)。这样就提高了微藻将不稳定碳(DIC)固定成惰性碳(有机碳)的能力，从而起到"碳汇效应"。因此，微藻胞外碳酸酐酶在"碳酸盐泵"和"生物泵"中都起到了重要的作用。

一、微藻生物溶蚀作用的种类特征

生物溶蚀是指矿物、岩石受生物生长及活动影响而发生的溶蚀作用，它又分为物理方式与化学方式两种形式。生物通过生命活动的黏着、穿插和剥离等机械活动使矿物颗粒分解，被认为是生物物理溶蚀作用；生物通过自身分泌及死后遗体析出的酸等物质，对岩石的腐蚀称为生物化学溶蚀(李莎等, 2007)。

目前，对生物风化的机理还没有一个较为统一的认识。但是大多数学者都认为，生物的生长对营养元素的需求是直接的动力。例如：①在进行微藻对碳酸盐岩的溶蚀实验过程中发现，随着藻类的急速生长，碳酸盐岩中阳离子的溶出速率也相应地急剧增加；②菌丝能直接将矿石与植物根系相连在一起，通过菌丝传输并将溶解出的元素直接供给植物利用(Aung and Ting，2005)。

在生物风化过程中，生物生长产生的代谢产物和分泌的化学物质对岩石的溶蚀分解起着重要的作用。各个研究者先后提出了酸解、络解、酶解、碱解，以及夹膜吸收、胞外多糖形成和氧化还原作用等多种观点(Li et al.，2006)。

1. 酸解

酸解，即生物代谢产生的无机酸(硫酸、盐酸、硝酸等)和有机酸(枸橼酸、草酸、葡萄糖酸、甲酸、乙酸、乳酸、琥珀酸、丙酮酸等)对岩石的溶解作用。大部分研究者认为，酸的溶解作用是生物促进岩石和矿物风化的重要途径(Cameselle et al.，2003；Jongmans et al.，1997；Puente et al.，2006)。Welch 等的研究表明在矿物性质及 pH 相同的情况下，草酸体系中矿物的溶解速率甚至是无机酸体系的 10 倍以上，这就说明有机酸比无机酸更易于溶解矿物(Welch and Ullman，1996)。黄黎英等的研究表明有机酸对石灰岩颗粒溶蚀作用大小的顺序为：乳酸＞乙酸＞枸橼酸＞甲酸＞草酸＞丙酮酸。同类型酸中如醇酸或羧酸中酸性较弱的酸溶蚀力反而较大。有机酸对石灰岩的溶蚀能力不仅与溶液中的 H^+ 和有机配位体浓度有关，而且与酸的类型、酸的强弱及反应后形成的盐的溶解度有关(黄黎英等，2006)。但也有一些学者认为有机酸只有在较低的 pH 条件下才能促进矿物溶解，有机酸溶解作用随着 pH 的升高而迅速减弱。当 pH＞5 时，释放金属离子的能力完全消失，但在有微生物体存在的情况下，即使体系 pH 不是很低，矿物依然能够溶解(Cameselle Claudio et al.，1997；盛下放和黄为一，2002)。

2. 碱解

碱解作用是异养微生物在生长的过程中形成有机碱类物质，这些碱类物质能提高矿物表面的 pH，从而促使二氧化硅溶解。Chen 等认为，长石的溶解速率与溶液的 pH 成 U 形关系，即在酸性区域随 pH 增大而减小、中性区域溶解速率低且受影响小、碱性区域随 pH 增大而增大(Chen et al.，1993)。Teng 的研究表明，长石在碱性条件下以化学计量比溶解，在酸性条件下的非化学计量比溶解只限于一个晶胞深度(Teng et al.，2001)。

3. 酶解

生物的生长过程中，将产生各种各样的酶，这些酶一方面影响生物的生长繁殖，一方面影响生物的新陈代谢产物。陈延伟等认为，硅酸盐细菌解钾的机理不是产酸，而是细菌与矿石接触并产生特殊的酶，破坏矿物晶格或是表面的物理化学接触所引起的(陈延伟和陈华癸，1960)。余龙江等的研究结果表明微生物碳酸酐酶能使灰岩溶出的导电离子总量和 Ca^{2+} 提高 40% 以上(余龙江等，2004)。刘再华等研究发现牛碳酸酐酶在不同的 CO_2 分压下能够使石灰岩溶蚀速率提高数倍(刘再华，2001)。

4. 其他

络解作用是在微生物作用下形成可溶出矿物元素(铁、铝、铜、锌、锰、钙、镁等)的螯合剂或络合剂的过程。微生物作用形成的螯合物中包括柠檬酸、草酸、酒石酸、乙酮-葡萄糖酸等有机酸和水杨酸、2,3-二羟基苯甲酸等石炭酸(Krumbein，1983)。铁和锰的氧化性或还原性风化作用物质主要由细菌产生，包括由嗜酸铁细菌产生的三价铁以及由硫酸盐还原细菌产生的硫化物，酸溶液中的三价铁可与 CuS 之类的硫化物矿物产生化学反应，而硫化物则可还原 MnO_2。

二、pH 与微藻生物溶蚀作用

水体的 pH 在碳酸盐岩的溶蚀过程中起到了重要的作用，pH 越低，碳酸盐岩的化学溶蚀作用越强。同时，pH 又会影响微藻的生理活动，改变微藻对碳酸盐岩的生物溶蚀作用。

在 250mL 含有 100mL 培养液的三角瓶中，25℃，$150\mu mol/(m^2 \cdot s)$ 的光照强度，光周期为 12h/12h，培养藻体，每天摇动培养物 5 次，共培养 7 天。三角瓶中添加不同的藻体(*Chlamydomonas reinhardtii*，C.R；*Chlorella pyrenoidosa*，C.P；*Microcystis aeruginosa*，M.A)、方解石矿粉(0.3g)，并用无菌封口膜封口，设置 6~7、7~8、8~9、9~10 四个 pH 梯度进行培养(所有处理都进行 3 个平行处理)，具体见表 5-1。每隔 6 个小时用 0.1mmol/L 或 1mmol/L 的 HCl 调节一次 pH，使 pH 维持在设定范围以内。其中培养液采用改进的 SE 培养液(不含 Ca，只含约 1/5 的 Mg)。分别在 6h、24h、48h、72h、120h 和 168h 的时候对藻液进行提取，并在 1600g 离心力条件下离心 10min。其中沉淀用于测定叶绿素 *a*，上清液用针式过滤器(0.22μm)过滤后测定其电导率、碱度、Ca^{2+}、Mg^{2+} 浓度。

表 5-1 实验处理
Tab. 5-1 Treatments of experiment

处理	藻	方解石	pH
1			6~7
2		方解石(ST1)*/方解石(ST2)**	7~8
3			8~9
4			9~10
5			6~7
6	C.R/C.P/M.A	方解石(ST1)*/方解石(ST2)**	7~8
7			8~9
8			9~10

注：$C.R$：莱茵衣藻，$C.P$：蛋白核小球藻，$M.A$：铜绿微囊藻。

*：添加 $\delta^{13}C$ 为 $-10.287‰$ 的方解石的处理；**：添加 $\delta^{13}C$ 为 $0.719‰$ 的方解石的处理。

Note：$C.R$：*Chlamydomonas reinhardtii*，$C.P$：*Chlorella pyrenoedosa*，$M.A$：*Microcystis aeruginosa*.

*：the treatment with added calcite with an $\delta^{13}C$ value of $-10.287‰$，**：the treatment with added calcite with an $\delta^{13}C$ value of $0.719‰$.

(一) pH 对微藻生物量的影响

在不同 pH 梯度处理下，不同微藻有着不同的生长规律，如图 5-1 所示。对于莱茵衣藻来说，在 pH=6~7 的时候，生长速率相对较小；而在 pH 大于 7 的情况下，生长速率相对较大，但是在 pH 为 7~10，莱茵衣藻的生长速率变化不明显。对于蛋白核小球藻来说，其生长速率受水体 pH 的影响不明显，各 pH 处理下的生物量只有微弱差异。这可能是因为在碱性环境更有利于空气中的 CO_2 转化为 HCO_3^-，作为微藻生长的无机碳源，而莱茵衣藻有较强的胞外碳酸酐酶活力，能够在高 pH 环境下将更多的 HCO_3^- 转化为 CO_2 来利用，蛋白核小球藻则只有微弱的胞外碳酸酐酶活力(Aizawa and Miyachi，1986；Roughton and Booth，1946)。而铜绿微囊藻的生长速率受水体 pH 的影响较大。可以发现，在 pH 为 7~8 时，铜绿微囊藻的生长速率最大；在 pH 为 9~10 时，其生长速率最小；在 pH 为 6~7 和 8~9 时，次之。这可能是因为，铜绿微囊藻没有胞外碳酸酐酶活力，其主要是通过直接途径和 CO_2 途径来利用水体中的 DIC 进行光合作用。在 pH 为 7~8 时，有较高的 HCO_3^- 和一定量的 CO_2 供铜绿微囊藻利用，其最接近自然淡水的 HCO_3^-/CO_2(Menéndez et al.，2001)；在 pH 为 9~10 时，DIC 主要以 HCO_3^- 和 CO_3^{2-} 的形式存在，无 CO_2，而 CO_3^{2-} 是无法被微藻利用，导致铜绿微囊藻没有充足的可利用的碳源。因此，对于有较高胞外碳酸酐酶的藻体，pH 主要是控制藻体的碳酸酐酶活力来控制藻体的生长，而对于无胞外碳酸酐酶的藻体，pH 主要控制 DIC 存在形式来控制藻体的生长。

在此基础上，对不同 pH 条件下各藻体的生物量建立随时间变化的方程。

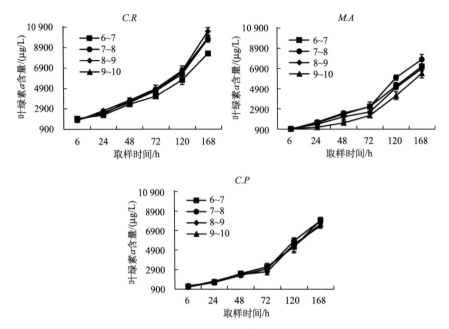

图 5-1 不同 pH 梯度处理下微藻的叶绿素 a 含量

$C.R$：莱茵衣藻；$C.P$：蛋白核小球藻；$M.A$：铜绿微囊藻；

6~7、7~8、8~9、9~10：分别为培养液 pH 为 6~7、7~8、8~9、9~10 的处理

Fig. 5-1 Chl-a content of microalgae under different pH treatments

$C.R$: *Chlamydomonas reinhardtii*; $C.P$: *Chlorella pyrenoedosa*; $M.A$: *Microcystis aeruginosa*; 6~7: the pH of medium is between 6 and 7; 7~8: the pH of medium is between 7 and 8; 8~9: the pH of medium is between 8 and 9; 9~10: the pH of medium is between 9 and 10

可以发现，随时间的迁移，各藻体呈指数增长，如表 5-2 所示。并对其求时间的积分，我们可以得到不同 pH 条件下各藻体的累计作用时间方程，如表 5-2 所示。

（二）pH 对方解石生物溶蚀过程中 Ca^{2+} 释放量的影响

从图 5-2 中，可以发现，Ca^{2+} 的总释放量随 pH 降低而逐渐增加。这主要是因为 pH 越低 $CaCO_3$ 的饱和度越大，并且 pH 越低，水体中的 H^+ 含量和 CO_2 比例越高，方解石的溶蚀越快（$CaCO_3 + 2H^+ \longleftrightarrow Ca^{2+} + 2HCO_3^-$，$CaCO_3 + CO_2 + H_2O \leftrightarrow Ca^{2+} + 2HCO_3^-$）。同时，我们还发现，在添加微藻的条件下，$Ca^{2+}$ 的释放量大于没有添加微藻的条件下（图 5-2）。说明微藻能够促进方解石 Ca^{2+} 的释放。此外，在 pH 为 6~8 的条件下，含有莱茵衣藻的培养液中 Ca^{2+} 离子释放量明显小于含有蛋白核小球藻和铜绿微囊藻的培养液，而在 pH 大于 8 的条

表 5-2 不同 pH 下生物增加曲线及生物累计作用时间模型
Tab. 5-2 Growth rate and biological cumulative action time model under different pH

藻体	处理	增长曲线	生物累计作用时间
C.P	6~7	$Chi = -4356 + 6133e^{0.0043T}$ $R^2 = 0.997$	$PTi = \int_0^T Chi(T) = -4357T + 1\,426\,511\,e^{0.0043T} - 1\,426\,511$
	7~8	$Chi = -4080 + 5867e^{0.005T}$ $R^2 = 0.994$	$PTi = \int_0^T Chi(T) = -4080T + 1\,173\,416\,e^{0.005T} - 1\,173\,416$
	8~9	$Chi = -3862 + 5695e^{0.0054T}$ $R^2 = 0.992$	$PTi = \int_0^T Chi(T) = -3862T + 1\,054\,661\,e^{0.0054T} - 1\,054\,661$
	9~10	$Chi = -6588 + 8276e^{0.004T}$ $R^2 = 0.996$	$PTi = \int_0^T Chi(T) = -6588T + 2\,068\,920\,e^{0.004T} - 2\,068\,920$
C.R	6~7	$Chi = -3301 + 4345e^{0.0057T}$ $R^2 = 0.999$	$PTi = \int_0^T Chi(T) = -3301T + 762\,324\,e^{0.0057T} - 762\,324$
	7~8	$Chi = -6173 + 6997e^{0.0042T}$ $R^2 = 0.99$	$PTi = \int_0^T Chi(T) = -6173T + 1\,665\,919\,e^{0.0042T} - 1\,665\,919$
	8~9	$Chi = -4274 + 5251e^{0.0048T}$ $R^2 = 0.985$	$PTi = \int_0^T Chi(T) = -4274T + 1\,093\,979\,e^{0.0048T} - 1\,093\,979$
	9~10	$Chi = -5366 + 6372e^{0.0042T}$ $R^2 = 0.999$	$PTi = \int_0^T Chi(T) = -5366T + 1\,517\,204\,e^{0.0042T} - 1\,517\,204$
M.A	6~7	$Chi = -16\,220 + 16\,953e^{0.0019T}$ $R^2 = 0.999$	$PTi = \int_0^T Chi(T) = -16\,220T + 8\,922\,778\,e^{0.0057T} - 8\,922\,778$
	7~8	$Chi = -20\,896 + 21\,516e^{0.0017T}$ $R^2 = 0.99$	$PTi = \int_0^T Chi(T) = -20\,896T + 12\,656\,464\,e^{0.0042T} - 12\,656\,464$
	8~9	$Chi = -5291 + 16\,953e^{0.0019T}$ $R^2 = 0.99$	$PTi = \int_0^T Chi(T) = -5291T + 1\,434\,995\,e^{0.0048T} - 1\,434\,995$
	9~10	$Chi = -1545 + 16\,953e^{0.0019T}$ $R^2 = 0.996$	$PTi = \int_0^T Chi(T) = -1546T + 293\,814\,e^{0.0042T} - 293\,814$

注:C.R:莱茵衣藻;C.P:蛋白核小球藻;M.A:铜绿微囊藻;6~7、7~8、8~9、9~10:分别为培养液 pH 为 6~7、7~8、8~9、9~10 的处理。

Note: C.R: *Chlamydomonas reinhardtii*; C.P: *Chlorella pyrenoedosa*; M.A: *Microcystis aeruginosa*; 6~7: the pH of mediumis between 6 and 7, 7~8: the pH of medium is between 7 and 8, 8~9: the pH of medium is between 8 and 9, 9~10: the pH of medium is between 9 and 10.

件下,Ca^{2+} 离子释放量差异较小,甚至释放大于蛋白核小球藻和铜绿微囊藻。这可能是由于,在 pH 为 6~8 的时候,莱茵衣藻的胞外碳酸酐酶活力最强(Williams and Colman,1996),而胞外碳酸酐酶在催化无机碳利用过程中会导致 $CaCO_3$ 的沉淀($Ca^{2+} + 2HCO_3^- \leftrightarrow CaCO_3 + CH_2O + O_2$)(Sharma and Bhattacharya,2010)。

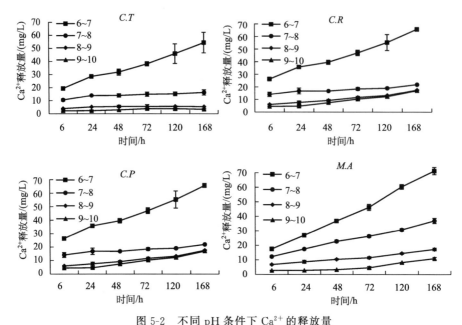

图 5-2 不同 pH 条件下 Ca²⁺ 的释放量

C.T: 空白; *C.R*: 莱茵衣藻; *C.P*: 蛋白核小球藻; *M.A*: 铜绿微囊藻; 6~7、7~8、8~9、9~10: 分别为培养液 pH 为 6~7、7~8、8~9、9~10 的处理

Fig. 5-2 Amount of Ca²⁺ released from calcite under different pH

C.T: control (without microalgae added); *C.R*: *Chlamydomonas reinhardtii*; *C.P*: *Chlorella-pyrenoedosa*; *M.A*: *Microcystis aeruginosa*; 6~7: the pH of medium is between 6 and 7, 7~8: the pH of medium is between 7 and 8, 8~9: the pH of medium is between 8 and 9, 9~10: the pH of medium is between 9 and 10

(三) pH 对方解石生物溶蚀过程中 Mg^{2+} 释放量的影响

如图 5-3 所示,在未添加微藻的培养液中,Mg^{2+} 释放量随 pH 的降低而增加,其与 Ca^{2+} 的释放规律相似。但是,在添加微藻的培养液中,不同 pH 条件下 Mg^{2+} 释放量的变化趋势与 Ca^{2+} 有所不同。在碱性 pH 条件下(pH>7),随 pH 的升高 Mg^{2+} 释放量逐渐减小。而在 pH 为 6~7 时,含有微藻的培养液中 Mg^{2+} 的释放量相对其他 pH 条件并没有增加,甚至下降。这可能是由于在酸性条件下,水体中的 DIC 以 CO_2 为主要存在形式,CO_2 供应量较为充足,诱导微藻主要以 CO_2 的方式利用 DIC 进行光合作用(Williams and Colman, 1996),特别是只有较小或者无胞外碳酸酐酶的 *C.P* 和 *M.A*,从而导致方解石溶解平衡向反方向进行($CaCO_3 + CO_2 + H_2O \leftrightarrow Ca^{2+} + 2HCO_3^-$)。

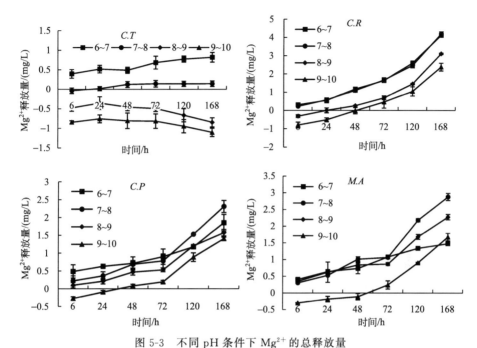

图 5-3 不同 pH 条件下 Mg^{2+} 的总释放量

$C.T$：空白；$C.R$：莱茵衣藻；$C.P$：蛋白核小球藻；$M.A$：铜绿微囊藻；6～7、7～8、8～9、9～10：分别为培养液 pH 为 6～7、7～8、8～9、9～10 的处理

Fig. 5-3 Amount of Mg^{2+} released from calcite under different pH

$C.T$: control (without microalgae added); $C.R$: *Chlamydomonas reinhardtii*; $C.P$: *Chlorella pyrenoedosa*; $M.A$: *Microcystis aeruginosa*; 6～7: the pH of medium is between 6 and 7, 7～8: the pH of medium is between 7 and 8, 8～9: the pH of medium is between 8 and 9, 9～10: the pH of medium is between 9 and 10

（四）不同 pH 条件下单位时间单位藻体的 Mg^{2+} 释放量

为了进一步研究不同 pH 条件下，各微藻的生物溶蚀作用。我们根据生物溶蚀作用模型（表 5-3），求得了不同 pH 条件下，各单位质量的微藻在单位时间的生物溶蚀量。

在不同 pH 培养条件下，各微藻对方解石的生物溶蚀作用是不同的。对于莱茵衣藻而言，随着 pH 的增加，Mg^{2+} 生物释放量逐渐减少；而对于含有铜绿微囊藻的处理来说，随着 pH 的增加，Mg^{2+} 生物释放量逐渐增加（图 5-4）。添加蛋白核小球藻的培养液，则表现居中，在 pH 为 7～8 时，较大幅的增加，随后下降并趋于稳定（图 5-4）。其可能是由于，莱茵衣藻有较强的胞外碳酸酐酶活力，能够通过 CAex 催化作用来快速利用 HCO_3^-，而胞外碳酸酐酶催化利用 HCO_3^-

表 5-3　不同 pH 条件下单位时间单位微藻的生物溶蚀定量模型（mg/μg-chl-h）

Tab. 5-3　Models of algae bio-dissolution unit algal biomass and unit time under different pH (mg/μg-chl-h)

藻体	处理	溶蚀模型	R^2
C.R	6～7	$N_i = 0.071 + 4.74 \times 10^{-6} PT_i$	0.988
	7～8	$N_i = 0.128 + 4.56 \times 10^{-6} PT_i$	0.988
	8～9	$N_i = -0.122 + 4.35 \times 10^{-6} PT_i$	0.991
	9～10	$N_i = 0.185 + 3.32 \times 10^{-6} PT_i$	0.972
C.P	6～7	$N_i = -0.0387 + 2.32 \times 10^{-6} PT_i$	0.982
	7～8	$N_i = -0.007 + 3.20 \times 10^{-6} PT_i$	0.996
	8～9	$N_i = -0.0206 + 2.86 \times 10^{-6} PT_i$	0.953
	9～10	$N_i = -0.103 + 2.90 \times 10^{-6} PT_i$	0.975
M.A	6～7	$N_i = 0.180 + 1.87 \times 10^{-6} PT_i$	0.969
	7～8	$N_i = 0.148 + 3.86 \times 10^{-6} PT_i$	0.978
	8～9	$N_i = 0.048 + 4.07 \times 10^{-6} PT_i$	0.983
	9～10	$N_i = -0.132 + 4.46 \times 10^{-6} PT_i$	0.990

注：C.R：莱茵衣藻；C.P：蛋白核小球藻；M.A：铜绿微囊藻；6～7、7～8、8～9、9～10：分别为培养液 pH 为 6～7、7～8、8～9、9～10 的处理。

Note：C.R：Chlamydomonas reinhardtii；C.P：Chlorella pyrenoedosa；M.A：Microcystis aeruginosa；6～7：the pH of mediumis between 6 and 7，7～8：the pH of medium is between 7 and 8，8～9：the pH of medium is between 8 and 9，9～10：the pH of medium is between 9 and 10.

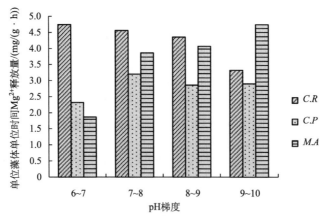

图 5-4　不同 pH 处理下单位时间单位叶绿素 a 微藻的 Mg^{2+} 释放量

C.R：莱茵衣藻；C.P：蛋白核小球藻；M.A：铜绿微囊藻；6～7、7～8、8～9、9～10：分别为培养液 pH 为 6～7、7～8、8～9、9～10 的处理

Fig. 5-4　Amounts of Mg^{2+} released per unit algal biomass and unit time under different pH

C.R：Chlamydomonas reinhardtii；C.P：Chlorella pyrenoedosa；M.A：Microcystis aeruginosa；6～7：the pH of medium is between 6 and 7；7～8：the pH of medium is between 7 and 8；8～9：the pH of medium is between 8 and 9；9～10：the pH of medium is between 9 and 10

的过程将促进碳酸盐岩的溶蚀(Xie and Wu,2014)。在 pH 为 6.5~7.5 时，其 CAex 活力最大，并随着 pH 的增加逐渐减小(Li et al.,2012)，使得二氧化碳途径与碳酸氢根途径的比例下降。铜绿微囊藻没有胞外碳酸酐酶活力，主要通过阴离子通道直接利用 HCO_3^-，随着 pH 增加，CO_2 比例逐渐减少，微藻为了维持无机碳的供应，可能提高对 HCO_3^- 的利用能力，从而使溶解平衡($MCO_3+CO_2+H_2O \leftrightarrow M+2HCO_3^-$)向正方向移动。

三、碳酸酐酶胞外酶在微藻生物溶蚀中的作用

在 500mL 含有 250mL 培养液的三角瓶中，25℃，150μmol/(m²·s)的光照强度，光周期为 12h/12h，培养藻体，每天摇动培养物 5 次，共培养 7 天。三角瓶中添加不同的藻体(*Chlamydomonas reinhardtii*，C.R；*Microcystis aeruginosa*，M.A)、方解石矿粉(1g)，并用封口膜(开放环境)或者碱石灰过气筛加 NaOH 吊瓶(隔离外界 CO_2 环境)封口，设置 0、1mmol/L、10mmol/L 三个乙酰唑胺(AZ)浓度梯度进行培养(所有处理都进行 3 个平行处理)，具体见表 5-4。其中培养液采用改进的 SE 培养液(不含 Ca，只含约 1/5 的 Mg)。分别在 6h、24h、48h、72h、120h 和 168h 时对藻液进行提取，并在 1600g 离心力条件下离心 10min。其中沉淀用于测定叶绿素 a，上清液用针式过滤器(0.22μm)过滤后测定其电导率，碱度，Ca^{2+}、Mg^{2+} 浓度。

表 5-4 实验处理

Tab. 5-4 Treatments of experiment

处理	环境	藻	方解石	AZ/(mmol/L)
1			方解石(ST1)*	0
2	开放或隔离	无	或方解石(ST2)**	1
3				10
5			方解石(ST1)*	0
6	开放或隔离	C.R/M.A	或方解石(ST2)**	1
7				10

注：C.R：莱茵衣藻；M.A：铜绿微囊藻。

*：添加 $\delta^{13}C$ 为 -10.287‰ 的方解石的处理；**：添加 $\delta^{13}C$ 为 0.719‰ 的方解石的处理。

Note：C.R：*Chlamydomonas reinhardtii*；M.A：*Microcystis aeruginosa*.

*：the treatment with added calcite with an $\delta^{13}C$ value of -10.287‰；**：the treatment with added calcite with an $\delta^{13}C$ value of 0.719‰.

(一) 不同 CO_2 供应下胞外碳酸酐酶抑制剂对微藻生物量的影响

藻体的叶绿素 a 含量能够有效的反映藻体生物量（Vörös and Padisák，1991）。因此，可以通过对培养液中藻体叶绿素 a 含量变化的观测来研究微藻生物量的变化。对于有较强胞外碳酸酐酶活力的莱茵衣藻来说，在开放环境下，未添加 AZ 培养液中的莱茵衣藻生长速率大于添加 AZ 培养液中的莱茵衣藻（图 5-5）。这说明胞外碳酸酐酶有助于莱茵衣藻的生长。然而，在隔离环境下，48h 以前未添加 AZ 的培养液中莱茵衣藻生长量大于添加 AZ 的培养液中，但是在培养 48h 以后添加 AZ 的培养液中莱茵衣藻生物量反而大于未添加 AZ 的培养液中（图 5-5）。这可能是由于隔离环境下无机碳含量的变化导致。在 48h 以前环境中的无机碳供应量较为充足，近似于开放环境。培养 48h 以后，随着水体中 DIC 被利用，又未能得到补给，使得微藻能利用的碳源较少。此时，胞外碳酸酐酶未受到抑制的处理，微藻由于缺少能够利用的碳源，而受到低碳源胁迫，生长受到抑制；但是，胞外碳酸酐酶受到抑制的处理，微藻的碳源利用途径减少，生

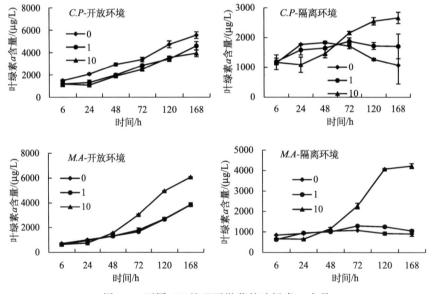

图 5-5 不同 AZ 处理下微藻的叶绿素 a 含量

0：未添加 AZ，1：添加 1mmol/L 的 AZ，10：添加 10mmol/L 的 AZ，$C.R$：莱茵衣藻，$M.A$：铜绿微囊藻

Fig. 5-5 Chl-a content of microalgae under different AZ treatments

0: without AZ, 1: 1mmol/L AZ, 10: 10mmol/L AZ, $C.R$: *Chlamydomonas reinhardtii*, $M.A$: *Microcystis aeruginosa*

长代谢放缓,对碳源的需求量降低,从而减少了低碳源胁迫带来的危害。同时,我们可以发现 AZ 浓度越大,莱茵衣藻生长受抑制的程度就越强。这主要是因为 AZ 具有浓度效应,AZ 浓度越大,胞外碳酸酐酶受抑制的程度越深,无机碳获取的能力越弱(Wu et al.,2011)。

此外,我们还发现,不管在开放还是隔离环境下,10mmol AZ 处理下铜绿微囊藻的生长量都大于 1mmol/L 或者没有 AZ 处理下的微藻生物量,同时 1mmol/L AZ 处理下的生物量和无 AZ 处理下的生物量非常接近,如图 5-5 所示。说明 AZ 并没有对铜绿微囊藻起到抑制作用。这与铜绿微囊藻不具有胞外碳酸酐酶活力有关。而 10mmol/L 处理下生物量较大的原因可能是由于 10mmol/L 的 AZ 处理下的培养液 pH 主要在 7~8.5,1mmol/L 或者没有 AZ 处理下的培养液 pH 会上升到 9~10.5(数据未列出),而 pH 为 7~8.5 的培养液最利于铜绿微囊藻的生长。

在此基础上,对不同处理条件下各藻体的生物量建立随时间变化的方程。发现在开放环境下,随时间的迁移,各藻体呈指数增长;在隔离环境下,各藻体由于处于限制性环境下,因此成对数正态分布(张怀强等,2014)。并对其求时间的积分,可以得到开放和隔离环境下不同 AZ 处理下各藻体的累计作用时间方程,如表 5-5 所示。

表 5-5 生物增加曲线及生物累计作用时间模型

Tab. 5-5 The mode between growth rate and biological cumulative action time

环境	藻体	AZ /(mmol/L)	增长曲线	生物累计作用时间
开放	C.R	0	$Chi = 1951.37e^{0.0065T}$ $R^2 = 0.937$	$PTi = \int_0^T 1951.37\,e^{0.0065T}d(T)$
		1	$Chi = 1370.72e^{0.0074T}$ $R^2 = 0.937$	$PTi = \int_0^T 1370.72\,e^{0.0074T}d(T)$
		10	$Chi = 1316.66e^{0.007T}$ $R^2 = 0.8981$	$PTi = \int_0^T 1316.66\,e^{0.007T}d(T)$
	M.A	0	$Chi = -2776 + 3415e^{0.0039T}$ $R^2 = 0.999$	$PTi = \int_0^T -2776 + 3415\,e^{0.0039T}d(T)$
		1	$Chi = -2503 + 3083e^{0.0043T}$ $R^2 = 0.999$	$PTi = \int_0^T -2503 + 3083\,e^{0.0043T}d(T)$
		10	$Chi = 1137e^{0.0104T}$ $R^2 = 0.903$	$PTi = \int_0^T 1137\,e^{0.0104T}d(T)$

续表

环境	藻体	AZ /(mmol/L)	增长曲线	生物累计作用时间
隔离	C.R	0	$Chi = 1051 + 877\, e^{-0.5\left(\frac{\ln\left(\frac{T}{37}\right)}{0.72}\right)^2}$ $R^2 = 0.964$	$PTi = \int_0^T 1051 + 877\, e^{-0.5\left(\frac{\ln\left(\frac{T}{37}\right)}{0.72}\right)^2} d(T)$
		1	$Chi = 1064 + 723\, e^{-0.5\left(\frac{\ln\left(\frac{T}{82}\right)}{1.42}\right)^2}$ $R^2 = 0.930$	$PTi = \int_0^T 1051 + 877\, e^{-0.5\left(\frac{\ln\left(\frac{T}{37}\right)}{0.72}\right)^2} d(T)$
		10	$Chi = 1102 + 1576\, e^{-0.5\left(\frac{\ln\left(\frac{T}{141}\right)}{0.67}\right)^2}$ $R^2 = 0.987$	$PTi = \int_0^T 1102 + 1576\, e^{-0.5\left(\frac{\ln\left(\frac{T}{141}\right)}{0.67}\right)^2} d(T)$
	M.A	0	$Chi = 854 + 218\, e^{-0.5\left(\frac{\ln\left(\frac{T}{58}\right)}{0.56}\right)^2}$ $R^2 = 0.958$	$PTi = \int_0^T 854 + 218\, e^{-0.5\left(\frac{\ln\left(\frac{T}{58}\right)}{0.56}\right)^2} d(T)$
		1	$Chi = 631 + 605\, e^{-0.5\left(\frac{\ln\left(\frac{T}{86}\right)}{0.95}\right)^2}$ $R^2 = 0.890$	$PTi = \int_0^T 631 + 605\, e^{-0.5\left(\frac{\ln\left(\frac{T}{86}\right)}{0.95}\right)^2} d(T)$
		10	$Chi = 651 + 3667\, e^{-0.5\left(\frac{\ln\left(\frac{T}{147}\right)}{0.56}\right)^2}$ $R^2 = 0.999$	$PTi = \int_0^T 651 + 3667\, e^{-0.5\left(\frac{\ln\left(\frac{T}{147}\right)}{0.56}\right)^2} d(T)$

注：C.R：莱茵衣藻，M.A：铜绿微囊藻。

Note: C.R: *Chlamydomonas reinhardtii*, M.A: *Microcystis aeruginosa*.

（二）不同 CO_2 分压下胞外碳酸酐酶抑制剂对微藻胞外碳酸酐酶的影响

AZ 能够特异性抑制胞外碳酸酐酶活力，广泛被用于针对性的研究胞外碳酸酐酶在微藻无机碳利用过程中的作用（Bozzo and Colman，2000；Wu et al.，2011）。从图 5-6 可以看出随着 AZ 浓度的增加，C.R 的胞外碳酸酐酶活力不断

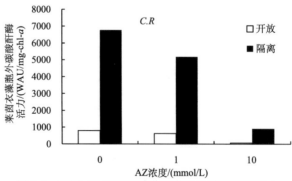

图 5-6 不同 AZ 处理下莱茵衣藻胞外碳酸酐酶活力
0：未添加 AZ，1：添加 1mmol/L 的 AZ，10：添加 10mmol/L 的 AZ，C.R：莱茵衣藻
Fig. 5-6 The activity of extracellular carbonic anhydrasein
C. reinhardtii s under different AZ treatments
0: without AZ, 1: 1mmol/L AZ, 10: 10mmol/L AZ, C.R: *Chlamydomonas reinhardtii*

减少,在10mmol/L 的 AZ 浓度处理下 $C.R$ 的 CAex 几乎被完全抑制。同时,我们还发现,隔离环境下的 $C.R$ 胞外碳酸酐酶活力明显高于开放环境下,平均高出 8 倍。这主要是由于,隔离环境下的 CO_2 分压较低,供应量较小,而低 CO_2 浓度环境下能诱导微藻胞外碳酸酐酶的表达(Aizawa and Miyachi,1986)。

(三) 不同 CO_2 分压下胞外碳酸酐酶抑制剂对 Ca^{2+} 释放量的影响

不同 CO_2 分压和 AZ 浓度处理下,方解石 Ca^{2+} 的释放量呈规律性变化。可以发现,在 10mmol/L AZ 浓度的处理下,含有微藻的培养液中 Ca^{2+} 的释放量大于不含有微藻的培养液中,说明微藻对方解石 Ca^{2+} 的释放有一定的促进作用。而在 0 或者 1mmol/L AZ 浓度的处理下未添加藻体的培养液中 Ca^{2+} 的释放量反而大于添加藻体的培养液(图 5-7)。这可能是由于方解石溶解释放的 HCO_3^- 被微藻所利用,而导致了 $CaCO_3$ 的沉淀和 pH 升高($CaCO_3 + CO_2 + H_2O \leftrightarrow Ca^{2+} + 2HCO_3^- \leftrightarrow CaCO_3 + CH_2O + O_2$)(Liu et al.,2010b)。从这个反应式可以看出,在有微藻的参与下,$CaCO_3$ 溶解($CaCO_3 + CO_2 + H_2O \leftrightarrow Ca^{2+} + 2HCO_3^-$)后会随着微藻对 HCO_3^- 的利用而再次沉淀($Ca^{2+} + 2HCO_3^- \leftrightarrow CaCO_3 + CH_2O + O_2$),但

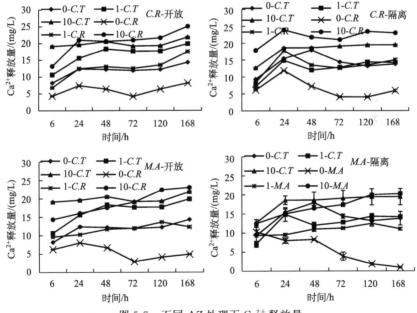

图 5-7 不同 AZ 处理下 Ca^{2+} 释放量

0:未添加 AZ,1:添加 1mmol/L 的 AZ,10:添加 10mmol/L 的 AZ;$C.T$:空白,$C.R$:莱茵衣藻,$M.A$:铜绿微囊藻

Fig. 5-7 Amount of Ca^{2+} released from calcite under different AZ treatments

0: without AZ, 1: 1mmol/L AZ, 10: 10mmol/L AZ; $C.T$: control (without microalgae added), $C.R$: *Chlamydomonas reinhardtii*, $M.A$: *Microcystis aeruginosa*

是水体却少了一份 CO_2，从而导致 pH 升高，降低 $CaCO_3$ 的饱和度，进一步促进 $CaCO_3$ 的沉淀。0 或者 1mmol/L AZ 处理下培养液的 pH 能很好地证明这一点（数据未列出来），其有较低的 Ca^{2+} 浓度，却有较快的 pH 升速。胞外碳酸酐酶由于能够快速催化 CO_2 水合反应，在此过程中起到关键的作用（Li et al., 2012a）。因此，在 10mmol/L 的 AZ 处理下情况相反，特别是莱茵衣藻。

开放环境下方解石中 Ca^{2+} 的释放量大于隔离环境下。这主要是由于开放环境下 CO_2 分压相对较高，使得 $CaCO_3+CO_2+H_2O\leftrightarrow Ca^{2+}+2HCO_3^-$ 朝着溶解的方向进行得相对较快。此外，我们还可以注意到，培养 24 小时以后在隔离环境下普遍出现 Ca^{2+} 释放量降低（图 5-7）。这可能是由于在隔离环境下随着微藻对 CO_2 的利用（$CO_2+H_2O\leftrightarrow CH_2O+O_2$）和方解石溶解对 CO_2 的吸收，使得水体中 CO_2 的浓度降低，而又得不到相应的补充，从而导致 Ca^{2+} 的再次沉淀。

（四）不同 CO_2 分压下胞外碳酸酐酶抑制剂对 Mg^{2+} 释放量的影响

从上述我们可以看出，由于 Ca^{2+} 受到 $CaCO_3$ 次生沉淀的影响，不能准确反映方解石的真实溶蚀情况。因此，我们选取饱和度较高的 Mg^{2+} 来进一步研究。从不同处理条件 Mg^{2+} 的释放量来看，含有微藻的培养液中 Mg^{2+} 的释放量明显高于不含有微藻的培养液中（图 5-8），说明微藻能够促进方解石中 Mg^{2+} 的释放。不过，培养后期（72h）在 0mmol/L AZ 处理的隔离环境下出现了 Mg^{2+} 释放量减少，并小于不含有微藻的培养液。这可能是在隔离环境的后期既存在微藻增生也存在死亡，由于缺少碳源导致死亡速率大于增生速率，从而使得我们对微藻吸收 Mg^{2+} 的量的估算偏小。这一点能够从微藻生长曲线得到印证（图 5-5）。对于开放环境来说，含有莱茵衣藻的培养液中 Mg^{2+} 的释放量随着 AZ 浓度的增加而减少，而铜绿微囊藻反之（图 5-8）。这主要是由于 AZ 抑制了莱茵衣藻胞外碳酸酐酶活力，导致 HCO_3^- 向 CO_2 的转化速率减小，从而降低了方解石的溶蚀；而铜绿微囊藻没有胞外碳酸酐酶活力，不受其影响。AZ 浓度越高，pH 相对越低，越利于铜绿微囊藻的生长，方解石的 Mg^{2+} 释放量越大。在隔离环境下，铜绿微囊藻有与开放环境下相似的变化规律；而莱茵衣藻却与开放环境下相反，特别是在不

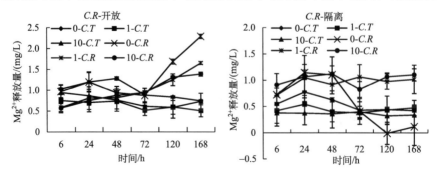

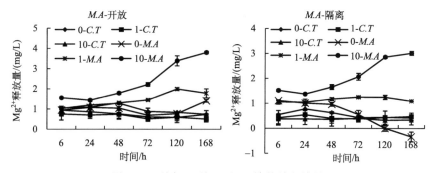

图 5-8 不同 AZ 处理下 Mg^{2+} 的总释放量

0：未添加 AZ，1：添加 1mmol/L 的 AZ，10：添加 10mmol/L 的 AZ；
C.R：莱茵衣藻，M.A：铜绿微囊藻

Fig. 5-8 Amount of Mg^{2+} released from calcite under different AZ treatments
0: without AZ, 1: 1mmol/L AZ, 10: 10mmol/L AZ; C.T: control (without microalgae added),
C.R: *Chlamydomonas reinhardtii*, M.A: *Microcystis aeruginosa*

含有 AZ 的处理条件下。

（五）不同 CO_2 分压下胞外碳酸酐酶抑制剂对单位时间单位藻体 Mg^{2+} 释放量的影响

为了进一步研究 AZ 和 P_{CO_2} 对微藻生物溶蚀作用的影响，通过本章第二节的方法确定了不同条件下单位时间单位质量微藻的生物溶蚀模型，如表 5-6 所示。

表 5-6 单位时间单位微藻的生物溶蚀定量模型

Tab. 5-6 Quantitative models of microalgae biological dissolution in unit algal biomass and unit time (mg/μg-chl-h)

环境	藻体	AZ/(mmol/L)	溶蚀模型	R^2
开放	C.R	0	$N_i = 0.1549 + 3.377 \times 10^{-6} PT_i$	0.993
		1	$N_i = 0.0436 + 3.180 \times 10^{-6} PT_i$	0.968
		10	$N_i = -0.0113 + 3.001 \times 10^{-6} PT_i$	0.865
	M.A	0	$N_i = 0.1191 + 3.357 \times 10^{-6} PT_i$	0.976
		1	$N_i = 0.4624 + 4.739 \times 10^{-6} PT_i$	0.951
		10	$N_i = 0.6856 + 6.041 \times 10^{-6} PT_i$	0.895
隔离	C.R	0	$N_i = 0.4061 - 2.276 \times 10^{-6} PT_i$	0.567
		1	$N_i = 0.4181 + 7.270 \times 10^{-7} PT_i$	0.335
		10	$N_i = 0.5126 + 9.862 \times 10^{-7} PT_i$	0.712
	M.A	0	$N_i = 0.6212 - 6.349 \times 10^{-6} PT_i$	0.951
		1	$N_i = 0.6586 + 6.091 \times 10^{-7} PT_i$	0.101
		10	$N_i = 1.083 + 4.336 \times 10^{-7} PT_i$	0.862

注：C.R：莱茵衣藻，M.A：铜绿微囊藻。

Note：C.R：*Chlamydomonas reinhardtii*，M.A：*Microcystis aeruginosa*.

不同环境下,单位时间单位藻体的 Mg^{2+} 释放量随着 AZ 浓度呈规律性变化。在开放环境下,莱茵衣藻的生物溶蚀量随着 AZ 浓度的增加而减少,这主要是由于 AZ 浓度越大,胞外碳酸酐酶受到抑制的程度越强,催化 HCO_3^- 与 CO_2 之间的转换的能力越低,$CaMgCO_3 + 2H_2O + 2CO_2 \leftrightarrow Ca^{2+} + Mg^{2+} + 4HCO_3^-$ 反应越慢。而铜绿微囊藻反之,这与其不具有胞外碳酸酐酶活力有关,并且其随 AZ 浓度增加生长速率增加,导致单位藻体与矿物交互作用小。在隔离环境下,各微藻随着 AZ 浓度的增加,对方解石 Mg^{2+} 释放效应不断增加(图 5-9)。这可能是由于隔离环境下各微藻随 AZ 浓度的增加,生长速率不断增加,单位藻体与矿物交互作用不断减小。其中,0mmol/L 的 AZ 处理下微藻对方解石的释放出现了负值。

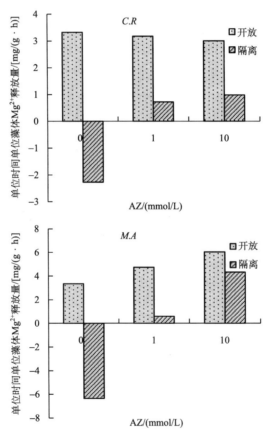

图 5-9 不同 AZ 处理下单位时间单位叶绿素 a 的微藻 Mg^{2+} 的释放量

C.R:莱茵衣藻,M.A:铜绿微囊藻

Fig. 5-9 Amount of Mg^{2+} released per unit algal biomass and unit time under different AZ treatments

C.R: *Chlamydomonas reinhardtii*, M.A: *Microcystis aeruginosa*

这可能是由于，0mmol/L AZ 处理下微藻的死亡速率大于生长速率(图 5-5)，对矿物的生物溶蚀作用极小，同时微藻吸收 Mg^{2+} 量又因具体生长量无法确定而被低估。

此外，我们还发现，各微藻在开放环境下单位藻体单位时间的 Mg^{2+} 释放量大于隔离环境(图 5-9)。这主要是由于在开放环境下，微藻利用水体的 CO_2 后能够很好地得到大气的补给，就如同一个"泵"使得大气中的 CO_2 源源不断地进入水体，从而促进 $CaMgCO_3 + 2H_2O + 2CO_2 \longleftrightarrow Ca^{2+} + Mg^{2+} + 4HCO_3^-$ 反应的进行，而隔离环境下，由于隔离大气中缺少 CO_2，当被藻体利用后无法得到相应的补给。

四、阴离子通道在微藻生物溶蚀中的作用

在 500mL 含有 250mL 培养液的三角瓶中，25℃，$150\mu mol/(m^2 \cdot s)$ 的光照强度，光周期为 12h/12h，培养微藻，每天摇动培养物 5 次，共培养 7 天。三角瓶中添加不同的藻体(*Chlamydomonas reinhardtii*，C.R；*Microcystis aeruginosa*，M.A)、方解石矿粉(1g)，并用封口膜(开放环境)或者碱石灰过气筛加 NaOH 吊瓶(隔离外界 CO_2 环境)封口，设置 0、0.1mmol/L、0.5mmol/L 三个 4,4'-二异硫氰-2,2'-二黄酸芑(DIDS)浓度梯度进行培养(所有处理都进行 3 个平行处理)，具体见表 5-7。其中培养液采用改进的 SE 培养液(不含 Ca，只含约 1/5 的 Mg)。分别在 6h、24h、48h、72h、120h 和 168h 时对藻液进行提取，并在 1600g 离心力条件下离心 10min。其中沉淀用于测定叶绿素 a，上清液用针式过滤器(0.22 μm)过滤后测定其电导率，碱度，Ca^{2+}、Mg^{2+} 浓度。

表 5-7 实验处理

Tab. 5-7 Treatments of experiment

处理	环境	藻	方解石	DIDS/(mmol/L)
1			方解石(ST1)*	0
2	开放或隔离	无		0.1
3			或方解石(ST1)**	0.5
5			方解石(ST1)*	0
6	开放或隔离	C.R/M.A		0.1
7			或方解石(ST1)**	0.5

注：C.R：莱茵衣藻，M.A：铜绿微囊藻。

*：添加 $\delta^{13}C$ 为 -10.287‰的方解石的处理，**：添加 $\delta^{13}C$ 为 0.719‰的方解石的处理。

Note: C.R: *Chlamydomonas reinhardtii*, M.A: *Microcystis aeruginosa*.

*: the treatment with added calcite with an $\delta^{13}C$ value of -10.287‰, **: the treatment with added calcite with an $\delta^{13}C$ value of 0.719‰.

(一) 不同 CO_2 供应下 DIDS 对微藻生物量的影响

从不同 DIDS 浓度处理下微藻的生物量变化来看, DIDS 对铜绿微囊藻起到明显的抑制作用, 其生物量随着 DIDS 浓度的增加而降低, 但是却未对莱茵衣藻的生物量起到抑制(图 5-10)。这可能是由于, 铜绿微囊藻没有胞外碳酸酐酶活力, 阴离子通道是利用 HCO_3^- 的主要途径, 在 DIDS 抑制下其碳源利用途径减少, 使生长受到抑制; 而莱茵衣藻有较强胞外碳酸酐酶活力, 当阴离子通道受到抑制时, 它能够提高胞外碳酸酐酶活力来利用碳源。其中, 莱茵衣藻的生物量在后期出现随着 DIDS 浓度增加而略有增加的趋势, 可能是由于随着 DIDS 浓度增加, pH 变化相对较小, 而当 pH 在 7 以上时, pH 较低更利于生长(图 5-1)。此外, 隔离环境下微藻的生物量明显小于开放环境下(图 5-10), 这主要是隔离环境下微藻缺少无机碳源的补给导致。特别是铜绿微囊藻表现得最为明显, 几乎一直处于下降趋势。

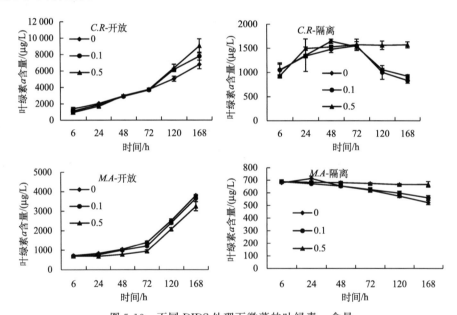

图 5-10 不同 DIDS 处理下微藻的叶绿素 a 含量

0: 未添加 DIDS, 0.1: 添加 0.1mmol/L 的 DIDS, 1: 添加 1mmol/L 的 DIDS;
$C.R$: 莱茵衣藻, $M.A$: 铜绿微囊藻

Fig. 5-10 Chl-a content of microalgae under different DIDS treatments

0: without DIDS, 0.1: 0.1mmol/L DIDS, 0.5: 0.5mmol/L DIDS;
$C.R$: *Chlamydomonas reinhardtii*, $M.A$: *Microcystis aeruginosa*

在此基础上，对不同处理条件下各藻体的生物量建立随时间变化的方程。可以发现，在开放环境下，随时间的迁移，各藻体呈指数增长；在隔离环境下，各藻体由于处于限制性环境下，因此成对数正态分布（张怀强等，2014）。并对其求时间的积分，得到开放和隔离环境下不同 DIDS 浓度处理下各藻体的累计作用时间方程，如表 5-8 所示。

表 5-8　生物增加曲线及生物累计作用时间模型

Tab. 5-8　The mode between growth rate and biological cumulative action time

环境	藻体	DIDS /(mmol/L)	增长曲线	生物累计作用时间
开放	C.R	0	$Chi = 1886.21 e^{0.0078T}$ $R^2 = 0.963$	$PTi = \int_0^T 1886.21 e^{0.0078T} d(T)$
		0.1	$Chi = 1788.65 e^{0.0091T}$ $R^2 = 0.949$	$PTi = \int_0^T 1788.65 e^{0.0091T} d(T)$
		0.5	$Chi = -18107 + 18828 e^{0.0022T}$ $R^2 = 0.998$	$PTi = \int_0^T -18107 + 18828 e^{0.0022T} d(T)$
	M.A	0	$Chi = -624 + 1220 e^{0.0077T}$ $R^2 = 0.999$	$PTi = \int_0^T -624 + 1220 e^{0.0077T} d(T)$
		0.1	$Chi = -215 + 788 e^{0.0095T}$ $R^2 = 0.993$	$PTi = \int_0^T -215 + 788 e^{0.0095T} d(T)$
		0.5	$Chi = 121 + 418 e^{0.0120T}$ $R^2 = 0.983$	$PTi = \int_0^T 121 + 418 e^{0.0120T} d(T)$
隔离	C.R	0	$Chi = 964 + 732 \exp\left(-0.5\left(\frac{\ln\left(\frac{T}{45}\right)}{0.54}\right)^2\right)$ $R^2 = 0.951$	$PTi = \int_0^T 964 + 732 \exp\left(-0.5\left(\frac{\ln\left(\frac{T}{45}\right)}{0.54}\right)^2\right) d(T)$
		0.1	$Chi = 964 + 732 \exp\left(-0.5\left(\frac{\ln\left(\frac{T}{44}\right)}{0.54}\right)^2\right)$ $R^2 = 0.951$	$PTi = \int_0^T 964 + 732 \exp\left(-0.5\left(\frac{\ln\left(\frac{T}{44}\right)}{0.54}\right)^2\right) d(T)$
		0.5	$Chi = -2920593 + 2922184 \exp\left(-0.5\left(\frac{\ln\left(\frac{T}{125}\right)}{83}\right)^2\right)$ $R^2 = 0.987$	$PTi = \int_0^T -2920593 + 2922184 \exp\left(-0.5\left(\frac{\ln\left(\frac{T}{125}\right)}{83}\right)^2\right) d(T)$
	M.A	0	$Chi = 352 + 365 \exp\left(-0.5\left(\frac{\ln\left(\frac{T}{15}\right)}{1.98}\right)^2\right)$ $R^2 = 0.992$	$PTi = \int_0^T 352 + 365 \exp\left(-0.5\left(\frac{\ln\left(\frac{T}{15}\right)}{1.98}\right)^2\right) d(T)$
		0.1	$Chi = 691 - 1191 \exp\left(-0.5\left(\frac{\ln\left(\frac{T}{36477}\right)}{2.56}\right)^2\right)$ $R^2 = 0.996$	$PTi = \int_0^T 691 - 1191 \exp\left(-0.5\left(\frac{\ln\left(\frac{T}{36477}\right)}{2.56}\right)^2\right) d(T)$
		0.5	$Chi = 686 - 21 \exp\left(-0.5\left(\frac{\ln\left(\frac{T}{145}\right)}{0.73}\right)^2\right)$ $R^2 = 0.999$	$PTi = \int_0^T 686 + 21 \exp\left(-0.5\left(\frac{\ln\left(\frac{T}{145}\right)}{0.73}\right)^2\right) d(T)$

注：C.R：莱茵衣藻，M.A：铜绿微囊藻。

Note：C.R：*Chlamydomonas reinhardtii*，M.A：*Microcystis aeruginosa*.

(二)不同CO_2供应下DIDS对微藻胞外碳酸酐酶的影响

DIDS是阴离子通道专一抑制剂,它广泛被用于研究阴离子通道在微藻无机碳利用过程中的作用(Larsson and Axelsson,1999)。然而,阴离子通道与碳酸酐酶关系密切。有学者报道,当胞内碳酸酐酶Ⅱ型被抑制时Cl^-/HCO_3^-离子通道的转换效率将降低50%~60%(Sterling et al.,2001)。Young等发现,具有较强胞外碳酸酐酶活力的 *Dunaliella tertiolecta* 存在一种不受DIDS影响的机制,但是其并没有查明是什么机制最终促使 *D. tertiolecta* 能够在DIDS作用下保持正常的生理功能(Young et al.,2001)。此外,其他学者还发现多种具有较强胞外碳酸酐酶活力的微藻在DIDS的作用下能够正常光合,如 *Chaetomorpha linum*、*Cladophora* sp.、*Spongomorpha arcta* 等(Larsson and Axelsson,1999)。

本小节研究了DIDS对 *C. reinhardtii* 胞外碳酸酐酶活力的影响。从图5-11可以看出,在开放环境下,随着DIDS浓度的增加,*C.R* 的胞外碳酸酐酶活力有微弱的增加,特别是在0.5mmol/L DIDS浓度的抑制作用下。这可能是由于胞外碳酸酐酶是一种诱导酶,微藻的可利用碳源不足时将被诱导表达,而当DIDS抑制 *C.R* 的阴离子通道后导致微藻的无机碳利用途径减少,可利用碳源减少。

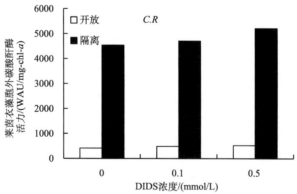

图5-11 不同DIDS处理下莱茵衣藻胞外碳酸酐酶活力
0:未添加DIDS,0.1:添加0.1mmol/L的DIDS,1:添加1mmol/L的DIDS;
C.R:莱茵衣藻,*M.A*:铜绿微囊藻

Fig. 5-11 The activity of extracellular carbonic anhydrase in
C. reinhardtii under different DIDS treatments
0: without DIDS, 0.1: 0.1mmol/L DIDS, 0.5: 0.5mmol/L DIDS;
C.R: *Chlamydomonas reinhardtii*, *M.A*: *Microcystis aeruginosa*

(三) 不同 CO_2 供应下 DIDS 对 Ca^{2+} 释放量的影响

从未添加微藻培养液中 Ca^{2+} 的释放情况来看，随着 DIDS 浓度的增加 Ca^{2+} 的释放量只有微弱的变化(图 5-12)。因此，DIDS 对方解石的溶蚀只有微弱的影响。同时，隔离环境下 Ca^{2+} 的释放量明显低于开放环境，这与隔离环境中大气 CO_2 分压较低有关。从总体来看，无论是添加莱茵衣藻还是铜绿微囊藻的培养液，Ca^{2+} 的释放量都随着 DIDS 浓度的增加而不断增加。由于 Ca^{2+} 极易产生 $CaCO_3$ 次生沉淀。因此，这主要受到两个重要因素的影响，分别是：方解石的溶蚀量和 $CaCO_3$ 的次生沉淀量。我们可以发现，在培养 48h 以后，出现了明显的下降，特别是 DIDS 处理下的莱茵衣藻。这可能是由于在培养 48h 后微藻进入对数生长期，无机碳利用增强，而微藻在利用 HCO_3^- 的过程中将导致 $CaCO_3$ 沉淀的产生，并且胞外碳酸酐酶在其中起到强烈促进作用(Sharma and Bhattacharya, 2010; Yates and Robbins, 1999)。

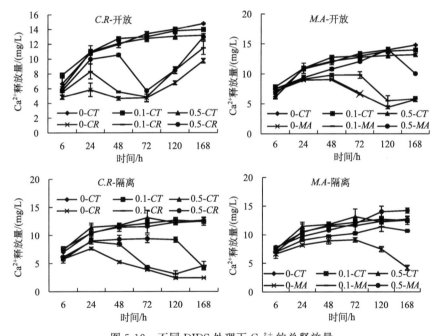

图 5-12 不同 DIDS 处理下 Ca^{2+} 的总释放量

0：未添加 DIDS，0.1：添加 0.1mmol/L 的 DIDS，1：添加 1mmol/L 的 DIDS；
C.R：莱茵衣藻，M.A：铜绿微囊藻

Fig. 5-12 Amount of Ca^{2+} released from calcite under different DIDS treatments

0: without DIDS, 0.1: 0.1mmol/L DIDS, 0.5: 0.5mmol/L DIDS;
C.R: *Chlamydomonas reinhardtii*, M.A: *Microcystis aeruginosa*

(四) 不同 CO_2 供应下 DIDS 对 Mg^{2+} 释放量的影响

为了进一步研究 DIDS 在微藻生物溶蚀过程中的作用，我们将从 Mg^{2+} 的释放特征来进行研究。从未添加微藻的空白对照组来看，DIDS 对方解石中 Mg^{2+} 的释放并没有规律性的影响。随着微藻的参与，在 DIDS 作用下方解石的生物溶蚀作用出现规律性的变化。从含有莱茵衣藻的培养液 Mg^{2+} 释放量来看，随着 DIDS 浓度的增加 Mg^{2+} 释放量不断增加。说明 DIDS 促进了莱茵衣藻参与下的方解石 Mg^{2+} 释放作用。对于铜绿微囊藻来说，随着 DIDS 浓度的增加 Mg^{2+} 释放量不断减少，说明 DIDS 抑制了 M.A 参与下的方解石 Mg^{2+} 释放作用。其与 Ca^{2+} 释放量的变化规律相反(图 5-12)，这可能是由于 Ca^{2+} 受化学作用的影响较强，而 Mg^{2+} 受生物溶蚀作用的影响较强。此外，隔离环境下 Mg^{2+} 的释放量明显小于开放环境(图 5-13)。这一方面由于隔离环境下 CO_2 分压较低，另一方面由于微藻生物量较低。

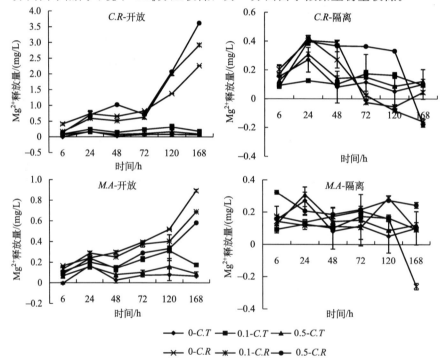

图 5-13 不同 DIDS 处理下 Mg^{2+} 的总释放量

0：未添加 DIDS，0.1：添加 0.1mmol/L 的 DIDS，1：添加 1mmol/L 的 DIDS；
C.R：莱茵衣藻，M.A：铜绿微囊藻

Fig. 5-13 Amount of Mg^{2+} released from calcite under different DIDS treatments
0：without DIDS, 0.1：0.1mmol/L DIDS, 0.5：0.5mmol/L DIDS；C.T：control
(without microalgae added)；
C.R：*Chlamydomonas reinhardtii*，M.A：*Microcystis aeruginosa*

（五）不同 CO_2 分压下 DIDS 对单位时间单位藻体 Mg^{2+} 释放量的影响

为了进一步研究 DIDS 和 P_{CO_2} 对微藻生物溶蚀作用的影响，通过本章第二节的方法确定了不同条件下单位时间单位质量微藻的生物溶蚀模型，如表 5-9 所示。

表 5-9　单位时间单位微藻的生物溶蚀定量模型
Tab. 5-9　Quantitative models of microalgae biological dissolution in unit algal biomass and unit time（mg/μg-chl-h）

环境	藻体	DIDS/(mmol/L)	溶蚀模型	R^2
开放	C.R	0	$N_i=0.3611+2.822\times10^{-6}PT_i$	0.992
		0.1	$N_i=0.0639+4.198\times10^{-6}PT_i$	0.971
		0.5	$N_i=0.2643+4.658\times10^{-6}PT_i$	0.969
	M.A	0	$N_i=0.1257+2.387\times10^{-6}PT_i$	0.967
		0.1	$N_i=0.0231+2.181\times10^{-6}PT_i$	0.983
		0.5	$N_i=0.0199+2.166\times10^{-6}PT_i$	0.977
隔离	C.R	0	$N_i=0.1687-1.678\times10^{-6}PT_i$	0.677
		0.1	$N_i=0.2226-1.660\times10^{-6}PT_i$	0.436
		0.5	$N_i=0.1830-7.858\times10^{-7}PT_i$	0.140
	M.A	0	$N_i=0.0666-4.0941\times10^{-6}PT_i$	0.161
		0.1	$N_i=0.0091+1.799\times10^{-7}PT_i$	0.101
		0.5	$N_i=0.0181+1.901\times10^{-7}PT_i$	0.228

注：C.R：莱茵衣藻，M.A：铜绿微囊藻。
Note：C.R：*Chlamydomonas reinhardtii*，M.A：*Microcystis aeruginosa*.

不同环境下，单位时间单位藻体的 Mg^{2+} 释放量随着 DIDS 浓度呈规律性变化。随着培养液中 DIDS 浓度的增加，单位时间单位铜绿微囊藻的 Mg^{2+} 释放量逐渐降低（图 5-14）。说明 DIDS 抑制了铜绿微囊藻对方解石中 Mg^{2+} 的释放。这是由于铜绿微囊藻主要通过阴离子通道途径来对水体中 HCO_3^- 进行利用，在 DIDS 的作用下，其对水体中 HCO_3^- 的利用受到抑制，而使 $CaMg(CO_3)_2 + 2CO_2 + 2H_2O \leftrightarrow Ca^{2+} + Mg^{2+} + 4HCO_3^-$ 朝正方向的反应速率减慢。但是，莱茵衣藻却表现为相反的变化规律（图 5-14），这主要由于莱茵衣藻对水体中 HCO_3^- 的利用途径不仅有阴离子通道，还有胞外碳酸酐酶，而胞外碳酸酐酶是一种诱导酶，当莱茵衣藻缺少碳源时将被诱导。因此，当 DIDS 抑制了莱茵衣藻的阴离子通道时，将诱导微藻胞外碳酸酐酶活力的升高（这一点能从生物量上获得证明），而胞外碳酸酐酶对方解石的生物溶蚀有重要的促进作用（Xie and Wu，2014）。

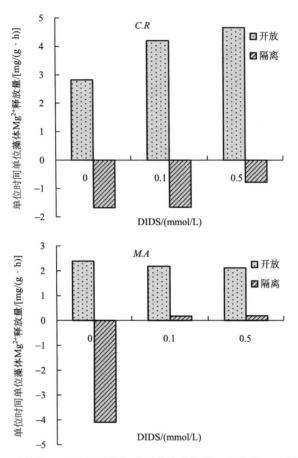

图 5-14 不同 DIDS 处理下单位时间单位叶绿素 a 藻体的 Mg^{2+} 释放量

0：未添加 DIDS，0.1：添加 0.1mmol/L 的 DIDS，1：添加 1mmol/L 的 DIDS；
$C.R$：莱茵衣藻，$M.A$：铜绿微囊藻

Fig. 5-14 Amount of Mg^{2+} released per unit algal biomass and unit time under different DIDS treatments

0: without DIDS, 0.1: 0.1mmol/L DIDS, 0.5: 0.5mmol/L DIDS;
$C.R$: *Chlamydomonas reinhardtii*, $M.A$: *Microcystis aeruginosa*

第四节 碳酸酐酶影响下的微藻对碳酸钙碳源的利用

长期以来，都认为碳酸盐岩的溶蚀对碳汇没有起到作用，因为碳酸钙溶蚀过程中（$CaCO_3 + CO_2 + H_2O \rightarrow Ca^{2+} + 2HCO_3^-$）消耗的 CO_2 又通过碳酸钙沉积而返回大气（Bricker，1988）。但是，其没有考虑到生物作用的参与，特别是微藻

这类光合生物。随着碳酸钙的溶蚀，其向水体释放出可溶性无机碳（DIC）。这些DIC一部分来自大气的无机碳，一部分来自于碳酸钙中的无机碳。而微藻可能利用水体中的这部分DIC，并进一步改变水体的DIC平衡，使碳酸钙溶解平衡发生变化。碳酸酐酶由于可以催化CO_2水合反应，促进微藻对水体DIC的利用和碳酸盐岩的溶蚀，在微藻与碳酸盐岩的交互系统中起到重要的作用。因此，本节将通过双向同位素示踪手段来研究微藻对碳酸钙碳源的利用，并进一步探讨碳酸酐酶在其中的作用。

一、酸碱度对微藻利用碳酸钙碳源的影响

在250mL含有100mL培养液的三角瓶中，25℃，$150\mu mol/(m^2 \cdot s)$的光照强度，光周期为12h/12h，培养微藻，每天摇动培养物5次，共培养7天。三角瓶中分别添加不同的藻体和两种碳同位素组成相差悬殊的方解石矿粉（0.3g）（$\delta^{13}C_{ST1}$：$-10.87‰$，$\delta^{13}C_{ST2}$：$0.719‰$），并用无菌封口膜封口，设置6~7、7~8、8~9、9~10四个pH梯度进行培养（所有处理都进行3个平行处理），每隔6h用稀HCl调节一次。在最后收获时，收集剩余的所有藻液进行碳氧同位素测定和NPOC测定。

（一）不同pH条件下微藻$\delta^{13}C$和$\delta^{18}O$值变化

从不同pH条件下微藻的碳同位素组成来看，藻体的$\delta^{13}C$值随着pH的增加逐渐降低（图5-15）。特别是在pH大于9时，下降最为明显。这主要是由于随着pH的增加，水体中CO_2与HCO_3^-的比例逐渐减少，诱导微藻利用HCO_3^-的比例逐渐增加，而微藻通过利用HCO_3^-途径这个相对CO_2途径的慢速过程，将产生一个趋于偏负的同位素分馏（Mook et al.，1974）。此外，我们可以发现，随着pH不断增加，$M.A$的$\delta^{13}C$值变化最大，从pH为6~7到pH为9~10，其$\delta^{13}C$值变化了约2‰；其次是$C.P$，变化了约1.5‰；变化最小的是$C.R$，约为1‰。这种现象的产生与这三种藻的胞外碳酸酐酶活力有关，$C.R$胞外碳酸酐酶活力最强，$C.P$次之，$M.A$没有胞外碳酸酐酶活力。而胞外碳酸酐酶催化利用HCO_3^-将减弱微藻利用HCO_3^-过程中产生的分馏效应（Wu et al.，2011）。

从不同pH条件下微藻的氧同位素组成来看，藻体的$\delta^{18}O$随着pH的增加逐渐增加，如图5-16所示。这主要是由于，微藻在进行光合作用（$CO_2 + H_2O \leftrightarrow CH_2O + O_2$）过程中，由于有气态氧分子的产生，将导致产生的有机物（CH_2O）的氧同位素产生一个相对偏正的氧分馏。而pH越低，水体中的DIC以CO_2的形式存在的比例越大，直接利用CO_2的途径能力越强；随着pH增加，HCO_3^-比例增加，导致微藻需要增加HCO_3^-途径来保证足够的碳源供应量。这个过程在无形中增强了微藻对氧同位素的分馏效应。

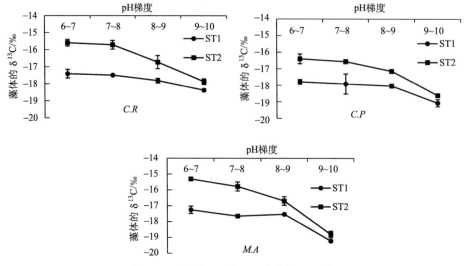

图 5-15　不同 pH 处理下微藻的 $\delta^{13}C$ 值

C.R：莱茵衣藻，C.P：蛋白核小球藻，M.A：铜绿微囊藻；ST1：添加 $\delta^{13}C$ 为 $-10.287‰$ 的方解石的培养组，ST2：添加 $\delta^{13}C$ 为 $0.719‰$ 的方解石的培养组；6~7、7~8、8~9、9~10：分别为培养液 pH 为 6~7、7~8、8~9、9~10 的处理

Fig. 5-15　The microalgal $\delta^{13}C$ value under different pH treatments

C.R：*Chlamydomonas reinhardtii*，C.P：*Chlorella pyrenoedosa*，M.A：*Microcystis aeruginosa*；ST1：the treatment with added calcite with an $\delta^{13}C$ value of $-10.287‰$，ST2：the treatment with added calcite with an $\delta^{13}C$ value of $0.719‰$；6~7：the pH of medium is between 6 and 7，7~8：the pH of medium is between 7 and 8，8~9：the pH of medium is between 8 and 9，9~10：the pH of medium is between 9 and 10

（二）不同 pH 条件下微藻对碳酸钙碳源和氧源的利用

无论是莱茵衣藻、蛋白核小球藻还是铜绿微囊藻，随着 pH 的增加，微藻对方解石碳源的利用比例都逐渐降低，而对水体固有碳源的利用比例增加（图 5-17）。这可能有两方面的原因，一方面是因为随着 pH 的增加方解石的溶蚀逐渐降低；另一方面是因为 pH 越高 CO_2 和 HCO_3^- 与 CO_3^{2-} 比例越低，而只有 CO_2 和 HCO_3^- 才能被藻体所利用，目前还没有明显的证据表明 CO_3^{2-} 能被微藻直接利用。在 pH 为 8~9 下降最为明显，也从侧面印证了这一点。因为，在 pH 为 8~9 的范围内，水体中各形态的 DIC 出现了最为明显的变化，当 pH 为 8.3 时水体中的 CO_2 将全部转换为 HCO_3^- 或者 CO_3^{2-}，以 CO_3^{2-} 形式存在的 DIC 由此出现。

随着 pH 的增加，微藻对方解石氧源的利用份额比例都逐渐降低，而对水体

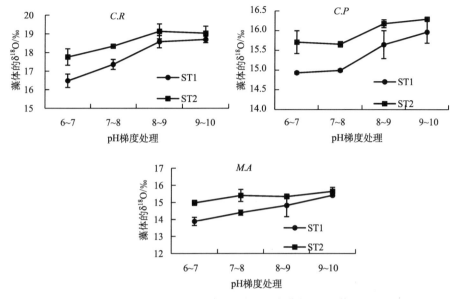

图 5-16 不 Note：同 pH 处理下微藻的 $\delta^{18}O$ 值

C.R：莱茵衣藻，C.P：蛋白核小球藻，M.A：铜绿微囊藻；ST1：添加 $\delta^{18}O$ 为 $-15.2‰$ 的方解石的培养组，ST2：添加 $\delta^{18}O$ 为 $-7.8‰$ 的方解石培养组；6~7、7~8、8~9、9~10：分别为培养液 pH 为 6~7、7~8、8~9、9~10 的处理

Fig. 5-16 The microalgal $\delta^{16}O$ under different pH treatments

C.R: *Chlamydomonas reinhardtii*, C.P: *Chlorella pyrenoedosa*, M.A: *Microcystis aeruginosa*; ST1: the treatment with added calcite with an $\delta^{18}O$ value of $-15.2‰$, ST2: the treatment with added calcite with an $\delta^{18}O$ value of $-7.8‰$; 6~7: the pH of medium is between 6 and 7, 7~8: the pH of medium is between 7 and 8, 8~9: the pH of medium is between 8 and 9, 9~10: the pH of medium is between 9 and 10

固有氧源的利用比例逐渐增加（图 5-18）。这一结果与微藻对方解石碳源的利用情况形成了很好的印证。

二、胞外碳酸酐酶对微藻利用碳酸钙碳源的影响

在 500mL 含有 250mL 培养液的三角瓶中，25℃，150μmol/(m²·s) 的光照强度，光周期为 12h/12h，培养微藻，每天摇动培养物 5 次，共培养 7 天。三角瓶中添加不同的藻体（*Chlamydomonas reinhardtii*，C.R；*Microcystis aeruginosa*，M.A）和两种碳同位素组成相差悬殊的方解石矿粉（0.3g）（$\delta^{13}C_{ST1}$：$-10.87‰$，$\delta^{13}C_{ST2}$：$0.719‰$），并用封口膜（开放环境）或者碱石灰过气筛加 NaOH 吊瓶（隔离外界 CO_2 环境）封口，设置 0、1mmol/L、10mmol/L 三个 AZ 浓度梯度进行培养（所有处理都进行 3 个平行处理）。在最后收获时，收集剩余的

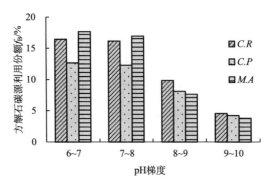

图 5-17　不同 pH 处理下微藻对方解石碳源的利用份额

C.R：莱茵衣藻，C.P：蛋白核小球藻，M.A：铜绿微囊藻；6~7、7~8、8~9、9~10：分别为培养液 pH 为 6~7、7~8、8~9、9~10 的处理

Fig. 5-17　Proportion of utilizationon carbon in calcite by microalgae under different pH treatments

C.R：*Chlamydomonas reinhardtii*，C.P：*Chlorella pyrenoedosa*，M.A：*Microcystis aeruginosa*；6~7：the pH of medium is between 6 and 7，7~8：the pH of medium is between 7 and 8，8~9：the pH of medium is between 8 and 9，9~10：the pH of medium is between 9 and 10

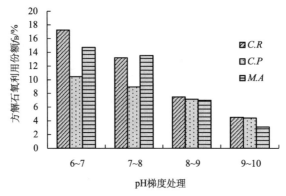

图 5-18　不同 pH 处理下微藻对方解石氧源的利用份额

C.R：莱茵衣藻，C.P：蛋白核小球藻，M.A：铜绿微囊藻；6~7、7~8、8~9、9~10：分别为培养液 pH 为 6~7、7~8、8~9、9~10 的处理

Fig. 5-18　Proportion of utilizationon oxygen in calcite by microalgae under different pH treatments

C.R：*Chlamydomonas reinhardtii*，C.P：*Chlorella pyrenoedosa*，M.A：*Microcystis aeruginosa*；6~7：the pH of medium is between 6 and 7，7~8：the pH of medium is between 7 and 8，8~9：the pH of medium is between 8 and 9，9~10：the pH of medium is between 9 and 10

所有藻液进行碳同位素测定和 NPOC 测定。

(一) 不同 CO_2 分压下胞外碳酸酐酶抑制剂对微藻 $\delta^{13}C$ 的影响

从不同 AZ 浓度处理下微藻的 $\delta^{13}C$ 值变化来看,随着 AZ 浓度的增加,微藻的 $\delta^{13}C$ 值不断降低,其中莱茵衣藻 $\delta^{13}C$ 值的下降程度大于铜绿微囊藻。这主要是由于随着 AZ 浓度增加微藻胞外碳酸酐酶受到抑制的程度增加,使得微藻通过 CAex 催化利用 HCO_3^- 的能力降低。而在无碳酸酐酶胞外酶催化的情况下,碳酸氢根离子的直接转运过程存在约 10‰ 的碳同位素分馏(Mook et al.,1974);而由碳酸酐酶胞外酶催化的碳酸氢根离子的间接转运过程只存在约 1.1‰ 的碳同位素分馏(Marlier and O'Leary,1984)。然而,对于没有胞外碳酸酐酶的铜绿微囊藻来说,其主要是受 pH 的影响,AZ 能够降低培养液的 pH,而 pH 越低微藻的 $\delta^{13}C$ 值越偏负(图 5-15)。此外,低 AZ 处理(0 或者 1mmol/L)的隔离环境下,莱茵衣藻和铜绿微囊藻的 $\delta^{13}C$ 值相对开放环境偏正;10mmol/L AZ 处理下反之(图 5-19)。这两种藻产生这一现象的原因是不同的。对于莱茵衣藻,其主要是由于在隔离环境下,CO_2 分压相对较低,胞外碳酸酐酶活力表达受到诱导

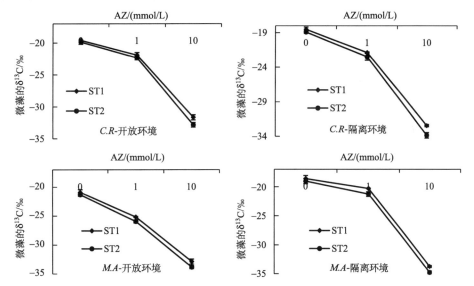

图 5-19 不同 AZ 处理下微藻的 $\delta^{13}C$

C.R:莱茵衣藻,M.A:铜绿微囊藻;ST1:添加 $\delta^{13}C$ 为 -10.287‰ 的方解石的处理,ST2:添加 $\delta^{13}C$ 为 0.719‰ 的方解石的处理

Fig. 5-19 The microalgal $\delta^{13}C$ value under different AZ treatments

C.R: *Chlamydomonas reinhardtii*, M.A: *Microcystis aeruginosa*; ST1: the treatment with added calcite with an $\delta^{13}C$ value of -10.287‰, ST2: the treatment with added calcite with an $\delta^{13}C$ value of 0.719‰

(Aizawa and Miyachi, 1986), 降低 DIC 利用过程中的分馏效应; 而在 10mmol/L AZ 处理下莱茵衣藻的胞外碳酸酐酶活力几乎完全被抑制, 隔离环境下的微藻由于缺少 CO_2, 只能增强直接利用 HCO_3^- 的途径来利用 DIC, 这个慢速反应将产生一个相对偏负的动力学同位素分馏。对于铜绿微囊藻, 其主要是由于低 AZ 处理下的隔离环境, 方解石溶蚀相对较慢, pH 相对较低, 因此 $\delta^{13}C$ 相对偏正。但是, 在 10mmol/L AZ 处理下的隔离与开放环境 pH 相似, 且较多的通过直接 HCO_3^- 途径利用 DIC, 因此 $\delta^{13}C$ 相对偏负。

(二) 不同 CO_2 分压下胞外碳酸酐酶抑制剂对微藻利用碳酸钙碳源的影响

不同 AZ 处理下, 微藻对方解石碳源的利用份额呈规律性变化。随着 AZ 浓度增加, 莱茵衣藻利用碳酸钙碳源的份额大幅度增加, 而铜绿微囊藻出现相对较小的增加(图 5-20)。这主要是由于, 水体中的 DIC 主要以水体固有无机碳为主, 莱茵衣藻主要利用固有的无机碳源, 在存在胞外碳酸酐酶抑制剂的情况下, 莱茵衣藻的胞外碳酸酐酶活性降低, 使得莱茵衣藻通过胞外碳酸酐酶的水合催化作用 ($HCO_3^- + H^+ \leftrightarrow CO_2 + H_2O$) 来利用固有无机碳源的能力降低, 从而降低了莱茵衣藻对固有无机碳源的利用份额。而铜绿微囊藻增幅较小主要是由于其不具有胞外碳酸酐酶活力, 其主要受不同 pH 条件下方解石的溶蚀差异影响。此外, 我们还可以看出, 隔离环境下, 微藻对方解石碳源的利用份额大于开放环境下(图 5-20)。这主要是因为在实验模拟的水体 pH 环境下 ($7 < pH < 10.3$), DIC 中的 HCO_3^- 比例大于 CO_2。在隔离环境下, 实验瓶空气中的 CO_2 浓度降低, 而水体中 HCO_3^- 的减少量大于 CO_2 的减少量, 从而使碳酸钙溶解平衡朝着溶解的方向移动, 使水体中由碳酸钙溶蚀形成的 DIC 相对较多。因此, 微藻利用碳酸钙溶

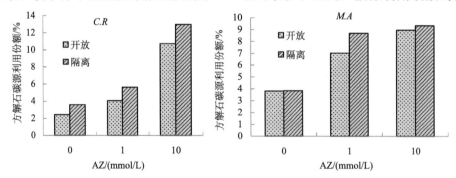

图 5-20 不同 AZ 处理下微藻对方解石碳源的利用份额

C.R: 莱茵衣藻, M.A: 铜绿微囊藻

Fig. 5-20 Proportion of utilizationon carbon in calcite by microalgae under different AZ treatments

C.R: *Chlamydomonas reinhardtii*, M.A: *Microcystis aeruginosa*

蚀所形成的 DIC 这种间接的方式来利用碳酸钙碳源的份额增加了，而在利用碳酸钙碳源的过程中很大程度上受到碳酸钙溶解平衡的影响。由此，可以推测，在湖泊、海洋等水体中，深度越深，固有无机碳碳源越少，微藻对碳酸钙碳源的利用量越大。

三、阴离子通道对微藻利用碳酸钙碳源的影响

在 500mL 含有 250mL 培养液的三角瓶中，25℃，150μmol/(m^2·s)的光照强度，光周期为 12h/12h，培养微藻，每天摇动培养物 5 次，共培养 7 天。三角瓶中添加不同的藻体（$Chlamydomonas\ reinhardtii$，C.R；$Microcystis\ aeruginosa$，M.A）和两种碳同位素组成相差悬殊的方解石矿粉（0.3g）（$\delta^{13}C_{ST1}$：$-10.87‰$，$\delta^{13}C_{ST2}$：0.719‰），并用封口膜（开放环境）或者碱石灰过气筛加 NaOH 吊瓶（隔离外界 CO_2 环境）封口，设置 0、0.1mmol/L、0.5mmol/L 三个 DIDS 浓度梯度进行培养（所有处理都进行 3 个平行处理）。在最后收获时，收集剩余的所有藻液进行碳同位素测定和不可吹灭有机碳（non-purgeable organic carbon，NPOC）测定。

（一）不同 CO_2 分压下 DIDS 对微藻 $\delta^{13}C$ 的影响

从处理后的同位素数据来看，微藻的同位素组成随着 DIDS 浓度的增加而普遍降低，但是开放环境下莱茵衣藻的同位素组成变化却相反（图 5-21）。这主要是由于，DIDS 对微藻的阴离子通道起到了抑制作用，而使微藻利用 HCO_3^- 的速率减慢，从而增加了动力学同位素分馏效应。然而，对于莱茵衣藻，主要是因为莱茵衣藻有较强的胞外碳酸酐酶活力。当阴离子通道受到抑制时，其可以提高胞外碳酸酐酶活力来弥补阴离子通道被抑制所带来碳源不足（图 5-11），而 CAex 的快速催化作用能降低动力学同位素分馏效应。而在隔离环境下，莱茵衣藻的$\delta^{13}C$值并没有在 DIDS 为 0.5mmol/L 的处理下上升。其原因可能是隔离环境下胞外碳酸酐酶已经被充分的诱导，而此时 DIDS 通过抑制阴离子通道所带来的偏负同位素分馏效应，大于胞外碳酸酐酶升高所带来同位素效应。从图 5-11 可以看出开放环境下莱茵衣藻在无 DIDS 处理下的 CAex 活力相比 0.5mmol/L 的 DIDS 处理下升高了约 25%，而隔离环境下仅升高 15%。

（二）不同 CO_2 分压下 DIDS 对微藻利用碳酸钙碳源的影响

不同 DIDS 处理下，微藻对方解石碳源的利用份额呈规律性变化。随着 DIDS 浓度的增加，莱茵衣藻利用方解石碳源的份额逐渐增加，而铜绿微囊藻反之（图 5-22）。这主要是受微藻对方解石的生物溶蚀作用所控制。随着培养液 DIDS 浓度的增加，莱茵衣藻对方解石的生物溶蚀作用逐渐增加（见本章第三

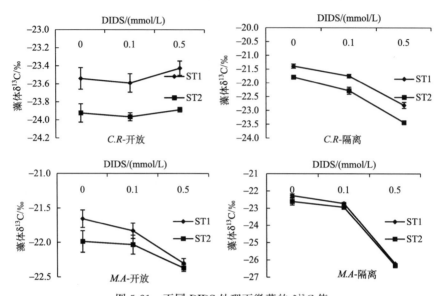

图 5-21 不同 DIDS 处理下微藻的 $\delta^{13}C$ 值

$C.R$：莱茵衣藻，$M.A$：铜绿微囊藻；ST1：添加 $\delta^{13}C$ 为 $-10.287‰$ 的方解石的处理，ST2：添加 $\delta^{13}C$ 为 $0.719‰$ 的方解石的处理

Fig. 5-21 The microalgal $\delta^{13}C$ value under different DIDS treatments

$C.R$: *Chlamydomonas reinhardtii*, $M.A$: *Microcystis aeruginosa*; ST1: the treatment with added calcite with an $\delta^{13}C$ value of $-10.287‰$, ST2: the treatment with added calcite with an $\delta^{13}C$ value of $0.719‰$

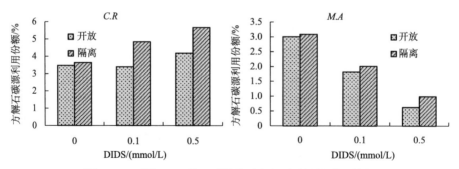

图 5-22 不同 DIDS 处理下微藻对方解石碳源的利用份额

$C.R$：莱茵衣藻，$M.A$：铜绿微囊藻

Fig. 5-22 Proportion of utilization on carbon in calcite by microalgae under different DIDS treatments

$C.R$: *Chlamydomonas reinhardtii*, $M.A$: *Microcystis aeruginosa*

节)，使得方解石释放的 DIC 占水体总 DIC 的比例逐渐增加。而铜绿微囊藻则是随着 DIDS 浓度的增加，生物溶蚀作用逐渐降低。

第五节 碳酸酐酶影响下的生物岩溶碳汇作用

水体中微藻参与下的碳酸盐岩溶蚀所释放的无机碳主要有三种归属，分别为：①以 DIC 的形式残留在水体(DIC)；②被藻体光合作用所利用形成有机碳(有机碳封存)；③再次沉淀形成 $CaCO_3$($CaCO_3$ 封存)。其中由于碳酸盐岩溶蚀($CaMg(CO_3)_2+2CO_2+2H_2O \leftrightarrow Ca^{2+}+Mg^{2+}+4HCO_3^-$)所吸收的一份 CO_2 会随着 $CaCO_3$ 的沉淀而释放出去($Ca^{2+}+2HCO_3^- \rightarrow CaCO_3+CO_2$)。因此，以 $CaCO_3$ 形式存在的无机碳不具有碳汇效应，只有以 DIC 或者有机碳形式存在的无机碳才具有实际的碳汇效应。碳酸酐酶一方面能够促进碳酸盐岩的溶蚀；另一方面能够促进 $CaCO_3$ 沉淀的产生，在微藻对碳酸盐岩生物溶蚀碳汇效应中起到重要作用。

一、酸碱度对生物岩溶碳汇的作用

从总体来看，在 pH 为 7～8 时，方解石在溶蚀过程中释放的无机碳最多，而其后随着 pH 的增加，方解石溶蚀所释放的无机碳逐渐减少。这可能是由于在 pH 为 7～8 时，生物溶蚀作用和化学溶蚀作用都处于相对高的水平，从而导致总体溶蚀作用较强；随着 pH 增加，化学溶蚀作用逐渐降低，从而使方解石的无机碳释放量减少。我们还发现，在 pH 为 6～7 时，含有蛋白核小球藻和铜绿微囊藻的培养液中方解石的无机碳释放量明显减少(图 5-23)。这主要是由于，在 pH 为 6～7 时，这两种微藻的生物溶蚀作用很小，这一点能够得到图 5-4 很好印证。此外，我们还发现，在 pH 为 7～8 时 $CaCO_3$ 的沉淀绝对量最大。这可能由于 pH 为 7～8 时方解石的总溶蚀量最大，同时又是微藻生长最适宜的 pH，DIC 利用量最大，而微藻在利用 HCO_3^- 的过程中将导致 $CaCO_3$ 的产生。

方解石溶蚀过程($MCO_3+CO_2+H_2O \leftrightarrow M+2HCO_3^-$)将吸收大气的 CO_2，形成碳汇效应。但是，其又可能会随着 $CaCO_3$ 的沉淀($Ca^{2+}+2HCO_3^- \leftrightarrow CaCO_3+CO_2+H_2O$)，又将 CO_2 释放出去，没有形成有效的碳汇效应。只有方解石溶蚀过程所吸收的大气 CO_2 被微藻利用形成有机碳(生物泵)或者残留在水体形成 DIC(碳酸盐泵)，才能形成有效的碳汇，则为岩溶碳汇。如图 5-23 所示，可以看出，随着 pH 降低，方解石溶蚀所释放的无机碳更多地被藻体所利用或形成 DIC。这主要是由于，pH 越低，方解石的化学溶蚀越强，微藻利用方解石碳源比例越大(图 5-18)，碳酸钙的饱和度越大。因此，在喀斯特湖泊中，pH 越小，其净的生物岩溶碳汇效应越强。

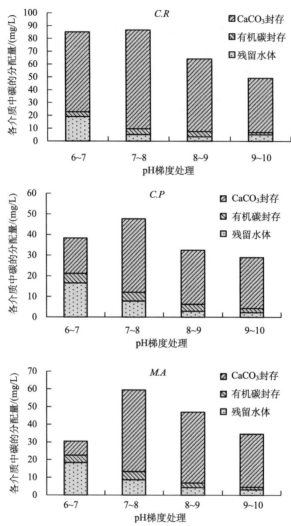

图 5-23　不同 pH 处理下方解石溶蚀碳在各介质中的分配

C.R：莱茵衣藻，C.P：蛋白核小球藻，M.A：铜绿微囊藻；6～7、7～8、8～9、9～10：分别为培养液 pH 为 6～7、7～8、8～9、9～10 的处理

Fig. 5-23　Distribution of carbon from calcite dissolution in each media under different pH treatments

C.R：*Chlamydomonas reinhardtii*，C.P：*Chlorella pyrenoedosa*，M.A：*Microcystis aeruginosa*；6～7：the pH of medium is between 6 and 7，7～8：the pH of medium is between 7 and 8，8～9：the pH of medium is between 8 and 9，9～10：the pH of medium is between 9 and 10

而对于岩溶体系的微藻来说不仅有生物岩溶碳汇效应，而且还具有自身的光合碳汇效应，其中生物岩溶碳汇效应与光合碳汇效应之和，则为微藻参与下的岩溶体系所产生的总碳汇。在此，我们选取具有较强胞外碳酸酐酶活力的 $C.R$，仅有较弱胞外碳酸酐酶活力的 $C.P$ 和没有胞外碳酸酐酶活力的 $M.A$ 来说明 pH 对微藻参与下的岩溶体系碳汇效应的影响。从表 5-10 可以看出，随着 pH 的增加，微藻参与下的岩溶体系中的岩溶碳汇强度不断减少，而此体系中微藻的光合碳汇不断增加。这主要是由于 pH 越低，更利于碳酸盐岩的溶蚀；而 pH 越高，微藻对大气碳源的利用份额越大。同时，像铜绿微囊藻此类的没有胞外碳酸酐酶活力的微藻，随着 pH 的增加将不断提高光合放氧速率和 DIC 亲和度（徐涛和宋立荣，2007），而对于有较强胞外碳酸酐酶活力的莱茵衣藻来说则可能随着 pH 的增加不断提高活细胞与死细胞的更新速度。此外，我们还发现，在 pH 为 7~8 的时候，此体系的总碳汇强度最小，而在 pH 为 6~7 和 9~10 时总碳汇相对较大，在酸性条件下岩溶碳汇能力大于光合碳汇能力，碱性条件下反之。这主要是由于，pH 为 6~7 时，岩溶碳汇最强，而在 9~10 时光合碳汇能力最强，pH 为 7~8 时，无论是岩溶碳汇能力还是光合碳汇能力都处于较低水平。

表 5-10 不同 pH 条件下生物岩溶体系的各碳汇能力

Tab. 5-10 The types of carbon sink in bio-karst system under different pH treatments

微藻	pH	岩溶碳汇/(mg/L)			光合碳汇/(mg/L)	总碳汇/(mg/L)
		DIC	NPOC	总和		
C.R	6~7	19.53	3.21	22.75	13.10	35.85
	7~8	5.57	4.10	9.67	17.18	26.85
	8~9	3.78	3.78	7.56	30.81	38.37
	9~10	5.19	1.67	6.85	33.30	40.15
M.A	6~7	18.91	3.55	22.46	13.00	35.46
	7~8	9.22	3.98	13.21	15.54	28.75
	8~9	4.51	2.34	6.86	25.90	32.75
	9~10	3.19	1.32	4.51	32.37	36.88
C.P	6~7	17.05	4.13	21.18	24.39	45.57
	7~8	6.57	5.51	12.07	21.41	33.48
	8~9	1.45	4.76	6.21	29.65	35.86
	9~10	1.48	2.77	4.25	39.86	44.11

注：$C.R$：莱茵衣藻，$C.P$：蛋白核小球藻，$M.A$：铜绿微囊藻；6~7、7~8、8~9、9~10：分别为培养液 pH 为 6~7、7~8、8~9、9~10 的处理。

Note：$C.R$：*Chlamydomonas reinhardtii*，$C.P$：*Chlorella pyrenoedosa*，$M.A$：*Microcystis aeruginosa*；6~7：the pH of medium is between 6 and 7，7~8：the pH of medium is between 7 and 8，8~9：the pH of medium is between 8 and 9，9~10：the pH of medium is between 9 and 10.

二、胞外碳酸酐酶对岩溶碳汇的作用

从总体来看，随着 AZ 浓度的增加，含有莱茵衣藻的培养液中方解石无机碳的释放量不断减少，而含有铜绿微囊藻的培养液反之（图 5-24）。说明胞外碳酸酐酶能够促进方解石中无机碳的释放。同时，隔离环境下方解石的无机碳释放量小于开放环境。

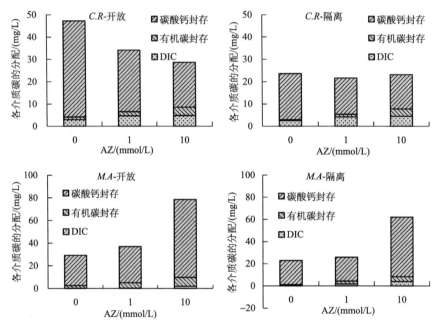

图 5-24 不同 AZ 处理下方解石溶蚀的碳在各介质中的分配
C.R：莱茵衣藻，M.A：铜绿微囊藻
Fig. 5-24 Distribution of carbon from calcite dissolution in each media under different AZ treatments
C.R：*Chlamydomonas reinhardtii*，M.A：*Microcystis aeruginosa*

不同 AZ 和 CO_2 分压下，微藻对方解石碳源的利用情况不一样，方解石的溶解和再次沉淀量也不一样。随着 AZ 浓度的增加，含有莱茵衣藻的培养液中 $CaCO_3$ 的沉淀量不断减少，而铜绿微囊藻反之（图 5-24）。这主要是由于莱茵衣藻通过胞外碳酸酐酶催化利用 HCO_3^- 的过程中将导致 $CaCO_3$ 沉淀的产生（Ca^{2+} + $2HCO_3^- \leftrightarrow CaCO_3 + CH_2O + O_2$）(Sharma and Bhattacharya, 2010)，随着 AZ 浓度增加微藻胞外碳酸酐酶活力降低，从而使得 $CaCO_3$ 的沉淀量减少。铜绿微囊藻不具有胞外碳酸酐酶，随着 AZ 浓度增加，方解石溶蚀量逐渐增加，Ca^{2+} 释放量变大，从而使得 $CaCO_3$ 再次沉淀的绝对值变大。

由于方解石溶解吸收的一份 CO_2，而后随之又形成 $CaCO_3$ 沉淀，再次释放出一份 CO_2，因此这一过程不具有碳汇效应。只有方解石溶蚀所吸收的 CO_2 残留在水体或者被微藻利用形成有机质，才具有碳汇效应。因此，可以发现随着 AZ 浓度的增加，残留在水体和被微藻吸收利用的无机碳不断增加（图 5-24）。特别是具有较强胞外碳酸酐酶活力的莱茵衣藻变化最为强烈。因此，由于胞外碳酸酐酶在催化过程中对 $CaCO_3$ 的沉淀作用大于溶蚀作用，其在岩溶碳汇效应过程中起到消极的影响。

微藻参与下的岩溶体系总碳汇不仅受到岩溶碳汇的影响，还受到微藻光合碳汇的影响。从表 5-11 可以看出，在开放环境下，对于没有胞外碳酸酐酶活力的铜绿微囊藻体系，随着 AZ 浓度的增加其岩溶碳汇和光合碳汇都不断增加，从而导致体系总碳汇不断增加。这说明，AZ 并没有对铜绿微囊藻参与下的岩溶体系碳汇效应起到抑制作用。然而，对于有较强胞外碳酸酐酶活力的莱茵衣藻来说，随着 AZ 浓度的增加岩溶碳汇不断增加，而光合碳汇不断减少。但是，由 AZ 作用引起的光合碳汇的减少量大于岩溶碳汇的增加量，从而使得整个体系的总碳汇不断减少。这一结果与铜绿微囊藻参与下的岩溶体系碳汇效应形成鲜明的对比，充分说明微藻胞外碳酸酐酶能够增强微藻参与下岩溶体系的碳汇效应。然而，在隔离 CO_2 环境下，其结果有所不同。如表 5-11 所示，随着 AZ 浓度的增加，无论是铜绿微囊藻还是莱茵衣藻，其岩溶碳汇和光合碳汇都不断地增加，从而使得整个岩溶体系总的碳汇不断增加。其中，莱茵衣藻主要是由于在隔离环境中可利用的碳源较少。因此，在 AZ 的抑制作用下，无机碳利用途径减少，从而导致低碳源胁迫作用降低，光合固碳量增加。

表 5-11 不同 AZ 处理下生物岩溶体系的各碳汇能力

Tab. 5-11 The types of carbon sink in bio-karst system under different AZ treatments

环境	微藻	处理 AZ/(mmol/L)	岩溶碳汇/(mg/L)			光合碳汇 /(mg/L)	总碳汇 /(mg/L)
			DIC	微藻	总和		
开放 CO_2	C.R	0	3.05	1.09	4.14	42.28	46.42
		1	4.94	1.54	6.49	35.32	41.81
		10	5.26	3.29	8.55	24.01	32.56
	M.A	0	0.41	2.10	2.51	50.76	53.27
		1	0.03	4.98	5.01	52.62	57.63
		10	3.55	6.16	6.66	57.63	64.29

续表

环境	微藻	处理 AZ/(mmol/L)	岩溶碳汇/(mg/L)			光合碳汇 /(mg/L)	总碳汇 /(mg/L)
			DIC	微藻	总和		
隔离 CO_2	C.R	0	2.70	0.22	2.92	5.73	8.65
		1	4.42	0.93	5.34	14.56	19.90
		10	5.05	2.67	7.73	15.23	22.96
	M.A	0	0.21	0.97	1.18	23.30	24.48
		1	2.23	2.12	4.35	20.08	24.43
		10	4.50	3.82	8.31	33.20	41.51

注：C.R：莱茵衣藻，M.A：铜绿微囊藻。
Note：C.R：*Chlamydomonas reinhardtii*，M.A：*Microcystis aeruginosa*.

三、阴离子通道对岩溶碳汇的作用

不同处理下，方解石生物溶蚀所释放的 DIC 在各介质中的分配比例不一样，而在不同介质中以不同形式存在的无机碳对碳汇的效应不同。方解石溶蚀所释放的 DIC 只有以有机碳或者 DIC 的形式存在才具有有效碳汇作用，而以 $CaCO_3$ 形式存在的无机碳不具有碳汇效应。随着 DIDS 浓度的增加方解石溶蚀的无机碳以有机质和 DIC 存在的形式不断增加（图 5-25）。因此，阴离子通道在生物岩溶碳汇过程中主要起到消极的作用。这主要是由于，微藻阴离子通道在对水体 HCO_3^- 的利用过程中对水体 $CaCO_3$ 的沉淀（$Ca^{2+} + 2HCO_3^- \leftrightarrow CaCO_3 + CH_2O + O_2$）作用大于对方解石的溶蚀（$CaMg(CO_3)_2 + 2CO_2 + 2H_2O \leftrightarrow Ca^{2+} + Mg^{2+} + 4HCO_3^-$）作用（Yates and Robbins，1999）。但是，随着 DIDS 浓度的增加，含有莱茵衣藻培养液中 $CaCO_3$ 的沉淀量不断增加。这主要是由于在 DIDS 作用过程中，莱茵衣藻的胞外碳酸酐酶活力将被诱导，而胞外碳酸酐酶在催化利用 DIC 过程中将导致 $CaCO_3$ 沉淀的产生。此外，隔离环境下方解石 DIC 释放量明显减少，但是以有机碳和 DIC 形式存在的比例却在增加。

从微藻参与下的岩溶体系总碳汇来看，在不同 DIDS 处理下，不同类型的微藻参与下的岩溶体系碳汇效应有所不同。在开放环境下，对于既有较强胞外碳酸酐酶活力又能够利用阴离子通道来利用水体中 HCO_3^- 的莱茵衣藻来说，随着 DIDS 浓度的增加，无论是岩溶碳汇还是光合碳汇都不断增加，从而导致整个体系的总碳汇不断增加（表 5-12）。这主要是由于在 DIDS 的作用下微藻的胞外碳酸酐酶表达受到诱导（图 5-11），从而增强微藻对方解石的溶蚀和大气无机碳源的利用。而对于没有胞外碳酸酐酶活力而主要利用阴离子通道来利用水体中 HCO_3^- 的铜绿微囊藻来说，随着 DIDS 浓度的增加，岩溶碳汇不断增加而光合碳

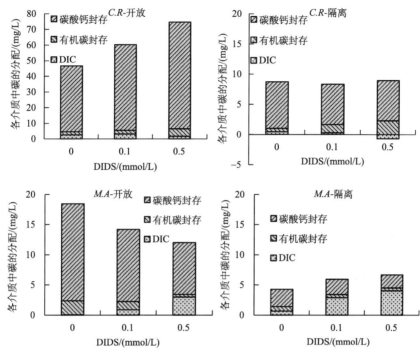

图 5-25 不同处理下方解石释放的碳在各介质中的分配

C.R：莱茵衣藻，M.A：铜绿微囊藻

Fig. 5-25 Distribution of carbon from calcite dissolution in each media under different DIDS treatments

C.R：*Chlamydomonas reinhardtii*，M.A：*Microcystis aeruginosa*

汇不断降低。但是，由 DIDS 引起的光合碳汇降低速率大于岩溶碳汇的增加速率，从而使得铜绿微囊藻参与下的岩溶体系碳汇效应不断降低。这一结果与莱茵衣藻参与下岩溶体系碳汇效应形成鲜明的对比，因此，阴离子通道在微藻参与下的岩溶体系总碳汇效应中起到明显的促进作用。然而，隔离 CO_2 环境下，此岩溶体系碳汇效应有所不同。无论是莱茵衣藻还是铜绿微囊藻，随着 DIDS 浓度的增加，其岩溶碳汇和光合碳汇都不断增加，从而导致整个体系的总碳汇不断增强（表 5-12）。这主要是由于，莱茵衣藻在 DIDS 的作用下能诱导胞外碳酸酐酶的表达，而铜绿微囊藻在 DIDS 的作用下能降低低碳源胁迫的影响，从而导致整个体系的碳汇效应不断增强。

表 5-12 不同 DIDS 处理下生物岩溶体系的各碳汇能力
Tab. 5-12 The types of carbon sink in bio-karst system under different DIDS treatments

环境	微藻	处理 DIDS /(mmol/L)	岩溶碳汇/(mg/L)			光合碳汇 /(mg/L)	总碳汇 /(mg/L)
			DIC	微藻	总和		
开放 CO_2	C.R	0	2.72	1.87	4.59	50.07	54.66
		0.1	3.25	2.34	5.60	64.07	69.67
		0.5	1.86	4.70	6.56	102.72	109.27
	M.A	0	0.21	2.15	2.37	67.32	69.68
		0.1	0.98	1.24	2.23	64.95	67.18
		0.5	3.06	0.38	3.44	56.90	60.34
隔离 CO_2	C.R	0	0.66	0.38	1.04	9.60	10.64
		0.1	0.54	1.13	1.67	21.15	22.82
		0.5	0	2.01	2.01	31.40	33.41
	M.A	0	0.84	0.57	1.42	17.40	18.82
		0.1	3.00	0.42	3.42	19.91	23.32
		0.5	4.07	0.43	4.50	42.48	46.98

注：C.R：莱茵衣藻，M.A：铜绿微囊藻。

Note：C.R：*Chlamydomonas reinhardtii*，M.A：*Microcystis aeruginosa*.

参 考 文 献

陈廷伟，陈华癸. 1960. 钾细菌的形态生理及其对磷钾矿物的分解能力. 微生物，2：104-112.
郭敏亮，高煜珠. 1989. 植物的碳酸酐酶. 植物生理学通讯，3：8.
黄黎英，曹建华，何寻阳，杨慧，李小方，申宏岗. 2006. 几种低分子量有机酸对石灰岩溶蚀作用的室内模拟试验. 地球与环境，34：44-50.
李冬玉，黄建华. 2009. 微生物成矿的研究现状. 科技信息，15-16.
李强，何媛媛，曹建华，梁建宏，朱敏洁. 2012. 植物碳酸酐酶对岩溶作用的影响及其生态效应. 生态环境学报，20：1867-1871.
李强，靳振江，孙海龙. 2005. 现代藻类碳酸钙沉积试验及其同位素不平衡现象. 中国岩溶，24(4)：261-264.
李莎，李福春，程良娟. 2007. 生物风化作用研究进展. 矿产与地质，20：577-582.
李为，刘丽萍，曹龙，余龙江. 2009. 碳酸盐生物沉积作用的研究现状与展望. 地球科学进展，24：597-605.
李为，曾宪东，栗茂腾，周蓬蓬，余龙江. 2011. 微藻及其碳酸酐酶对石灰岩土壤系统中钙元素迁移的驱动作用实验研究. 矿物岩石地球化学通报，30：261-264.
刘再华. 2000. 碳酸盐岩岩溶作用对大气 CO_2 沉降的贡献 α. 中国岩溶，19(4)：293-300.
刘再华. 2001. 碳酸酐酶对碳酸盐岩溶解的催化作用及其在大气 CO_2 沉降中的意义. 地球学报，

22: 477-480.

刘志礼, 刘雪娴. 1999. 藻类及其有机质的成矿作用试验. 沉积学报, 17: 9-18.

盛下放, 黄为一. 2002. 硅酸盐细菌 NB 丁菌株释钾条件的研究. 中国农业科学, 35: 673-677.

徐涛, 宋立荣. 2007. 三株铜绿微囊藻对外源无机碳利用的研究. 水生生物学报, 31: 245-250.

余龙江, 吴云, 李为, 曾宪东, 付春华. 2004. 微生物碳酸酐酶对石灰岩的溶蚀驱动作用研究. 中国岩溶, 23: 225-228.

袁道先. 1993. 碳循环与全球岩溶. 第四纪研究, 1: 6.

袁道先. 1995. 岩溶与全球变化研究. 地球科学进展, 10: 471-474.

张怀强, 公维丽, 赵越, 高培基. 2014. 构建限制性培养条件下细菌群体生长模型的探讨. 中国科学: 生命科学, 44: 185-196.

Aizawa K, Miyachi S. 1986. Carbonic anhydrase and CO_2 concentrating mechanisms in microalgae and cyanobacteria. FEMS Microbiology Letters, 39: 215-233.

Aung K M M, Ting Y P. 2005. Bioleaching of spent fluid catalytic cracking catalyst using *Aspergillus niger*. Journal of Biotechnology, 116: 159-170.

Bozzo G, Colman B. 2000. The induction of inorganic carbon transport and external carbonic anhydrase in *Chlamydomonas reinhardtii* is regulated by external CO_2 concentration. Plant, Cell and Environment, 23: 1137-1144.

Bricker O P. 1988. The global water cycle geochemistry and environment. Eos, Transactions American Geophysical Union, 69: 51.

Cameselle C, Núñez M J, Lema J M. 1997. Leaching of kaolin iron-oxides with organic acids. Journal of Chemical Technology and Biotechnology, 70: 349-354.

Cameselle C, Ricart M, Nunez M, Lema J. 2003. Iron removal from kaolin. Comparison between "in situ" and "two-stage" bioleaching processes. Hydrometallurgy, 68: 97-105.

Chen C, Mei B, Mao Z. 1993. The initial experimental study for dissolving silicate mineral by dicarboxylic acid in aqueous systems. Journal of Mineralogy and Petrology, 13: 103-107.

Chen T. 2011. Microbial carbon pump: additional considerations. Nature Reviews Microbiology, 9(7): 555.

Chen Z, Cheng H, Chen X. 2009. Effect of Cl^- on photosynthetic bicarbonate uptake in two cyanobacteria *Microcystis aeruginosa* and *Synechocystis* PCC6803. Chinese Science Bulletin, 54: 1197-1203.

Dreybrodt W, Lauckner J, Zaihua L, Svensson U, Buhmann D. 1996. The kinetics of the reaction $CO_2 + H_2O \leftrightarrow H^+ + HCO_3^-$ as one of the rate limiting steps for the dissolution of calcite in the system H_2O-CO_2-$CaCO_3$. Geochimica et Cosmochimica Acta, 60: 3375-3381.

Hammes F, Verstraete W. 2002. Key roles of pH and calcium metabolism in microbial carbonate precipitation. Reviews in Environmental Science and Biotechnology, 1: 3-7.

Hewett-Emmett D, Tashian R E. 1996. Functional diversity, conservation, and convergence in the evolution of the α-, β-, and γ-carbonic anhydrase gene families. Molecular Phylogenetics And Evolution, 5: 50-77.

Houghton R A, Woodwell G M. 1989. Global climate change. Scientific American, 260: 36-40.

Huang C Y L, Schulte E. 1985. Digestion of plant tissue for analysis by ICP emission spectroscopy. Communications in Soil Science & Plant Analysis, 16: 943-958.

Jongmans A, Van Breemen N, Lundström U, et al. 1997. Rock-eating fungi. Nature, 389: 682-683.

Kaufmann G, Dreybrodt W. 2007. Calcite dissolution kinetics in the system $CaCO_3 — H_2O — CO_2$ at high undersaturation. Geochimica et Cosmochimica Acta, 71: 1398-1410.

Krumbein W E. 1983. Microbial geochemistry. Boston: Blackwell Scientific Publications.

Larsson C, Axelsson L. 1999. Bicarbonate uptake and utilization in marine macroalgae. European Journal of Phycology, 34: 79-86.

Li F, Li S, Yang Y, Cheng L. 2006. Advances in the study of weathering products of primary silicate minerals, exemplified by mica and feldspar. Acta Petrol Mineral, 25: 440-448.

Li L, Fu M L, Zhao Y H, Zhu Y T. 2012. Characterization of carbonic anhydrase II from *Chlorella vulgaris* in bio-CO_2 capture. Environmental Science and Pollution Research, 19: 4227-4232.

Liu Y, Liu Z, Zhang J, He Y, Sun H. 2010a. Experimental study on the utilization of DIC by *Oocystis solitaria* Wittr and its influence on the precipitation of calcium carbonate in karst and non-karst waters. Carbonates and Evaporites, 25: 21-26.

Liu Z, Dreybrodt W, Wang H. 2010b. A new direction in effective accounting for the atmospheric CO_2 budget: Considering the combined action of carbonate dissolution, the global water cycle and photosynthetic uptake of DIC by aquatic organisms. Earth-Science Reviews, 99: 162-172.

Liu Z, Dreybrod W. 1997. Dissolution kinetics of calcium carbonate minerals in H_2O-CO_2 solutions in turbulent flow: The role of the diffusion boundary layer and the slow reaction $H_2O + CO_2 \rightarrow H^+ + HCO_3^-$. Geochimica et Cosmochimica Acta, 61: 2879-2889.

Liu Z, Yuan D, Dreybrodt W. 2005. Comparative study of dissolution rate-determining mechanisms of limestone and dolomite. Environmental Geology, 49: 274-279.

Li W, Yu L, He Q, Wu Y, Yuan D, Cao J. 2005. Effects of microbes and their carbonic anhydrase on Ca^{2+} and Mg^{2+} migration in column-built leached soil-limestone karst systems. Applied Soil Ecology, 29: 274-281.

Marlier J F, O'Leary M H. 1984. Carbon kinetic isotope effects on the hydration of carbon dioxide and the dehydration of bicarbonate ion. Journal of the American Chemical Society, 106: 5054-5057.

Menéndez M, Martínez M, Comín F A. 2001. A comparative study of the effect of pH and inorganic carbon resources on the photosynthesis of three floating macroalgae species of a Mediterranean coastal lagoon. Journal of Experimental Marine Biology and Ecology, 256: 123-136.

Miyamoto H, Miyashita T, Okushima M, Nakano S, Morita T, Matsushiro A. 1996. A carbonic anhydrase from the nacreous layer in oyster pearls. Proceedings of the National Academy of Sciences, 93: 9657-9660.

Mook W, Bommerson J, Staverman W. 1974. Carbon isotope fractionation between dissolved bicarbonate and gaseous carbon dioxide. Earth and Planetary Science Letters, 22: 169-176.

Puente M E, Rodriguez-Jaramillo M C, Li C Y, Bashan Y. 2006. Image analysis for quantification of bacterial rock weathering. Journal of Microbiological Methods, 64: 275-286.

Roughton F, Booth V. 1946. The effect of substrate concentration, pH and other factors upon the activity of carbonic anhydrase. Biochemical Journal, 40: 319.

Saarnio J, Parkkila S, Parkkila A K, et al. 1998. Immunohistochemical study of colorectal tumors for expression of a novel transmembrane carbonic anhydrase, MN/CA IX, with potential value as a marker of cell proliferation. The American journal of pathology, 153: 279-285.

Sharma A, Bhattacharya A. 2010. Enhanced biomimetic sequestration of CO_2 into $CaCO_3$ using purified carbonic anhydrase from indigenous bacterial strains. Journal of Molecular Catalysis B-Enzymatic, 67: 122-128.

Sterling D, Reithmeier R A, Casey J R. 2001. A transport metabolon functional interaction of carbonic anhydrase II and chloride/bicarbonate exchangers. Journal of Biological Chemistry, 276: 47886-47894.

Teng H H, Fenter P, Cheng L, Sturchio N C. 2001. Resolving orthoclase dissolution processes with atomic force microscopy and X-ray reflectivity. Geochimica et Cosmochimica Acta, 65: 3459-3474.

Topper K, Kotuby-Amacher J. 1990. Evaluation of a closed vessel acid digestion method for plant analyses using inductively coupled plasma spectrometry. Communications in Soil Science & Plant Analysis, 21: 1437-1455.

Vörös L, Padisák J. 1991. Phytoplankton biomass and chlorophyll-a in some shallow lakes in central Europe. Hydrobiologia, 215: 111-119.

Wallin M, Bjerle I. 1989. A mass transfer model for limestone dissolution from a rotating cylinder. Chemical Engineering Science, 44: 61-67.

Welch S, Ullman W. 1996. Feldspar dissolution in acidic and organic solutions: Compositional and pH dependence of dissolution rate. Geochimica et Cosmochimica Acta, 60: 2939-2948.

Williams T, Colman B. 1996. The effects of pH and dissolved inorganic carbon on external carbonic anhydrase activity in *Chlorella saccharophila*. Plant, Cell and Environment, 19: 485-489.

Wu Y, Xu Y, Li H, Xing D. 2011. Effect of acetazolamide on stable carbon isotope fractionation in *Chlamydomonas reinhardtii* and *Chlorella vulgaris*. Chinese Science Bulletin, 57: 786-789.

Xie T, Wu Y. 2014. The role of microalgae and their carbonic anhydrase on the biological dissolution of limestone. Environmental Earth Sciences, 71: 5231-5239.

Yadav S, Chakrapani G, Gupta M. 2008. An experimental study of dissolution kinetics of calcite, dolomite, leucogranite and gneiss in buffered solutions at temperature 25℃ and 5℃. Environmental Geology, 53: 1683-1694.

Yates K K, Robbins L L. 1999. Radioisotope tracer studies of inorganic carbon and Ca in microbially derived $CaCO_3$. Geochimica et Cosmochimica Acta, 63: 129-136.

Young E, Beardall J, Giordano M. 2001. Inorganic carbon acquisition by *Dunaliella tertiolecta* (Chlorophyta) involves external carbonic anhydrase and direct HCO_3^- utilization insensitive to the anion exchange inhibitor DIDS. European Journal of Phycology, 36: 81-88.

第六章 无机元素与微藻碳酸酐酶的相互作用

摘　要

无机元素既可以促进碳酸酐酶活性，也可以抑制碳酸酐酶活性。大多数阴离子，由于降低酶与底物的亲和力或与酶的底物发生竞争，降低了酶的活性。有些种类的重金属在低浓度下对碳酸酐酶活力起促进作用，在高浓度下对碳酸酐酶活力起抑制作用，表现出明显的"激素效应"。研究无机元素与微藻碳酸酐酶的相互作用，有助于阐明微藻碳酸酐酶生物地球化学作用机制。

乙酰唑胺（acetazolamide，AZ）是胞外碳酸酐酶的专一抑制剂。AZ 对莱茵衣藻（*Chlamydomonas reinhardtii*）氮、磷、钾的吸收具有抑制作用；对蛋白核小球藻（*Chlorella pyrenoidosa*）氮、钾吸收也表现出抑制作用，而对磷的吸收影响不大。碳酸酐酶对微藻无机元素的吸收起着重要的作用。

无机元素氟对莱茵衣藻的碳酸酐酶和光合净氧的释放有显著的影响。3 天的短期处理，$0\sim 2$mmol/L 的氟对碳酸酐酶活力和净光合氧的释放具有促进作用，而高浓度对这些指标具有抑制作用。碳酸酐酶活力最大和净光合氧的释放最多的处理是 2mmol/L。莱茵衣藻碳酸酐酶活力与净光合氧的释放有显著的正相关。

10 天的中长期处理，在低浓度 F^-（$0\sim 10$mmol/L）下，莱茵衣藻的叶绿素 a 和叶绿素 b 含量随着 F^- 浓度升高呈现出下降趋势，而蛋白核小球藻的叶绿素 a 和叶绿素 b 含量在此浓度范围内则没有变化，说明莱茵衣藻的对环境变化的响应更灵敏。

在相同 pH 条件和相同生长阶段，莱茵衣藻叶绿素 a 合成量明显高于蛋白核小球藻，表明莱茵衣藻具有比蛋白核小球藻更强的生长繁殖能力。当 $F^- < 1$mmol/L 时，两种微藻叶绿素 a 的变化量并不明显，微藻胞外碳酸酐酶活性主要受 pH 变化的影响；当 $F^- > 1$mmol/L 时，随着 F^- 浓度的升高，叶绿素 a 含量迅速降低，说明高浓度的 F^- 抑制莱茵衣藻和蛋白核小球藻的生长。两种微藻的胞外碳酸酐酶活性受 F^- 浓度和 pH 的共同影响，但 F^- 的影响更大。在不同 pH 条件下，莱茵衣藻和蛋白核小球藻表达出最高碳酸酐酶活性时的 F^- 浓度不同：pH 分别为 5、7、9 时，莱茵衣藻分别在 F^- 浓度为 10mmol/L、50mmol/L、100mmol/L，蛋白核小球藻分别为 50mmol/L、100mmol/L、100mmol/L 时有最高的胞外碳酸酐酶活性。此外，氟对微藻的生长和碳酸酐酶的影响所呈现的剂量效应存在着对处理时间的依存性。

低浓度的 Fe、Mn、Co、Ni 和 Cd 促进莱茵衣藻碳酸酐酶胞外酶的活性，高浓度的各种金属抑制它的活性。100μmol/L 的 Cu、Zn、Cd、Mn、Fe、Co 和 Ni 对莱茵衣藻碳酸酐酶胞外酶的活性的抑制作用随这些微量金属相对原子质量的增加而增大。Cu 对莱茵衣藻和红枫湖中的微藻的碳酸酐酶胞外酶活性的抑制相似，但对阿哈湖中的微藻的碳酸酐酶胞外酶的抑制极为强烈。Pb 对莱茵衣藻和阿哈湖中的微藻的碳酸酐酶胞外酶活性的抑制相似，但 Pb 对红枫湖中的微藻的碳酸酐酶胞外酶的活性在低浓度下还具有促进作用。金属元素对碳酸酐酶胞外酶活力的影响在不同的湖泊中表现的差异，与不同湖泊的微藻种类差异有关。

Chapter 6 The interaction between inorganic elements and carbonic anhydrase in microalgae

Abstract

Inorganic elements can not only promote carbonic anhydrase (CA) activity, but also inhibit its activity. For most anions, they reduce the affinity between enzyme and substrates or compete with substrates, which leads to the decrease enzyme activity. A few species of heavy metal elements promote CA activity at low concentrations, while inhibit CA activity in high concentration. It shows a significant "hormone effect". The interaction between inorganic elements and microalgae will helps to clarify the mechanism of biogeochemistry action with microalgae.

Acetazolamide(AZ) is a specific inhibiter on extracellular carbonic anhydrase. AZ inhibitsthe uptake on nitrogen, phosphorus, and potassium by *Chlamydomonas reinhardtii* and the uptake on nitrogen, and potassium by *Chlorella pyrenoidosa*, but AZ has no effect on the uptake on phosphorus by *Chlorella pyrenoidosa*. Extracellular carbonic anhydrase plays a critical role in inorganic elements uptake in microalgae.

Inorganic element, fluorine (F), has a significant effect on extracellular carbonic anhydrase and net photosynthetic O_2 evolution in *Chlamydomonas reinhardtii*. When the treatment time in the short term (3 days), 0～2mmol/L F promotes the activity extracellular CA as well as net photosynthetic O_2 evolution, while high concentration of F inhibits these parameters. The treatment resulted in the greatest activity of CA and net photosynthetic O_2 evolution is that of 2mmol/L. The activity of CA has a significant positive relationship with net photosynthetic O_2 evolution.

When the treatment time in the mid-long term (10 days), the content of chlorophyll *a* and *b* in *Chlamydomonas reinhardtii* decreases with increasing of F^- under low concentration of F^- (0~10mmol/L), while no change in *Chlorella pyrenoidosa* at the range of F^-. It suggests that *Chlamydomonas reinhardtii* is more sensitive to response to environment change.

In the same condition of pH and growth phases, the synthetic amount of chlorophyll *a* in *Chlamydomonas reinhardtii* is obvious more than that in *Chlorella pyrenoidosa*, which reveals that *Chlamydomonas reinhardtii* has greater propagation abilities than *Chlorella pyrenoidosa*. The content of chlorophyll *a* in these two species of algae changes a little and the CA activity is mainly influenced by pH change when F^- less than 1mmol/L. The content of chlorophyll *a* decreases with the concentration of F^- increasing, quickly, when F^- greater than 1mmol/L. It illustrates that high concentration of F^- inhibits the growth of *Chlamydomonas reinhardtii* and *Chlorella pyrenoidosa*. The activity of CA in two species of algae was influenced by both concentration of F^- and pH, however, the F^- has a larger effect than pH. The concentration of F^-, which resulted in the highest activity of CA in *Chlamydomonas reinhardtii* and *Chlorella pyrenoidosa*, was different with different pH. Under pH at 5, 7 and 9, the concentration of F^-, which demonstrated the highest CA activity in *Chlamydomonas reinhardtii*, was 10mmol/L, 50mmol/L and 100mmol/L, respectively, that in *Chlorella pyrenoidosa* 50mmol/L, 100mmol/L and 100mmol/L, respectively. In addition, the dosage effect, which resulted from the influence of fluorine on microalgal growth and carbonic anhydrase, shows the treatment time-dependence.

Low concentration of Fe, Mn, Co, Ni and Cd promote extracellular CA activity, while inhibit its activity in *Chlamydomonas reinhardtii* at high concentration. Inhibition effect on the activity of extracellular CA in *Chlamydomonas reinhardtii*, which was induced from 100μmol/L of Cu, Zn, Cd, Mn, Fe, Co and Ni, increases with increasing of the atoms in these trace metal species. Inhibition of Cu on the activity of extracellular CA in *Chlamydomonas reinhardtii* and microalgae in Hongfeng Lake are similar, however, it strongly inhibits the activity of extracellular CA in microalgae in Aha Lake. Inhibition of Pb on the activity of extracellular CA in *Chlamydomonas reinhardtii* is similar to microalgae in Aha Lake, but it promotes extracellular CA activity of microalgae in Hongfeng Lake under low concentration. The effect of metallic elements on the activity in extracellular CA differs in the two lakes, which is due to the differences of algal species in the two lakes.

第一节 无机元素对碳酸酐酶影响的"激素"效应

一、无机元素对碳酸酐酶影响的剂量效应

所谓的激素效应是指激素对效应物既可以起促进作用，也可以起抑制作用。对于无机元素对碳酸酐酶作用的"激素"效应是指无机元素既可以促进碳酸酐酶活性，也可以抑制碳酸酐酶活性。

大多数阴离子，如 F^-、Cl^-、HCO_3^-、Br^-、I^-、SO_4^{2-}、NO_3^- 等，由于降低酶与底物的亲和力或与酶的底物发生竞争，降低了酶的活性(Innocenti et al., 2004；Nishimori et al., 2007；Nishimori et al., 2009；Lopez Mananes et al., 1993a, b；Wieth, 1979)。但有些种类的阴离子在低浓度下对碳酸酐酶活力起促进作用。在 2.0 mmol/L 以下，F^- 对莱因衣藻的胞外碳酸酐酶有明显的促进作用(Wu et al., 2007)；小于 20 mmol/L 的低浓度 Cl^-、Br^- 以及磷酸盐对人体碳酸酐酶Ⅳ具有明显的活化作用(Baird et al., 1997)。NaCl 对两种岩生苔藓植物的碳酸酐酶活力的影响也表现为"激素"效应，当浓度分别小于 2%(w/v) 和 1%(w/v) 时，NaCl 对中华墙藓(*Tortula sinensis* (Mull. Hal.) Broth)和尖叶扭口藓(*Barbula convoluta* Hedw.)的碳酸酐酶具有促进作用，大于此浓度则呈现出抑制作用(Wu et al., 2006)。小于 0.5 mmol/L 的低浓度碳酸氢根离子对莱因衣藻(*Chlamydomonas reinhardtii*)的胞外碳酸酐酶有促进作用，超过这个浓度，随着浓度的增加，抑制作用增加。Co(Ⅱ)替代的流感嗜血杆菌(*Haemophilus influenzae*)β-碳酸酐酶也受 pH 和重碳酸盐的变构调节(Hoffmann et al., 2011)。因此，我们可以看出，F^-、重碳酸盐、Cl^- 以及 Br^- 等对碳酸酐酶的作用都具有明显的激素效应。

很多重金属元素(Co(Ⅱ)、Cu(Ⅱ)、Zn(Ⅱ)、Ag(Ⅰ)、Cd(Ⅱ)、Pb、Hg)，由于它们或者能够替换碳酸酐酶作用中心的锌，或者能直接改变生物活性，或者间接地改变碳酸酐酶的代谢，从而降低了酶的活性(Sas et al., 2006；Lionetto et al., 1998；Soyut et al., 2008；Gilbert et al., 2001)。二价金属离子 Co、Mn、Ni、Cu、Fe、Cd 可以代替碳酸酐酶锌表现出不同的活性(Lindskog, 1963)。但有些种类的重金属在低浓度下对碳酸酐酶活力起促进作用，在高浓度下对碳酸酐酶活力起抑制作用，表现出明显的激素效应。添加低浓度的锌、铁培养鸭红细胞可以提高碳酸酐酶的活力，而高浓度则抑制碳酸酐酶活力(Wu et al., 2007)。低浓度的镉增加莱因衣藻胞外碳酸酐酶活力，在高浓度下镉同样对该酶起抑制作用(Wang et al., 2005)。添加镉来培养海洋浮游植物，碳酸酐酶活力得到加强(Cullen et al., 1999)。在 0~50μmol/L 范围内，Zn/Cd 超富集植物滇苦菜(*Picris divaricata*)的碳酸酐酶活力与地上部镉的含量成明显地正相关(Ying et al., 2010)。重金属超富集植物天蓝遏蓝菜(*Thlaspi caerulescens*)暴露在镉下，碳酸酐酶活力增强(Liu et al., 2008)。1mmol/L 锌、铁、镉增加假单胞菌(*Pseudomonas fragi*)胞外碳酸酐酶活性(Sharma et al., 2009)。2μmol/L 的锰、铁、钴、镍和镉能提高莱因衣藻(*Chlamydomonas reinhardtii*)胞外碳酸酐酶活性，浓度为 100μmol/L 时，起抑制作用(王宝利等，2006)。因此，可以看出，镉、铁、锌、锰、钴和镍对碳酸酐酶的作用具有明显的激素效应。

二、"激素"对碳酸酐酶的影响

迄今为止，六个从系统发生上已确定为独立基因家族编码的不同类型的碳酸酐酶(α、β、γ、δ、ε 和 ζ 类型)都没有明显的序列同源性(Hewett-Emmett and Tashian, 1996; So et al., 2004, Bertucci et al., 2009)。碳酸酐酶具有多种同工酶，不同的同工酶具有组织器官甚至细胞器的特异性(Smith and Ferry, 2000; Aspatwar et al., 2010)。无论是植物还是动物，激素也同样有几种类型。每种类型的激素在生物体的不同组织器官中的分布和含量是不同的。激素在生物体中这种分布特征非常相似于碳酸酐酶在生物体中分布的特征。

激素对碳酸酐酶的影响是多方面的。它们既可以对碳酸酐酶活力起促进作用，也可以起抑制作用。动物激素影响碳酸酐酶活力的报道极多。睾丸激素降低肾脏碳酸酐酶活力，并存在剂量效应。胞液的碳酸酐酶活力和血清睾酮浓度之间具有明显的负相关性。睾丸激素和雌二醇影响雄性老鼠肾脏碳酸酐酶活力，给雄性老鼠服用雌二醇能激活其肾脏的碳酸酐酶(Suzuki et al., 1996)。甲状旁腺素能激活破骨细胞中的碳酸酐酶活力(Anderson et al., 1985)。红细胞的甲状腺素刺激 CAⅡ基因的转录(Rascle et al., 1994)。此外，甲状腺素还能调节老鼠的碳酸酐酶Ⅲ蛋白质(Fremont et al., 1987)。降血钙素能使人红细胞的 CAⅡ活力增加 2 倍，使甲状旁腺素下降 50%(Arlot-Bonnemains et al., 1985)。β-雌二醇和黄体酮可以诱导大鼠肝细胞碳酸酐酶活力(Garg, 1975)。褪黑激素抑制虹鳟红细胞碳酸酐酶(Hisar et al., 2005)。随着体外培养时间的不同，苯甲酸雌二醇和孕酮对鸭红细胞碳酸酐酶活力影响也显著不同。当鸭红细胞培养 2 个小时和 14 小时时，苯甲酸雌二醇增加碳酸酐酶活力。当鸭红细胞培养三十小时时，添加高浓度的苯甲酸雌二醇(高于 200 pg/mL)并没有促进碳酸酐酶活力。当鸭红细胞培养十四个小时时，添加 0.5～50 ng/mL 孕酮能提高碳酸酐酶活力；当鸭红细胞只培养两个小时时，添加 1～10 ng/mL 孕酮却降低碳酸酐酶活力；当鸭红细胞培养三十小时时，添加 0.5 ng/mL 孕酮就能抑制碳酸酐酶活力(Wu et al., 2010)。由此可以看出，动物碳酸酐酶受到了激素的双向影响。

植物激素影响碳酸酐酶活力的报道较少。10^{-6} mol/L 吲哚丁酸(IAA)、赤霉素(GA_3)和激动素(Kin)以及 10^{-8} mol/L 的表高油菜素内酯(HBR)能够提高芥菜型油菜的碳酸酐酶活力。上述激素对碳酸酐酶刺激作用的顺序为表高油菜素内酯＞赤霉素＞吲哚丁酸＞激动素。脱落酸(ABA)能使碳酸酐酶活力下降(Hayat et al.，2001)。10^{-10} mol/L至 10^{-6} mol/L的吲哚丁酸和氯代生长素(4-Cl-IAA)能提高叶中碳酸酐酶的活力。10^{-8} mol/L氯代生长素能最大地促进碳酸酐酶活力(Ali et al.，2008)。低浓度6-氨苄嘌呤和赤霉素能提高芥菜型油菜碳酸酐酶的活力，高浓度则抑制碳酸酐酶的活力，最适6-氨苄嘌呤和赤霉素的浓度是5 μmol/L。同样，在低浓度下萘乙酸提高了碳酸酐酶活力，高浓度抑制其活性，最适浓度是0.5 μmol/L(李西腾，2007)。因此，可以看出，植物激素对碳酸酐酶的影响也是双重的。

三、无机元素对碳酸酐酶作用的"激素"效应的可能机制

激素的作用牵涉环腺苷酸的积累和特定的基因表达(Sutherland，1972)。碳酸酐酶失活和激活也与环腺苷酸有关。蛋白激酶的激活与磷酸化的酶蛋白有关，而碳酸酐酶可以被依赖蛋白激酶的环腺苷酸所活化(Narumi and Miyamoto，1974)。碳酸酐酶抑制剂乙酰唑胺能够增加大鼠体外肾皮质片的环腺苷酸的产生(Rodriguez et al.，1974)。破骨细胞的碳酸酐酶的作用需要环腺苷酸参与(Hall and Kenny，1986)。在4~5℃下，青蛙皮肤上皮提取物的碳酸酐酶活性受到3-异丁基-1-甲基黄嘌呤(IBMX)的抑制，在0℃时，长时间暴露于环腺苷酸下也抑制了该酶的活性(Galar and Marroquin，1990)。碳酸酐酶和 $Mg^{2+}-HCO_3^--$ ATP酶共同调节小肠黏膜醛固酮的作用(Suzuki et al.，1985)。碳酸酐酶Ⅱ被认为是雌激素响应基因。子宫和肝脏体内碳酸酐酶Ⅱ表达被雌激素下调，染料木黄酮使子宫碳酸酐酶Ⅱ的基因表达略有下调，使肝脏碳酸酐酶Ⅱ的基因表达下调到60%(Caldarelli et al.，2005)。甲状腺素也能改变碳酸酐酶基因表达(Conrad et al.，2006)。

碳酸酐酶基因表达调控受多种因素影响。甲状腺激素在调节胚胎神经视网膜碳酸酐酶基因表达中发挥着重要作用(Peterson et al.，1996)。在人类乳房肿瘤细胞中，导致癌细胞微环境酸化的碳酸酐酶ⅩⅡ(CA12)的基因表达与雌激素的α受体高度相关(Barnett et al.，2008)。编码水稻叶子和根碳酸酐酶基因的表达响应于盐和渗透胁迫(Yu et al.，2007)。寒冻逆境会严重抑制绿豆幼苗编码β-碳酸酐酶基因的表达(Yang et al.，2005)。生长在20℃/50 μmol·m^{-2}·s^{-1}环境下的黑麦叶片编码β-碳酸酐酶基因表达水平明显低于其他(水或光)逆境下的(Ndong et al.，2001)。在碱性环境中(pH=9)编码秀丽隐杆线虫(Caenorhabditis elegans)α-碳酸酐酶 CAH-4b 基因 cah-4b 的表达水平是中性环境(pH=7)下的6倍(Hall et al.，2008)。添加Cd到Zn限制培养的硅藻中可以提高Cd专一性的碳酸酐酶基因表达(Lane and Morel，2000)。将斑节对虾(Penaeus monodon)从正常盐度移至低盐条件下，虾鳃碳酸酐酶的mRNA和活

力显著增加(Pongsomboon et al.，2009)。盐度还可刺激蓝蟹的两个碳酸酐酶同工酶的表达(Serrano et al.，2007)。脱水和聚乙二醇模拟干旱对拟南芥不同品系的 CA1 和 CA2 影响也显著不同(Wu et al.，2012)。莱茵衣藻和蛋白核小球藻中碳酸酐酶胞外酶(CAH1)基因在低浓度碳酸氢钠条件下表达旺盛，而高浓度碳酸氢钠则被下调(见第二章)。氟对莱因衣藻的 CAH3、CAH5 和 CAG1 基因也具有明显的激素效应(见第二章)。从上面的事实可以看出，碳酸酐酶的基因表达极易受到外界环境的调节。很可能，无机元素也是同样通过调节碳酸酐酶的基因表达来影响碳酸酐酶活力的。今后，希望获得这方面的证据。

第二节 微藻碳酸酐酶对氮磷钾吸收的影响

一、植物离子吸收动力学

离子吸收动力学参数 V_{max}、K_m 和 C_{min} 可用来表征浮游植物对限制性营养盐离子吸收的动力学特征。其中，V_{max} 表示离子吸收所能达到的最大速率，其值越大，离子吸收的能力越强。在不考虑底物浓度的情况下，载体介导的运输速率不会超过 V_{max}，当载体的底物结合位点饱和或通道的流量达最大时，运输速率将接近 V_{max} (Francois and Morel，1987)。运输体的密度是限制运输速率的主要因素，因此 V_{max} 能表征细胞膜上特异功能蛋白质分子的数目，即 V_{max} 的大小取决于细胞膜上某种离子转运载体蛋白的数量。K_m 则表示吸收位点(转运载体)和所吸收离子间的亲和性，其值越小，亲和性越好(谢少平和倪晋山，1990)。K_m 在数值上等于运输速率达 $1/2V_{max}$ 时的底物浓度，它是载体的特征参数，只和载体性质有关，而与载体数量无关。C_{min} 是净吸收速率为 0 时的底物浓度，其大小能表征藻类耐低营养盐环境的能力(Spilling et al.，2010)。但是，植物对某一种离子吸收的动力学参数，并非常数，而是受诸多因素影响(韩振海等，1994)。卡盾藻在光照条件下吸收氮、磷的 V_{max} 要大于黑暗条件下的(Nakamura and Watanabe，1983)。温度和光照强度等环境因子也影响着 V_{max} 和 K_m 值(刘静雯等，2001；Carter and Lathwell，1967；Naldi and Viaroli，2002)。铜离子会影响柑橘幼苗对氮、磷、钾、钙四种营养元素的吸收(黎耿碧等，1996)。NH_4^+ 能显著抑制 K^+ 的吸收，使 V_{max} 显著减小，但 K_m 值变化不大(孙小茗等，2007；Pettersson and Jensen，1983)；NO_3^- 和稀土元素能使小麦显著提高 K^+ 的亲和力，从而促进 K^+ 吸收(黄建国等，1995)。

二、胞外碳酸酐酶抑制剂对微藻氮磷钾吸收的影响

为了研究微藻碳酸酐酶对微藻吸收无机大量营养元素氮、磷、钾的影响，选取蛋白核小球藻和莱茵衣藻作为研究对象，在缺氮、缺磷和缺钾胁迫下进行饥饿

处理，通过添加碳酸酐酶的专一抑制剂——乙酰唑胺(acetazolamide，AZ)，采用常规耗竭法研究了微藻碳酸酐酶对其无机元素吸收的影响。

图6-1、图6-2、图6-3和图6-4分别为在AZ影响下两种藻培养液中无机元素氮、磷、钾的消耗曲线。由图可知，加AZ组莱茵衣藻培养液中氮、磷、钾浓度下降曲线在对照组的上方，表明AZ对莱茵衣藻氮、磷、钾的吸收具有抑制作用；加AZ组蛋白核小球藻培养液中氮、钾浓度下降曲线在对照组上方，而磷浓度下降曲线则位于对照组下方，且紧挨对照组，表明AZ对蛋白核小球藻氮、钾吸收表现出抑制作用，而对磷的吸收影响不大。

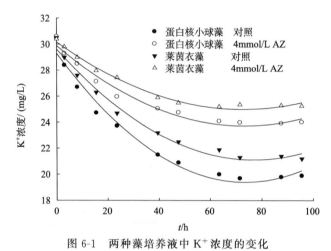

图 6-1 两种藻培养液中 K^+ 浓度的变化

Fig. 6-1 The change in concentration of K^+ in culture medium of two algal species

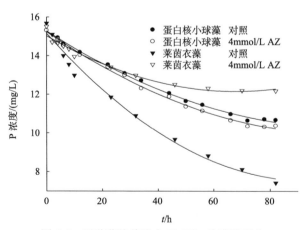

图 6-2 两种藻培养液中 $H_2PO_4^-$ 浓度的变化

Fig. 6-2 The change in concentration of $H_2PO_4^-$ in culture medium of two algal species

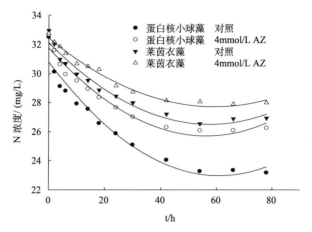

图 6-3 两种藻培养液中氨氮浓度的变化

Fig. 6-3 Variation in concentration of ammonia nitrogen in culture medium of two algal species

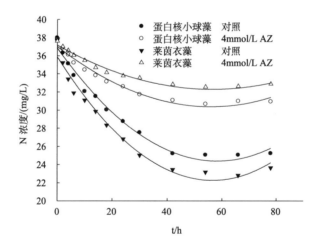

图 6-4 两种藻培养液中硝态氮浓度的变化

Fig. 6-4 Variation in concentration of nitrate nitrogen in culture medium of two algal species

表 6-1 为 AZ 影响下的两种微藻吸收氮、磷、钾无机盐的动力学参数。由表可知，不管是蛋白核小球藻，还是莱茵衣藻，不加 AZ 组的 V_{max} 均大于加 AZ 组（蛋白核小球藻 $H_2PO_4^-$ 的吸收除外），加 AZ 组的 K_m 与 C_{min} 均大于未加 AZ 组，表明在微藻碳酸酐酶活性受抑制的情况下，微藻对无机元素氮、磷、钾的亲和力均下降，耐贫无机营养盐环境的能力下降，对无机盐的吸收能力下降。由此进一

步可知，微藻碳酸酐酶对其无机营养盐的吸收起着重要作用，其活性强，微藻吸收利用无机元素的能力强。这主要与微藻体内需要无机元素氮、磷、钾参与的生命活动过程大多与碳酸酐酶的催化活动有关。

表 6-1　AZ 影响下的两种微藻无机营养吸收动力学参数
Tab. 6-1　The kinetic parameter of inorganic nutrient uptake by two algal species under the influence of AZ

微藻	无机离子	无 AZ			加 AZ		
		V_{max}/(fg·cell^{-1}·h^{-1})	K_m/(mg/L)	C_{min}/(mg/L)	V_{max}/(fg·cell^{-1}·h^{-1})	K_m/(mg/L)	C_{min}/(mg/L)
蛋白核小球藻	NH_4^+	75.02	24.92	22.96	63.79	27.21	25.70
	NO_3^-	169.68	27.60	24.50	96.87	32.00	30.33
	$H_2PO_4^-$	6.83	11.29	9.96	7.12	11.70	10.52
	K^+	15.23	21.74	19.22	9.67	25.06	23.45
莱茵衣藻	NH_4^+	89.81	27.91	26.51	49.06	28.77	27.54
	NO_3^-	267.39	25.57	22.13	93.21	33.45	32.19
	$H_2PO_4^-$	66.79	9.66	7.83	35.09	12.78	12.03
	K^+	75.71	23.06	20.87	49.75	26.36	25.09

藻的种类不同，其体内碳酸酐酶的活性不同，受碳酸酐酶抑制剂的抑制程度不同，其无机元素吸收所受影响亦不同。表 6-2 为 AZ 处理下的两种藻无机盐吸收动力学参数相对对照的变化幅度。由表 6-2 可知，AZ 对两种藻的硝态氮和钾的吸收影响均较大，且莱茵衣藻铵态氮吸收所受影响程度更大，但二者的钾吸收所受影响程度差别不大。在四种无机离子中，蛋白核小球藻对 $H_2PO_4^-$ 的吸收受 AZ 的影响最小，而莱茵衣藻的 $H_2PO_4^-$ 吸收所受影响较大，其对铵态氮的吸收受 AZ 的影响最小，但蛋白核小球藻的铵态氮吸收所受影响却较大。表明碳酸酐酶对蛋白核小球藻铵态氮、硝态氮和钾营养吸收具有重要作用，但对 $H_2PO_4^-$ 的吸收影响较小；碳酸酐酶对莱茵衣藻硝态氮、钾和 $H_2PO_4^-$ 的吸收起着重要作用，而对铵态氮的吸收影响较小。此外，在三个动力学参数中，两种藻的 V_{max} 的变化幅度均最大，而 K_m 和 C_{min} 的变化幅度相对较小，且二者的差别较小，两种藻的动力学参数受 AZ 的影响程度总体呈以下顺序：$V_{max} > C_{min} > K_m$。

表 6-2 AZ 处理下两种藻的无机离子吸收动力学参数相对于对照的变化幅度

Tab. 6-2 Difference of inorganic nutrient uptake kinetic parameter of two algal spceies treated with AZ compared with the control

无机离子	蛋白核小球藻			莱茵衣藻		
	V_{max} 降幅/%	K_m 增幅/%	C_{min} 增幅/%	V_{max} 降幅/%	K_m 增幅/%	C_{min} 增幅/%
NH_4^+	14.97	9.19	11.93	45.37	3.08	3.89
NO_3^-	42.91	15.94	23.80	65.14	30.82	45.46
$H_2PO_4^-$	-4.25	3.63	5.62	47.46	32.30	53.64
K^+	36.51	15.27	22.01	34.29	14.31	20.22

综上所述，碳酸酐酶对微藻无机元素的吸收起着重要的作用，酶活性高，微藻无机元素吸收能力强，耐贫无机营养盐环境的能力强，微藻对无机元素的亲和力强。但碳酸酐酶对微藻无机元素吸收的影响情况与藻种及无机元素的种类有关。

三、微藻碳酸酐酶影响无机元素吸收的可能机制

碳酸酐酶广泛存在于动物和植物中，可以催化 CO_2 可逆水合反应：

$$CO_2 + H_2O \leftrightarrow H_2CO_3 \leftrightarrow H^+ + HCO_3^-$$

碳酸酐酶影响藻类的离子吸收，可能与碳酸酐酶催化上述水合反应形成的质子和 HCO_3^- 有关。中国螃蟹的碳酸酐酶通过提供碳酸氢根离子和质子给 V-型 H^+-ATPase、Na^+/H^+ 运输蛋白以及 Cl^-/HCO_3^- 转运蛋白等在致电离子吸收中发挥着重要作用(Gilles and Péqueux，1986；Onken et al.，1991；Onken and Putzenlechner，1995；Genovese et al.，2005)。*Dilocarcinus pagei* 鳃的 Cl^- 吸收与 V-型 H^+-ATPase、Na^+/H^+ 运输蛋白以及 Cl^-/HCO_3^- 转运蛋白有关(Onken and McNamara，2002；Weihrauch et al.，2004)。乙酰唑胺处理蛋壳腺匀浆使 Ca^{2+}-Mg^{2+}-ATP 酶活力可降到 42%，相应地也同样使碳酸酐酶活力降到了 36%(Lundholm，1990)。

碳酸酐酶通过影响酸碱平衡来影响离子吸收。斑马鱼的碳酸酐酶通过影响 H^+-ATP 酶来影响体内的酸碱平衡从而影响钠离子的吸收(Lin et al.，2008)。大鼠肝脏溶酶体的 ATP 酶和碳酸酐酶系统共同作用维持了溶酶体之间的质子梯度(Iritani and Wells，1974)，碳酸酐酶抑制剂使天竺鼠的大脑和肾脏皮质片钠增加、钾减少(Davies et al.，1955)。鼠胃黏膜中碳酸酐酶与 Mg^{2+}-ATP 酶也具有功能上的关系，胃黏膜线粒体的碳酸酐酶和 Mg^{2+}-ATP 酶共同影响着胃酸的分泌(Narumi and Kanno，1973)。碳酸酐酶抑制剂可以减少水网藻(*Hydrodictyon reticulatum*)质子和碳酸氢根离子的产生，从而影响离子吸收，降低了光合作用(Rybova and Slavikova，1974)。

第三节　无机元素对微藻碳酸酐酶的影响

一、氟对微藻碳酸酐酶的影响

氟是元素周期表中电负性最大的元素，具有高度的化学活泼性，几乎可以与周期表中所有元素作用形成化合物。常温下氟为淡黄色带刺激性臭味的气体，相对原子质量 18.9985，其固体密度 1.900g/cm³。液体密度 1.633g/cm³（−207℃），1.512 g/cm³（−188℃），气体密度 1.696g/L（标）。

氟在生物圈中的循环是通过水中的溶解态氟的流动性来完成的。水体中氟浓度和特定的地理环境如气候、水化学环境、地表物质、地貌等密切相关，除此之外还受人类活动的影响。人为污染也能导致局部地区地表水和浅层地下水氟含量升高。随着国民经济建设高速发展，工业三废排放量日益增加，导致局部地区浅层地下水氟含量逐渐升高。地貌条件决定地表水和地下水的流向、流速，影响氟的迁移和聚集。氟元素在迁移过程中，通常顺应地势向低处迁移，而富集于地势低洼地（许晓路和申秀英，1998）。

（一）氟对植物的影响

氟作为生物活性元素能被生物吸收，并在体内某些器官中富集。水、土壤和大气是生物摄取氟的重要场所，对于高等植物而言根系从土壤中吸收氟是植物体内氟积累的重要途径，对于水生植物而言水中的氟是植物体氟的来源。

Arnesen 认为，植物体内氟的背景浓度一般＜10g/kg，最大允许浓度为 30g/kg，超过这一阈限，植物就可能中毒，并通过食物链危及动物和人类的健康（Arnesen，1997）。植物出现氟毒害症状的氟临界浓度受包括植物品种、土壤性质、氟的存在形态以及毒害时间等多种因素影响。氟污染对植物生长的影响与植物种类有很大关系，或反映在产量上，或反映在植物的生理功能作用和营养成分上。

氟在植物体内的积累对植物的影响或表现为品质的改变，或表现为器官的畸变。氟可使佛手叶的光合作用及叶绿素含量下降，可溶性总糖及蔗糖含量下降，淀粉及果糖含量上升，使佛手花粉畸形率增加，萌发率及花粉管长度下降（徐丽珊等，2003）。大气氟污染对小麦的产量有较大的影响（勾晓华等，1999）。此外，氟还通过抑制 SOD 活性降低了叶绿素的含量，通过抑制 SOD 活性合成大量乙烯参与叶片脱落的纤维素酶的活性诱导（孟范平等，2002）。

尽管天然水体中含氟量较低，但由于氟是已知元素中电负性最高的，化学性质十分活泼，可与所有的金属形成氟化物，也可以与大多数非金属直接发生反应，因此氟对水生生物的毒性是不容忽视的。氟化物较强的水溶迁移性能够使其广泛存在于水体和水生生物体内,当其浓度超过一定的临界浓度时,就会成为对生物的有毒元素。并且由于氟的本底含量较低以及生物富集的特性,持续摄入少量的氟化物就能对水生生物特别是微藻产生显著的影响。

氟化物对水生植物尤其是藻类的影响研究相对较少。氟化物对藻类的影响并不呈现出一致的表现。氟对有些藻类(如 Chaetoceros gracilis)的生长起促进作用,对有些藻类(如 Amphidinim carteri)的生长起抑制作用(Antia and Klut, 1981),甚至在一定的浓度范围内对有些藻类(如:Nannochloris oculata)没有影响(Oliveira et al., 1978)。氟化物通过降低希尔反应和耗竭细胞内的 ATP 来抑制藻类光合放氧、增加二氧化碳的释放(Camargo, 2003)。但是这种情况并不是在所有藻类中都存在。在一些藻类如 Chlorella、Scenedesmus obliquus、Ankistrodesmus braunii 和 Chlamydomonas moewusii 并没有发现氟引起的光合放氧的减少(Bhatnagar and Bhatnagar, 2000)。氟化物对微藻的效应或抑制或促进,这取决于氟化物的浓度、在氟环境中的暴露时间以及藻的种类等(Camargo, 2003)。

(二) 氟对微藻碳酸酐酶活力的急性影响

以莱茵衣藻(Chlamydomonas reinhardtii)为实验材料,将其培养在 SE 培养基内。分别在培养基中添加 0、0.5mol/L、2mol/L、8mol/L 和 32mmol/L 的氟化钠,在温度 $25.0℃±1.0℃$,光强为 $150~\mu mol \cdot m^{-2} \cdot s^{-1}$ 的 16h/8h 的光周期下培养 3 天后,测定光合放氧、碳酸酐酶活力等指标。

氟对莱茵衣藻的碳酸酐酶和光合净氧的释放有显著的影响(表 6-3),0~2mmol/L 的氟对碳酸酐酶活力和净光合氧的释放具有促进作用,而高浓度对这些指标具有抑制作用,表现出明显的"激素"效应。碳酸酐酶活力最大和净光合氧的释放最多的处理是 2mmol/L。

表 6-3 衣藻不同氟浓度处理下的碳酸酐酶活力和净光合放氧速率

Tab. 6-3 CA activity and net photosynthetic oxygen evolution rate of *C. reinhardtii* under the treatment with different concentration of F

氟浓度 mmol/L	碳酸酐酶活力 (WAU/10^9cells)	净光合放氧速率 (nmol $O_2 \cdot h^{-1} \cdot \mu gchl^{-1}$)
0	8.82±0.83 b	57.28±8.50b
0.5	10.65±0.96c	71.08±6.14c
2	13.58±0.72d	84.48±6.31d
8	7.93±0.62b	45.17±2.31b
32	6.40±0.96a	34.25±4.55a

注:平均值±SD, $n=3$,平均值±SD 后面的字母代表同一列的数据差异性,ANOVA, $P<0.05$。

Note: Mean±SD, $n=3$, alphabet stands for data differentiation, ANOVA, $P<0.05$.

一些研究(如 Maren et al.,1976)表明氟能抑制碳酸酐酶活力。但是,我们的研究表明,低浓度的氟可以促进莱茵衣藻的胞外碳酸酐酶的活力。造成这种状况的可能原因是氟影响碳酸酐酶的作用有环腺苷酸参与,这样使氟对碳酸酐酶的影响呈现出"激素"效应。实际上,氟的"激素"效应在许多情况下也发生过(Bohatyrewicz et al.,2001;Machalinska et al.,2002;Burgstahler,2002)。这只是一种可能的解释。更合理的解释需要更多的证据。

莱茵衣藻碳酸酐酶活力与净光合氧的释放有显著的正相关($P<0.01$,图 6-5)。这个结果与一些植物如烟草和油菜等的结果相似(Williams et al.,1996;Ahmad et al.,2001)。这似乎是莱茵衣藻受到氟对碳酸酐酶的影响而影响其生理和生长的另一种机制。

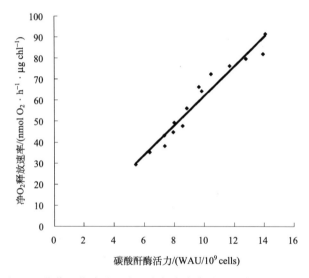

图 6-5 莱茵衣藻碳酸酐酶活力与净光合放氧速率之间的关系
Fig. 6-5 Relation between CA activity and net photosynthetic oxygen evolution rate of *C. reinhardtii*

(三)氟对莱茵衣藻和蛋白核小球藻胞外碳酸酐酶的影响

以莱茵衣藻(*Chlamydomonas reinhardtii*)和蛋白核小球藻(*Chlorella pyrenoidosa*)为实验材料,将其培养在 SE 培养基内。分别在培养基中添加 0.1mmol/L、1mmol/L、10mmol/L、50mmol/L、100mmol/L、200mmol/L 的氟化钠,在温度 25.0℃±1.0℃,pH 为 6.5±0.2,光照强度为 40μmol·m^{-2}·s^{-1}的 16h/8h 的光周期下培养。每天人工摇瓶 2~3 次。接种前将处于对数生长期的藻液离心(3000 r/min,5min)浓缩,用浓缩的藻液接种于已灭菌的培养基中,初始接种吸光值约为 0.5A(波长:680nm)。每个处理设置 3 个平行样。培养 10 天后,测定藻体叶绿素含量和胞外碳酸酐酶活力。

1. 氟对莱茵衣藻和蛋白核小球藻叶绿素含量的影响

由图 6-6 和图 6-7 可见在 0~1mmol/L 浓度范围内的，F^- 对莱茵衣藻和蛋白核小球藻的叶绿素 a、叶绿素 b 和叶绿素 c 含量没有显著影响（$n=9$，$P>0.05$），在 $[F^-]>1$mmol/L 时，三种叶绿素含量均随 F^- 浓度的升高而显著下降（$n=9$，$P<0.05$）。F^- 可能抑制了两种微藻叶绿素的合成，且造成叶绿素分解破坏。同时发现莱茵衣藻的叶绿素 c 含量极低，而蛋白核小球藻则含有一定比例的叶绿素 c，表明同为绿藻的莱茵衣藻和蛋白核小球藻在细胞组成成分上不同，莱茵衣藻的叶绿素组成更接近于高等植物，蛋白核小球藻的叶绿素组成与低等植物类似。

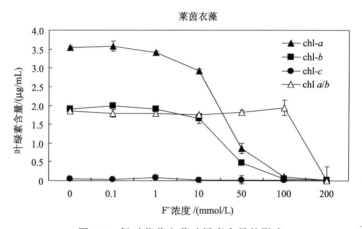

图 6-6　氟对莱茵衣藻叶绿素含量的影响

Fig. 6-6　Effect of F on chlorophyll content of *C. reinhardtii*

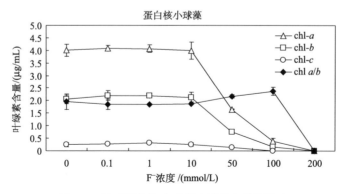

图 6-7　氟对蛋白核小球藻叶绿素含量的影响

Fig. 6-7　Effect of F on chlorophyll content of *Chlorella pyrenoidosa*

在培养基中 F^- 浓度为 1~10mmol/L 时，随着 F^- 浓度升高莱茵衣藻的叶绿素 a 和叶绿素 b 含量呈现出下降趋势，而蛋白核小球藻的叶绿素 a 和叶绿素 b 含量在此浓度范围内则没有变化，说明莱茵衣藻对环境变化的响应更灵敏，较低浓度的 F^- 即能够胁迫莱茵衣藻，而蛋白核小球藻则表现出更强的适应性，受环境变化的影响较小。

莱茵衣藻的叶绿素 a 和叶绿素 b 的比值在 0~100mmol/L 范围内没有变化，说明 F^- 对莱茵衣藻吸收利用光能的机制没有影响；蛋白核小球藻叶绿素 a 和叶绿素 b 的比值在 50mmol/L 和 100mmol/L 处理组中高于其他处理组，说明在高氟胁迫情况下，蛋白核小球藻吸收利用光能的能力增强，另外的可能原因是高氟胁迫导致叶绿素 b 合成量少于叶绿素 a 的合成量，使叶绿素 a 和叶绿素 b 的比值升高。

2. 氟对莱茵衣藻和蛋白核小球藻的胞外碳酸酐酶活性的影响

氟对莱茵衣藻和蛋白核小球藻的胞外碳酸酐酶活性有显著影响（图 6-8）。低浓度的 F^- 对莱茵衣藻和蛋白核小球藻胞外碳酸酐酶活性影响甚微，F^- 浓度为 0.1mmol/L、1mmol/L 和 10mmol/L 时与对照组没有显著性差异（$n=12$，$P>0.05$），较高浓度的 F^- 则能够促进胞外碳酸酐酶的表达，F^- 为 50mmol/L 和 100mmol/L 与其他浓度均有显著差异（$n=18$，$P<0.05$），当 $[F^-]<10$mmol/L 时，莱茵衣藻的胞外碳酸酐酶活性随 F^- 浓度的增加略有升高，蛋白核小球藻胞外碳酸酐酶活性基本没有变化；当 $[F^-]>10$mmol/L 时，胞外碳酸酐酶活性随着 F^- 浓度的升高而逐渐增强，在 F^- 浓度为 100mmol/L 左右时达到峰值后，胞外碳酸酐酶活性开始降低直至检测不出。由图 6-8 中还可看出，在受到高氟（50~100mmol/L）胁迫时，莱茵衣藻表达出的胞外碳酸酐酶活性（1423.77 ± 25.35 EU/mg）显著高于蛋白核小球藻（166.25 ± 18.84 EU/mg），说明在受到高氟逆境胁迫时，莱茵衣藻存在更为灵敏的响应机制应对不利环境的胁迫。

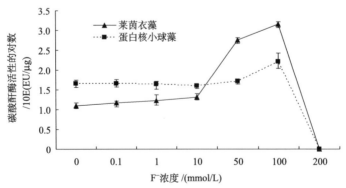

图 6-8　F^- 对莱茵衣藻和蛋白核小球藻胞外碳酸酐酶活性的影响

Fig. 6-8　Effect of F^- on CA_{ex} of *C. reinhardtii* and *Chlorella pyrenoidosa*

胞外碳酸酐酶活性与叶绿素含量有一定的相关性。在 F^- 为 10~100mmol/L 的处理组中，胞外碳酸酐酶活性和叶绿素含量变化呈相反趋势，胞外碳酸酐酶活性升高的同时叶绿素含量降低。不同浓度的 F^- 对莱茵衣藻和蛋白核小球藻表现出了相同的作用效果。低浓度的 F^-（<10mmol/L）对两种微藻的生长没有影响；高浓度的 F^-（>10mmol/L）则抑制微藻的生长，相应叶绿素含量降低。

（四）不同 pH 环境下氟对莱茵衣藻和蛋白核小球藻胞外碳酸酐酶的影响

F^- 超标水体的 pH 往往不稳定，已有研究证明 F^- 对微藻的生长会产生影响（Camargo，2003），同时 pH 也是微藻生长繁殖的重要影响因素之一（Bhatti and Colman，2008），pH 对微藻光合作用有显著影响（Semesi et al.，2009），生长在水体中的微藻受到 pH 和 F^- 的共同作用。微藻的胞外碳酸酐酶通过催化 CO_2 的可逆水合反应，促进水环境中的无机碳向微藻细胞内流动（Coleman，2004），为光合作用固定无机碳源源不断地提供原料，胞外碳酸酐酶还与微藻的抗逆性有关，在逆境下其活性会升高。目前，有关 pH 影响重金属对微藻毒性影响研究是国内外研究者感兴趣的内容（Boullemant，2009；赵娜等，2010），而 pH 和 F^- 对微藻的生理作用的交互影响尚未见报道。本实验以莱茵衣藻为实验材料，同时以蛋白核小球藻作参照，研究 pH 和 F^- 对莱茵衣藻胞外碳酸酐酶活性的交互作用，初步探讨两种因子对微藻抗逆性的影响，并为微藻除氟的可行性提供理论依据。

按需要用 HCl 或 NaOH 溶液调节培养基 pH 为 5.0、7.0、9.0。添加 NaF 使培养基中 F^- 终浓度分别为：0.1mmol/L、1mmol/L、10mmol/L、50mmol/L、100mmol/L 和 200mmol/L。在温度 25.0℃±1.0℃，pH 为 6.5±0.2，光照强度为 40μmol·m^{-2}·s^{-1} 的 16h/8h 的光周期下培养。每天人工摇瓶 2~3 次。接种前将处于对数生长期的藻液离心（3000 r/min，5min）浓缩，用浓缩的藻液接种于已灭菌的培养基中，初始接种吸光值约为 0.5A（波长：680nm）。每个处理设置 3 个平行样。培养 10 天后，测定藻体叶绿素含量和胞外碳酸酐酶活力。

不同 pH 环境下氟对两种微藻叶绿素 a 合成量的影响。叶绿素 a 的含量变化一定程度上反映微藻叶绿体的含量变化，同时叶绿素 a 作为一种色素，其净增加量能表现微藻合成有机物量的能力。在相同 pH 条件和相同生长时间下，莱茵衣藻叶绿素 a 合成量明显高于蛋白核小球藻，表明莱茵衣藻具有比蛋白核小球藻更强的生长繁殖能力。从图 6-9 可以看出，pH 影响微藻叶绿素 a 合成，碱性和中性环境中两种微藻叶绿素 a 的合成量比酸性环境中高，说明酸性环境不利于微藻的生长和有机物的积累。当[F^-]<1mmol/L 时，两种微藻叶绿素 a 的变化量并不明显，可见低浓度的氟对两种微藻的生长没有显著影响（$n=9$，

$P > 0.05$)。当[F⁻]>1mmol/L 时，随着 F⁻ 浓度的升高，叶绿素 a 含量迅速降低，说明高浓度的 F⁻ 抑制莱茵衣藻和蛋白核小球藻的生长。在受到 F⁻ 和 pH 共同作用时，F⁻ 浓度的升高和 pH 的下降都导致叶绿素 a 含量减少。

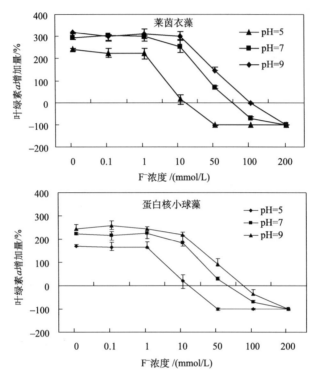

图 6-9　pH 和 F⁻ 对莱茵衣藻和蛋白核小球藻叶绿素 a 合成量的影响
Fig. 6-9　Effect of pH and fluoride on synthesis of Chlorophyll a in *Chlamydomonas reinhardtii* and *Chlorella pyrenoidosa*

不同 pH 环境下氟对两种微藻胞外碳酸酐酶活性的影响。pH 和 F⁻ 对莱茵衣藻和蛋白核小球藻胞外碳酸酐酶活性的影响如图 6-10 所示。两种微藻在同一 pH 且[F⁻]<1mmol/L 培养基中培养时，两者的胞外碳酸酐酶活性变化甚微，而在相同 F⁻ 浓度(0.1mmol/L 和 1mmol/L)但不同 pH 条件时，它们各自的胞外碳酸酐酶活性随 pH 下降而降低。由此可见，当 F⁻ 浓度为 0.1mmol/L 和 1mmol/L 时，微藻胞外碳酸酐酶活性主要受 pH 变化的影响。当 F⁻>1mmol/L 时，两种微藻的胞外碳酸酐酶活性受 F⁻ 浓度和 pH 的共同影响，但 F⁻ 浓度的影响更大。当 pH 一定时，胞外碳酸酐酶活性随着 F⁻ 浓度的增加而逐渐升高，到一定峰值后活性降低直至检测不出。一定范围内，较高浓度的 F⁻ 能够显著提高莱茵衣藻和蛋白核小球藻胞外碳酸酐酶的活性($n=18$, $P<0.01$)。在不同 pH 条件下，莱

莱茵衣藻和蛋白核小球藻表达出最高碳酸酐酶活性时的 F^- 浓度不同：pH 分别为 5、7、9 时，莱茵衣藻分别在 F^- 浓度为 10mmol/L、50mmol/L、100mmol/L 时有最高的胞外碳酸酐酶活性；蛋白核小球藻在 F^- 浓度为 50mmol/L、100mmol/L、100mmol/L 时有最高的胞外碳酸酐酶活性。

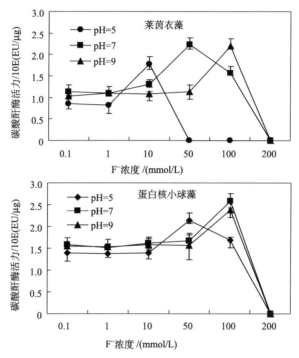

图 6-10 pH 和 F^- 对莱茵衣藻和蛋白核小球藻胞外碳酸酐酶活性的影响

Fig. 6-10 Effect of pH and fluoride on extracelluar carbonic anhydrase activity of *Chlamydomonas reinhardtii* and *Chlorella pyrenoidosa*

莱茵衣藻和蛋白核小球藻的胞外碳酸酐酶活性和叶绿素 a 合成量呈现出相似的变化趋势，总的来说，在 $[F^-]<1mmol/L$ 情况下，莱茵衣藻和蛋白核小球藻几乎不受到 F^- 浓度变化的影响，而受 pH 变化影响较大，表现为胞外碳酸酐酶活性在酸性环境中低于中性和碱性环境中，叶绿素 a 合成量随着 pH 的升高而增加。在 $[F^-]>1mmol/L$ 时，两种微藻受到 F^- 和 pH 共同影响，表现为胞外碳酸酐酶活性先升高后降低，叶绿素 a 合成量则随着 F^- 浓度增加而迅速下降。在研究高 F^- 浓度造成的逆境中，胞外碳酸酐酶活性随 F^- 浓度变化波动较大，在 10～100mmol/L 范围内酶活显著高于 0～10mmol/L 范围（$n=18$，$P<0.05$），表明胞外碳酸酐酶活性能够作为指示微藻处于高 F^- 浓度逆境的指标。叶绿素 a 合成量在逆境中下降，且与 F^- 浓度有一定的对应关系，可见叶绿

素 a 合成量也可以作为指示微藻处于高 F^- 浓度逆境的指标。

交互作用指在一定条件下，两个或多个元素的结合生理效应小于或超过它们各自效应之和(祖艳群等，2008)。在本实验中，pH=5 时较低浓度 F^- 的作用下两种微藻的胞外碳酸酐酶活性即达到最高值，而在 pH=7 和 pH=9 时，诱导出两种微藻最高胞外碳酸酐酶活性所需的 F^- 浓度更高。导致这一现象的可能原因是，在酸性环境下，微藻生长状况较差，H^+ 和 F^- 的协同作用下 F^- 的效应被放大，故微藻胞外碳酸酐酶活性最高点出现在较低 F^- 浓度水平。LeBlanc 的研究也揭示 pH 6.0 条件下生长的普通小球藻能够耐受的 F^- 浓度低于 pH 6.8 时的(LeBlanc，1984)。偏碱性环境对微藻的生长有利，OH^- 部分抵消了 F^- 的胁迫作用，表现为在更高浓度 F^- 作用下两种微藻胞外碳酸酐酶活性才达到最高。在偏碱性环境中，微藻生长状况最好，叶绿素的合成量和总无机碳固定量大于中性和酸性环境中。叶绿素含量的多少影响了光合效率的高低。酸性环境和高 F^- 逆境下，两种微藻叶绿素合成量减少。碱性环境中莱茵衣藻和蛋白核小球藻叶绿素 a 合成量明显增加，表明碱性环境有利于叶绿素的合成，即莱茵衣藻和蛋白核小球藻更适于生长在碱性环境下。

总而言之，胞外碳酸酐酶活性和叶绿素 a 的变化情况都表明碱性环境对莱茵衣藻和蛋白核小球藻的生长有利，能够耐受的 F^- 浓度也较中性环境和酸性环境下高。低于 1mmol/L 的 F^- 对莱茵衣藻和蛋白核小球藻的胞外碳酸酐酶活性影响甚微，对叶绿素 a 的合成也没有显著影响。自然水体中 F^- 浓度变化很大但均远低于 1mmol/L(Amini et al.，2008)，说明两种微藻在自然界高氟水体内均能正常生长。水生植物体内氟含量随环境中 F^- 升高以及培养时间的增加而上升(Camargo，2003)，可见微藻可以作为除去水体中低浓度超标 F^- 的材料。

(五) 氟对微藻影响的剂量效应的依存性

氟对微藻的生长和碳酸酐酶的影响所呈现的剂量效应存在着时间的依存性。3 天的短期处理，0~2mmol/L 的氟就可以促进光合能力和碳酸酐酶表达的增加，大于 8mmol/L 的氟才对这些指标有抑制作用。而 10 天的中长期处理，低于 10mmol/L 的氟对两种微藻的生长没有影响；只有大于 10mmol/L 的氟才表现出对生长的抑制，而 50~100mmol/L 的氟对碳酸酐酶具有促进作用；也就是说，对碳酸酐酶的作用中长期培养使其效应值提高 25~50 倍。这可能与中长期培养微藻经过了适应性的变化耐受性增强有关(Li et al.，2013)。

氟对微藻的生长和碳酸酐酶的影响所呈现的剂量效应也与环境的 pH 有关。在酸性环境下，微藻碳酸酐酶更易受到氟的影响，偏碱性环境下，对微藻碳酸酐酶活力具有最大促进作用的氟浓度在升高。因此，为了获得较好的除氟效果，预先必须调节废水的 pH，通过设置较长的处理时间，来达到最佳的微藻除氟效果。

二、重金属污染物对微藻碳酸酐酶的影响

微量金属元素,是指在环境、生物或地质样品中以痕量出现的各种元素。微量金属元素虽然含量甚微,但一部分却是生物生长所必需的微量营养元素,比如元素 Fe、Mn、Zn、Cu、B、Mo、Co、V 是藻生长所必需的(Hutchinson,1967)。这部分微量元素具有两面性:一方面,作为生物生化酶的辅助因子,必需微量金属元素是生物新陈代谢的必需要素,它们的生物有效利用率直接影响着生物量的变化;另一方面,当必需微量金属元素的供给超过了生物的需要时,就会对生物产生毒害作用。对生物而言,非必需微量金属元素(Cd、Pb、Hg、As、Tl)是有毒性的,这些微量金属元素进入生物细胞后,竞争并取代必需微量金属元素,造成细胞新陈代谢紊乱,从而产生毒害。微量金属元素一直是环境地球化学研究的重要内容。碳酸酐酶为含 Zn 金属酶,它催化可逆的 CO_2 水和反应。碳酸酐酶胞外酶又为诱导酶,它一部分在细胞内,一部分在细胞表面。当外界 CO_2 供应受到限制时,碳酸酐酶胞外酶使得藻细胞可以利用水中的 HCO_3^- 作为无机碳源(Tsuzuki and Miyachi,1989),是某些藻类无机碳浓缩机制中的一个重要组成成分(Aizawa and Miyachi,1986),在全球碳的生物地球化学循环中具有重要的作用。工业革命以来,人为活动强烈地改变了地球表层的局部乃至全球的微量金属元素的含量(e.g. Hg),对生物正常生长乃至人类的健康造成了潜在的威胁。研究微量金属元素对微藻碳酸酐酶的影响,可了解微量金属元素的生态环境效应,明确微藻碳酸酐酶生物地球化学作用的影响因素。

在此,以实验室内 SE 培养基纯培养的莱茵衣藻(*Chlamydomonas reinhardtii*)及阿哈湖和红枫湖收集的微藻为材料,研究了 Mn^{2+}、Fe^{2+}、Co^{2+}、Ni^{2+}、Cu^{2+}、Zn^{2+}、Cd^{2+}、Pb^{2+} 对微藻碳酸酐酶胞外酶(CA_{ex})的影响。阿哈湖和红枫湖是中国西南贵阳市附近的人工水库。莱茵衣藻、阿哈湖和红枫湖微藻 CA_{ex} 活性分别为 5.27 ± 0.66、10.39 ± 0.17、10.15 ± 0.23($\times 10^{-3}$ Wilbur-Anderson units · (μg chla^{-1}),平均值$\pm$SD);阿哈湖和红枫湖微藻 CA_{ex} 活性明显高于莱茵衣藻的 CA_{ex} 活性,而阿哈湖和红枫湖的微藻 CA_{ex} 的活性基本相同;HCO_3^- 是阿哈湖和红枫湖中的主要阴离子,而 SE 培养液中 HCO_3^- 的浓度几乎为零;可见 CA_{ex} 的活性与环境中 HCO_3^- 的浓度有关。

随着微量金属浓度的增加,它们对碳酸酐酶胞外酶(CA_{ex})活性的抑制越来越强烈(表 6-4)。然而,低浓度的 Cd 却刺激了 CA_{ex} 的活性;室内培养实验研究表明:Cd 的添加增加了碳酸酐酶的活性(Morel et al.,1994;Lee et al.,1995);显然,Cd 对碳酸酐酶比较特殊,这可能与碳酸酐酶为 Zn 金属酶,而且在元素周期表上 Cd 和 Zn 又处于同一族有关;然而随着 Cd 浓度的继续升高,Cd 像 Zn、Cu、Pb 一样,开始抑制 CA_{ex} 的活性。研究发现,Cd、Zn 和 Cu

对河口生长的螃蟹的碳酸酐酶的活性具有抑制效应(Vitale et al., 1999; Skaggs and Henry, 2002)。Co(Ⅱ)对虹鳟鱼肌肉体外碳酸酐酶活性具有竞争性抑制,而Cu(Ⅱ)、Zn(Ⅱ)和Ag(Ⅰ)对虹鳟鱼肌肉体外碳酸酐酶活性具有非竞争性抑制(Soyut et al., 2012)。本研究结果也表明,Cd、Zn、Cu和Pb对CA_{ex}的活性的抑制效应因金属浓度的不同而不同,而且抑制作用因微藻种类的不同而不同;并不是所有的微藻都含有碳酸酐酶胞外酶,每种微藻忍受微量金属元素的门槛也不一样。

表6-4a Cu、Zn、Cd、Pb对莱茵衣藻CA_{ex}活性的影响

Tab. 6-4a Effect of Cu、Zn、Cd、Pb on the activity of CA_{ex} in *C. reinhardtii*

处理浓度/(mg/L)	相对CA_{ex}的活性			
	Cu	Zn	Cd	Pb
0	1.00 ± 0.13[①]	1.00 ± 0.13	1.00 ± 0.13	1.00 ± 0.13
0.5	1.17 ± 0.11	0.88 ± 0.17	1.16 ± 0.42	0.61 ± 0.15[②]
1.0	0.92 ± 0.06	0.82 ± 0.09	1.03 ± 0.18	0.73 ± 0.02[②]
1.5	0.77 ± 0.11	0.54 ± 0.21[②]	1.10 ± 0.11	0.84 ± 0.07
2.0	0.49 ± 0.08[②]	0.31 ± 0.09[②]	1.05 ± 0.09	0.77 ± 0.11

①正常生长情况。
②和正常情况相比存在显著性差异(Tukey, $P<0.05$;平均值 ± SD)。
①Normal culture condition.
② The difference is significant between metal-excessive condition and normal condition (Tukey, $P<0.05$; values are means ± SD).

表6-4b Cu、Zn、Cd、Pb对阿哈湖微藻CA_{ex}活性的影响

Tab. 6-4b Effect of Cu、Zn、Cd、Pb on the activity of microalgal CA_{ex} in Aha Lake

处理浓度/(mg/L)	相对CA_{ex}的活性			
	Cu	Zn	Cd	Pb
0	1.00 ± 0.08[①]	1.00 ± 0.08	1.00 ± 0.08	1.00 ± 0.08
0.5	0.18 ± 0.06[②]	0.73 ± 0.08[②]	1.25 ± 0.10	0.78 ± 0.05
1.0	0.15 ± 0.05[②]	0.65 ± 0.09[②]	0.66 ± 0.14[②]	0.57 ± 0.09[②]
1.5	0.11 ± 0.03[②]	0.40 ± 0.09[②]	0.34 ± 0.13[②]	0.23 ± 0.05[②]
2.0	0.03 ± 0.04[②]	0.06 ± 0.02[②]	0.23 ± 0.07[②]	0.23 ± 0.14[②]

①正常生长情况。
②和正常情况相比存在显著性差异(Tukey, $P<0.05$;平均值 ± SD)。
①Normal culture condition.
② The difference is significant between metal-excessive condition and normal condition (Tukey, $P<0.05$; values are means ± SD).

表 6-4c　Cu、Zn、Cd、Pb 对红枫湖微藻 CA_{ex} 活性的影响

Tab. 6-4c　Effect of Cu、Zn、Cd、Pb on the activity of microalgal CA_{ex} in Hongfeng Lake

处理浓度/(mg/L)	相对 CA_{ex} 的活性			
	Cu	Zn	Cd	Pb
0	1.00 ± 0.09[①]	1.00 ± 0.09	1.00 ± 0.09	1.00 ± 0.09
0.5	0.87 ± 0.13	0.87 ± 0.11	1.37 ± 0.12[②]	1.07 ± 0.07
1.0	0.90 ± 0.12	0.74 ± 0.06[②]	1.18 ± 0.09	0.94 ± 0.11
1.5	0.83 ± 0.11	0.69 ± 0.05[②]	1.17 ± 0.07	0.87 ± 0.06
2.0	0.56 ± 0.08[②]	0.46 ± 0.13[②]	0.88 ± 0.11	0.82 ± 0.06

①正常生长情况。

②和正常情况相比存在显著性差异(Tukey, $P<0.05$；平均值 ± SD)。

①Normal culture condition.

② The difference is significant between metal-excessive condition and normal condition (Tukey, $P<0.05$; values are means ± SD).

微量金属元素对 CA_{ex} 活性的影响主要通过三个途径：①直接与 CA_{ex} 这一含 Zn 金属酶作用；②间接影响藻细胞内 CA_{ex} 的新陈代谢；③直接改变 CA_{ex} 的生物活性。为了深入了解微量金属元素对微藻碳酸酐酶胞外酶的作用机理，对莱茵衣藻碳酸酐酶胞外酶提取前后分别加入微量金属进行了对比研究。结果表明：Cd、Cu 和 Pb 在这两种情况下对 CA_{ex} 活性的影响基本一致，只有 Zn 的不同(表 6-5)。这表明 Cd、Cu 和 Pb 可能直接通过与 CA_{ex} 作用来影响其活性，而 Zn 还可能通过影响藻细胞 CA_{ex} 的代谢来影响它的活性。由于蛋白质的金属结合位点不具有专一性，它们也可以结合离子半径和配位结构相似的其他金属离子，这可能是 Cd 和 Cu 抑制 CA_{ex} 的活性的原因。除了竞争生化酶上的金属结合位点，Pb 还可以直接改变蛋白质的生物活性。

表 6-5　不同处理下莱茵衣藻(*C. reinhardtii*)的相对 CA_{ex} 的活性

Tab. 6-5　Relative activity of CA_{ex} in *C. reinhardtii* under different treatment

	Cu	Zn	Cd	Pb
After	0.57 ± 0.06	0.93 ± 0.19	1.02 ± 0.15	0.68 ± 0.10
Before	0.47 ± 0.08	0.31 ± 0.09	1.05 ± 0.09	0.78 ± 0.12

注："Before"和"After"之间存在显著性差异(Tukey, $P<0.05$；平均值±SD)。"Before"指在提取 CA_{ex} 前已经用 2.0mg/L 四种金属处理了 24 小时，"After"指在提取 CA_{ex} 后暴露在 2.0mg/L 的四种金属中，再测定的 CA_{ex} 的活性。

Note: The difference is significant between"Before"and"After"(Tukey, $P<0.05$; values are means ± SD). "Before" refers to the relative activity of CAex from *Chlamydomonas reinhardtii* that was exposed to 2.0 mg/L heavy metals for 24 hours before extracting, "After" refers to the relative activity of CAex that was exposed to 2.0 mg/L heavy metals for 24 hours after being extracted from *Chlamydomonas reinhardtii*.

为了更好地理解不同种微量金属对 CA_{ex} 活性影响的差异，我们又研究了 Mn、Fe、Co 和 Ni 对 CA_{ex} 活性的影响。$2\mu mol/L$ 的 Mn、Fe、Co、Ni 和 Cd 处理下，莱茵衣藻 CA_{ex} 活性比正常情况下都高，说明 $2\mu mol/L$ 的这些金属刺激了 CA_{ex} 的活性；但当浓度升高到 $100\mu mol/L$，Mn、Co、Ni、Cu、Zn 和 Cd 都抑制了莱茵衣藻 CA_{ex} 的活性，而且抑制作用随这些微量金属原子质量的增加而增大的迹象，但不是很明显，这时 Cu、Zn 和 Cd 对 CA_{ex} 活性的影响程度基本一致（图 6-11）。

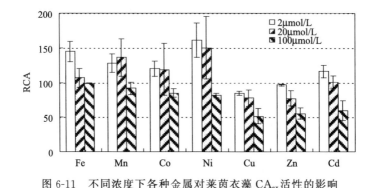

图 6-11　不同浓度下各种金属对莱茵衣藻 CA_{ex} 活性的影响

RCA：相对 CA_{ex} 活性

Fig. 6-11　Effect of different metal species with different concentration on the activity of CA_{ex} in *Chlamydomonas reinhardtii*

RCA：Relative CA_{ex} Activity

$20\mu mol/L$ 的 Fe、Mn 对莱茵衣藻 CA_{ex} 提取前后活性的影响情况和 $100\mu mol/L$ 的情况截然相反（图 6-12）。$20\mu mol/L$ 时，经过 24 小时的处理，Fe 和 Mn 对莱茵衣藻 CA_{ex} 活性稍有刺激作用，CA_{ex} 活性和正常情况相差不大；莱茵衣藻 CA_{ex} 提取后直接分别加入 Fe 和 Mn，它们对 CA_{ex} 有强烈的刺激作用，Fe 和 Mn 最高分别使 CA_{ex} 的活性提高到原来的 2 倍和 2.5 倍。$100\mu mol/L$ 时，经过 24 小时的处理，Fe 和 Mn 对莱茵衣藻 CA_{ex} 活性的影响不大，Mn 的 CAex 活性比正常情况的稍低一点；莱茵衣藻 CA_{ex} 提取后直接分别加入同浓度的 Fe 和 Mn，它们对 CA_{ex} 有强烈的抑制作用，Fe 和 Mn 使 CA_{ex} 的活性降低到原来的 50% 左右。由此可见：①Fe 和 Mn 不仅直接同 CA_{ex} 作用，还通过影响莱茵衣藻 CA_{ex} 的新陈代谢来影响它的活性；②莱茵衣藻对 Fe 和 Mn 影响的生物调节作用明显，24 小时后，经过莱茵衣藻的生物调节，$20\mu mol/L$ 和 $100\mu mol/L$ 的 Fe 和 Mn 对莱茵衣藻 CA_{ex} 的活性影响相差不大，而莱茵衣藻 CA_{ex} 提取后直接加入 $20\mu mol/L$ 和 $100\mu mol/L$ 的 Fe 和 Mn 时 CA_{ex} 的活性相差很大。

Co 和 Ni 分别对莱茵衣藻 CA_{ex} 提取前后活性的影响基本相同，它们在

20μmol/L 的两种处理下分别都刺激了 CA_{ex} 的活性，Ni 的刺激作用比 Co 的明显；它们在 100μmol/L 的两种处理下分别都抑制了 CA_{ex} 的活性；从实验结果看，莱茵衣藻对 Co 和 Ni 也具有生物调节作用，但效果不是很明显（图 6-12）。可见 Co 和 Ni 直接同 CA_{ex} 作用影响其活性，随着浓度的升高，它们的抑制作用越强烈。

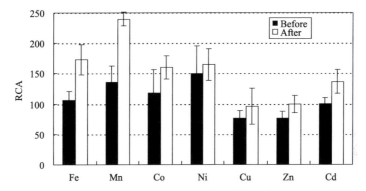

图 6-12a　20μmol/L 的各种金属对提取前后莱茵衣藻 CA_{ex} 活性的影响

Fig. 6-12a　Effect of different metal species (20μmol/L) on the activity of CA_{ex} in *Chlamydomonas reinhardtii* before and after extracting

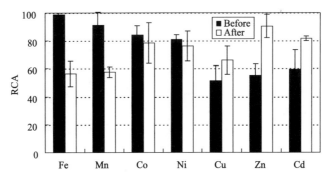

图 6-12b　100μmol/L 的各种金属对提取前后莱茵衣藻 CA_{ex} 活性的影响

Fig. 6-12b　Effect of different metal species (100μmol/L) on the activity of CA_{ex} in *Chlamydomonas reinhardtii* before and after extracting

Cu 在这些金属中对 CA_{ex} 的抑制最强，它在 20μmol/L 和 100μmol/L 的各种处理下对莱茵衣藻 CA_{ex} 活性的影响基本相同，100μmol/L 的抑制作用比 20μmol/L 的强烈（图 6-12）。可见 Cu 直接同 CA_{ex} 作用影响其活性，浓度越高抑制越强。

Zn 和 Cd 分别对莱茵衣藻 CA_{ex} 提取前后活性的影响不同(图 6-12)。20μmol/L 和 100μmol/L 时,经过 24 小时的处理,Zn 对 CA_{ex} 具有抑制作用,但 CA_{ex} 提取后加入 Zn 时对其活性影响不大;可见 Zn 主要通过影响莱茵衣藻 CA_{ex} 的新陈代谢来影响它的活性。20μmol/L 的 Cd 对莱茵衣藻经过 24 小时的处理后,莱茵衣藻 CA_{ex} 的活性变化不大,CA_{ex} 提取后加入 Cd 时刺激了 CA_{ex} 的活性;100μmol/L 时,Cd 在两种处理下都对莱茵衣藻 CA_{ex} 的活性具有抑制作用,经过 24 小时处理的抑制作用要强一些;可见 Cd 不仅直接同 CA_{ex} 作用,还通过影响莱茵衣藻 CA_{ex} 的新陈代谢来影响它的活性。

参 考 文 献

勾晓华,王勋陵,陈发虎. 1999. 氟化氢熏气对植物的伤害研究. 兰州大学学报(自然科学版),35(2):141-145.

韩振海,王永章,王倩. 1994. 植物的离子吸收动力学研究的现状和前景. 北京农业大学学报,20(4):381-387.

黄建国,杨邦俊,袁铃. 1995. 小麦不同品种吸收钾离子的动力学研究. 植物营养与肥料学报,1(1):38-42.

黎耿碧,陈二钦,Alva A K. 1996. 外界铜离子浓度对柑橘小苗常量元素吸收特性的影响. 广西农业大学学报,15(3):195-201.

李西腾. 2007. 生长调节物质对油菜碳酸酐酶活性的影响. 安徽农业科学,35:5354-5355.

刘静雯,董双林,马生生. 2001. 温度和盐度对几种大型海藻生长率和 NH_4-N 吸收的影响. 海洋学报,23(2):109-116.

孟范平,李桂芳,吴方正. 2002. 氟害大豆超氧化物歧化酶活性与叶绿素含量及叶片脱落的关系. 农村生态环境,18(2):34-38.

孙小茗,封克,汪晓丽. 2007. K^+ 高亲和转运系统吸收动力学特征及其受 NH_4^+ 影响的研究. 植物营养与肥料学报,13(2):208-212.

王宝利,刘丛强,吴沿友. 2006. 藻碳酸酐酶胞外酶活性对过渡金属响应的实验研究. 矿物岩石地球化学通报,25(Z1):42-44.

谢少平,倪晋山. 1990. 水稻(威优 49)幼苗根系 K^+($80Rb^+$)吸收的调节. 植物生理学报,16(1):63-691.

徐丽珊,申秀英,许晓路,郑珊波. 2003. 氟化物对金华佛手碳代谢及花粉的影响. 农业环境科学学报,22(4):474-476.

许晓路,申秀英. 1998. 金华地区氟污染成因危害及对策. 农业环境保护,17(1):11-14.

赵娜,冯鸣凤,朱琳. 2010. 不同 pH 值条件下 Cr^{6+} 对小球藻和斜生栅藻的毒性效应. 东南大学学报(医学版),29(4):382-386.

祖艳群,李元,Bock L,Schvartz C,Colinet G,胡文友. 2008. 重金属与植物 N 素营养之间的交互作用及其生态学效应. 农业环境科学学报,27(1):7-14.

Ahmad A, Hayat S, Fariduddin Q, Ahmad I. 2001. Photosynthetic efficiency of plants of Brassica juncea, treated with chlorosubstituted auxins. Photosynthetica, 39: 565-568.

Aizawa K, Miyachi S. 1986. Carbonic anhydrase and carbon concentrating mechanisms in microalgae and cyanobacteria. FEMS Microbiology Reviews, 39: 215-233.

Ali B, Hayat S, Hasan S A, Ahmad A. 2008. IAA and 4-Cl-IAA increases the growth and nitrogen fixation in mung bean. Communications in Soil Science and Plant Analysis, 39: 2695-2705.

Amini M, Mueller K, Abbaspour K C, et al. 2008. Statistical modeling of global geogenic fluoride contamination in groundwaters. Environmental Science and Technology, 42(10): 3662-3668.

Anderson R E, Jee W S, Woodbury D M. 1985. Stimulation of carbonic anhydrase in osteoclasts by parathyroid hormone. Calcified Tissue International, 37: 646-650.

Antia N J, Klut M E. 1981. Floride addition effects on euryhaline phytoplankter growth in nutrient-enriched seawater at an estuarine level of salinity. Botanica Marina, 24, 147-152.

Arlot-Bonnemains Y, Fouchereau-Peron M, Moukhtar M S, Benson A A, Milhaud G. 1985. Calcium-regulating hormones modulate carbonic anhydrase II in the human erythrocyte. Proceedings of the National Academy of Sciences of the U. S. A., 82: 8832-8834.

Arnesen A K M. 1997. Availability of fluoride to plants grown in contaminated soils. Plant and Soil, 191: 13-25.

Aspatwar A, Tolvanen M E E, Parkkila S. 2010. Phylogeny and expression of carbonic anhydrase-related proteins. BMC Molecular Biology, 11: 25-43.

Baird T T, Waheed A, Okuyama T, Sly W S, Fierke C A. 1997. Catalysis and inhibition of human carbonic anhydrase IV. Biochemistry, 36: 2669-2678.

Barnett D H, Sheng S, Charn T H, et al. 2008. Estrogen receptor regulation of carbonic anhydrase XII through a distal enhancer in breast cancer. Cancer Research, 68: 3505-3515.

Bertucci A, Innocenti A, Zoccola D, Scozzafava A, Tambutte S, Supuran C T. 2009. Carbonic anhydrase inhibitors. Inhibition studies of a coral secretory isoform by sulfonamides. Bioorganic and Medical Chemistry, 17: 5054-5058.

Bhatnagar M, Bhatnagar A. 2000. Algal and cyanobacterial responses to fluoride. Fluoride, 33: 55-65.

Bhatti S, Colman B. 2008. Inorganic carbon acquisition in some synurophyte algae. Physiologia Plantarum, 133(1): 33-40.

Bohatyrewicz A, Białecki P, Larysz D, Gusta A. 2001. Compressive strength of cancellous bone in fluoride-treated rats. Fluoride, 34: 236-241.

Boullemant A. 2009. Uptake of hydrophobic metal complexes by three freshwater algae: unexpected influence of pH. Environmental Science and Technology, 43(9): 3308.

Burgstahler A W. 2002. Paradoxical Dose-response effects of fluoride. Fluoride, 35: 143-7.

Caldarelli A, Diel P, Vollmer G. 2005. Effect of phytoestrogens on gene expression of carbonic anhydrase II in rat uterus and liver. Journal of Steroid Biochemistry and Molecular Biology,

97(3): 251-256.

Camargo J A. 2003. Fluoride toxicity to aquatic organisms: a review. Chemosphere, 50: 251-264.

Carter O G, Lathwell D V. 1967. Effects of temperature on orthophosphate absorption by excised corn roots. Plant Physiology, 42: 1407-1412.

Coleman J. 2004. Carbonic anhydrase and its role in photosynthesis. Photosynthesis, 9: 353-367.

Conrad A H, Zhang Y T, Walker A R, et al. 2006. Thyroxine affects expression of KSPG-related genes, the carbonic anhydrase II gene, and KS sulfation in the embryonic chicken cornea. Investigative Ophthalmology and Visual Science, 47(1): 120-132.

Cullen J T, Lane T W, Morel F M M, Sheerell R M. 1999. Modulation of cadmium uptake in phytoplankton by seawater CO_2 concentration. Nature, 402: 165-167.

Davies R E, Galston A W, Whittam R. 1955. The effects of an inhibitor of carbonic anhydrase on sodium and potassium movements in brain and kidney cortex slices. Biochimica et Biophysica Acta, 17: 434-439.

Francois M, Morel M. 1987. Kinetics of nutrient uptake and growth in phytoplankton. Journal of Phycology, 23: 137-150.

Fremont P, Lazure C, Tremblay R R, Chretien M, Rogers P A. 1987. Regulation of carbonic anhydrase III by thyroid hormone: opposite modulation in slow- and fast-twitch skeletal muscle. Biochemistry and Cell Biology, 65: 790-797.

Galar I, Marroquin M C. 1990. Cyclic AMP effects on chloride transport and carbonic anhydrase activity in frog skin. Canadian Journal of Physiology and Pharmacology, 68: 1269-1274.

Garg L C. 1975. Induction of hepatic carbonic anhydrase by estrogen. Journal of Pharmacology and Experimental Therapeutics, 192: 297-302.

Genovese G, Ortiz N, Urcola M R, Luquet C M. 2005. Possible role of carbonic anhydrase, V^- H^+-ATPase, and Cl^-/HCO_3^- exchanger in electrogenic ion transport across the gills of the euryhaline crab Chasmagnathus granulates. Comparative Biochemistry and Physiology Part A: Molecular and Integrative Physiology, 142(3): 362-369.

Gilbert A L, Guzman H M. 2001. Bioindication potential of carbonic anhydrase activity in anemones and corals. Marine Pollution Bulletin, 42: 742-744.

Gilles R, Péqueux A. 1986. Physiological and ultrastructural studies of NaCl transport in crustacean gills. Bolletino di Zoology, 53: 173-182.

Hall G E, Kenny A D. 1986. Bone resorption induced by parathyroid hormone and dibutyryl cyclic AMP: role of carbonic anhydrase. Journal of Pharmacology and Experimental Therapeutics, 238: 778-782.

Hall R A, Vullo D, Innocenti A, et al. 2008. External pH influences the transcriptional profile of the carbonic anhydrase, CAH-4b in Caenorhabditis elegans. Molecular and Biochemical Parasitology, 161: 140-149.

Hayat S, Ahmad A, Mobin M, Fariduddin Q, Azam Z M. 2001. Carbonic anhydrase, photosynthesis, and seed yield in mustard plants treated with phytohormones. Photosynthetica, 39: 111-114.

Hewett-Emmett D, Tashian R E. 1996. Functional diversity, conservation and convergence in the evolution of the α-, β- and γ-carbonic anhydrase gene families. Molecular Phylogenetics and Evolution, 5: 50-77.

Hisar O, Beydemir S, Gülçin I, Aras Hisar S, Yanik T, Küfrevioglu O I. 2005. The effects of melatonin hormone on carbonic anhydrase enzyme activity in rainbow trout (Oncorhynchus mykiss) erythrocytes in vitro and in vivo. Turkish Journal of Veterinary and Animal Sciences, 29: 841-845.

Hoffmann K M, Samardzic D, Heever K, Rowlett R S. 2011. Co(II)-substituted Haemophilus influenzae β-carbonic anhydrase: Spectral evidence for allosteric regulation by pH and bicarbonate ion. Archives of Biochemistry and Biophysics, 511(1-2): 80-87.

Hutchinson G E. 1967. A treatise on limnology. New York: Wiley, 2: 1115.

Innocenti A, Zimmerman S, Ferry J G, Scozzafava A, Supuran C T. 2004. Carbonic anhydrase inhibitors. Inhibition of the beta-class enzyme from the methanoarchaeon Methanobacterium thermoautotrophicum (Cab) with anions. Bioorganic and Medical Chemistry Letter, 14: 4563-4567.

Iritani N, Wells W W. 1974. Studies on a HCO_3^- stimulated ATPase and carbonic anhydrase system in rat liver lysosomes. Archives of Biochemistry and Biophysics, 164(1): 357-366.

Lane T W, Morel F M. 2000. A biological function for cadmium in marine diatoms. Proceedings of the National Academy of Sciences of the U. S. A., 97: 4627-4631.

LeBlanc G A. 1984. Interspecies relationships in acute toxicity of chemicals to aquatic organisms. Ecotoxicology and Environmental Safety, 3: 47-60.

Li Q, Wu Y Y, Wu Y D. 2013. Effects of fluoride and chloride on the growth of Chlorella pyrenoidosa. Water Science and Technology, 68(3): 722-727.

Lin T Y, Liao B K, Horng J L, Yan J J, Hsiao C D, Hwang P P. 2008. Carbonic anhydrase 2-like a and 15a are involved in acid-base regulation and Na^+ uptake in zebrafish H^+-ATPase-rich cells. American Journal of Physiology-Cell Physiology, 294(5): C1250-C1260.

Lindskog S. 1963. Effects of pH and inhibitors on some properties related to metal binding in bovine carbonic anhydrase. Journal of Biological Chemistry, 238: 945-951.

Lionetto M G, Maffia M, Cappello M S, Giordano M E, Storelli C, Schettino T. 1998. Effect of cadmium on carbonic anhydrase and Na^+-K^+-ATPase in eel, Anguilla anguilla, intestine and gills. Comparative Biochemistry and Physiology Part A: Molecular and Integrative Physiology, 120: 89-91.

Liu M Q, Yanai J T, Jiang R F, Zhang F, McGrath S P, Zhao F J. 2008. Does cadmium play a physiological role in the hyperaccumulator Thlaspi caerulescens? Chemosphere, 71: 1276-1283.

Lundholm C E. 1990. The eggshell thinning action of acetazolamide: Relation to the binding of

Ca^{2+} and carbonic anhydrase activity of the shell gland homogenate. Comparative Biochemistry and Physiology Part C: Comparative Pharmacology, 95(1): 85-89.

Machalinska A, Machoy-Mokrzynska A, Marlicz W, Stecewiecz I, Machalinski B. 2001. NaF-induced apoptosis in human bone marrow and cord blood CD34 positive cells. Fluoride, 34: 258-263.

Maren T H, Rayburn C S, Liddell N E. 1976. Inhibition by anions of human red cell carbonic anhydrase B: physiological and biochemical implications. Science, 191: 469-472.

Mañanes A A L, Daleo G R, Vega F V. 1993. pH-dependent association of carbonic anhydrase (CA) with gastric light microsomal membranes isolated from bovine abomasum. Partial characterization of membrane-associated activity. Comparative Biochemistry and Physiology B, 105: 175-82.

Nakamura Y, Watanabe M M. 1983. Nitrate and phosphate uptake kinetics of Chattonella antiqua grown in light/dark cycles. Journal of the Oceanographical Society of Japan, 39(4): 167-170.

Naldi M, Viaroli P. 2002. Nitrate uptake and storage in the seaweed Ulva rigida C. Agardh in relation to nitrate availability and thallus nitrate content in a eutrophic coastal lagoon (Sacca di Go ro, Po River Delta, Italy). Journal of Experimental Marine Biology and Ecology, 269: 65-83.

Narumi S, Kanno M. 1973. Effects of gastric acid stimulants and inhibitors on the activities of HCO_3^--stimulated, Mg^{2+}-dependent ATPase and carbonic anhydrase in rat gastric mucosa. Biochimica et Biophysica Acta (BBA) - Biomembranes, 311(1): 80-89.

Narumi S, Miyamoto E. 1974. Activation and phosphorylation of carbonic anhydrase by adenosine 3′, 5′ monophosphate dependent protein kinases. Biochimica Biophysica Acta, 350: 215-224.

Ndong C, Danyluk J, Huner N P A, Sarhan F. 2001. Survey of gene expression in winter rye during changes in growth temperature, irradiance or excitation pressure. Plant Molecular Biology, 45: 691-703.

Nishimori I, Innocenti A, Vullo D, Scozzafava A, Supuran C T. 2007. Carbonic anhydrase inhibitors. Inhibition studies of the human secretory isoform VI with anions. Bioorganic and Medical Chemistry Letters, 17: 1037-1042.

Nishimori I, Minakuchi T, Onishi S, Vullo D, Cecchi A, Scozzafava A, Supuran C T. 2009. Carbonic anhydrase inhibitors. cloning, characterization and inhibition studies of the cytosolic isozyme III with anions. Journal of Enzyme Inhibition and Medicinal Chemistry, 24: 70-76.

Oliveira L, Antia N J, Bisalputra T. 1978. Culture studies on the effects from fluoride pollution on the growth of marine phytoplankters. Journal of the Fisheries Research Board of Canada, 35: 1500-1504.

Onken H, Graszynski K, Zeiske W. 1991. Na^+ independent, electrogenic Cl^- uptake across the posterior gills of the Chinese crab (*Eriocheir sinensis*) voltage clamp and microelectrode studies. Journal of Comparative Physiology, 161: 293-301.

Onken H, McNamara J C. 2002. Hyperosmoregulation in the red freshwater crab *Dilocarcinus pagei* (Brachyura, Trichodactylidae): structural and functional asymmetries of the posterior gills. Journal of Experimental Biology, 205: 167-175.

Onken H, Putzenlechner M. 1995. A V-ATPase drives active, electrogenic and Na^+-independent Cl^- absorption across the gills of Eriocheir sinensis. Journal of Experimental Biology, 198: 767-774.

Peterson R E, Orozco A, Linser P J. 1996. Thyroid hormone and its receptors and metabolizing enzymes contribute to the regulation of carbonic anhydrase expression in neural retina. Investigation Ophthalmology and Visual Science, 37: 1-1154.

Pettersson S, Jensen P. 1983. Variation among species an varieties in uptake and utilization of potassium. Plant Soil, 72: 231-237.

Pongsomboon S, Udomlertpreecha S, Amparyup P, Wuthisuthimethavee S, Tassanakajon A. 2009. Gene expression and activity of carbonic anhydrase in salinity stressed *Penaeus monodon*. Comparative Biochemistry and Physiology a-Molecular and Integrative Physiology, 152(2): 225-233.

Rascle A, Ghysdael J, Samarut J. 1994. c-ErbA, but not v-ErbA, competes with a putative erythroid repressor for binding to the carbonic anhydrase II promoter. Oncogene, 9: 2853-2867.

Rodriguez H J, Walls J, Yates J, Klahr S. 1974. Effects of acetazolamide on the urinary excretion of cyclic AMP and on the activity of renal adenyl cyclase. Journal of Clinical Investigation, 53: 122-130.

Rybova R, Slavikova M. 1974. Ion transport in the Alga *Hydrodictyon reticulatum* under conditions of carbonic anhydrase inhibition. Zeitschrift fur Pflanzenphysiologie, 72(4): 287-296.

Sas K N, Kovacs L, Zsiros O, et al. 2006. Fast cadmium inhibition of photosynthesis in cyanobacteria in vivo and in vitro studies using perturbed angular correlation of gamma-rays. Journal of Biological Inorganic Chemistry, 11: 725-734.

Semesi I S, Juma K, Mats B. 2009. Alterations in seawater pH and CO_2 affect calcification and photosynthesis in the tropical coralline alga, Hydrolithon sp. (Rhodophyta). Estuarine, Coastal and Shelf Science, 84(3): 337-341.

Serrano L, Halanych K M, Henry R P. 2007. Salinity-stimulated changes in expression and activity of two carbonic anhydrase isoforms in the blue crab *Callinectes sapidus*. Journal of Experimental Biology, 210: 2320-2332.

Sharma A, Bhattacharya A, Singh S. 2009. Purification and characterization of an extracellular carbonic anhydrase from Pseudomonas fragi. Process Biochemistry, 44: 1293-1297.

Skaggs H S, Henry R P. 2002. Inhibition of carbonic anhydrase in the gills of two euryhaline crabs, *Callinectes sapidus* and *Carcinus maenas*, by heavy metals. Comparative Biochemistry Physiology Part C: Toxicology and Pharmacology, 133: 605-612.

Smith K S, Ferry J G. 2000. Prokaryotic carbonic anhydrases. FEMS Microbiology Reviews, 24: 335-366.

So A K C, Espie G S, Williams E B, Shively J M, Heinhorst S, Cannon G C. 2004. A novel evolutionary lineage of carbonic anhydrase (ε class) is a component of the carboxysome shell. Journal of Bacteriology, 186: 623-630.

Soyut H, Beydemir S, Ceyhun S B, Erdogan O, Kaya E D. 2012. Changes in carbonic anhydrase activity and gene expression of Hsp70 in rainbow trout (*Oncorhynchus mykiss*) muscle after exposure to some metals. Turkish Journal of Veterinary and Animal Sciences, 36(5): 499-508.

Soyut H, Beydemir S. 2008. Purification and some kinetic properties of carbonic anhydrase from rainbow trout (*Oncorhynchus mykiss*) liver and metal inhibition. Protein Peptide Letters, 5: 528-535.

Spilling K, Tamminen T, Andersen T, Kremp A. 2010. Nutrient kinetics modeled from time series of substrate depletion and growth: dissolved silicate uptake of Baltic Sea spring diatoms. Marine biology, 157(2): 427-436.

Sutherland E W. 1972. Studies on the mechanism of hormone action. Science, 177: 401-408.

Suzuki S, Yoshida J, Takahashi T. 1996. Effect of testosterone on carbonic anhydrase and Mg^{2+}-dependent HCO_3^- stimulated ATPase activities in rat kidney: comparison with estradiol effect. Comparative Biochemistry and Physiology Part C: Pharmacology, Toxicology and Endocrinology, 114: 105-112.

Suzuki S, Takamura S, Yoshida J, Ozaki N. 1985. Effect of aldosterone antagonists on aldosterone-induced activation of Mg^{2+}-HCO_3^--ATPase and carbonic anhydrase in rat intestinal mucosa. Journal of Steroid Biochemistry, 23(1): 57-66.

Tsuzuki M, Miyachi S. 1989. The function of carbonic anhydrase in aquatic photosynthesis. Aquatic Botany, 34: 85-104.

Vitale A M, Monserrat J M, Castilho P, Rodriguez E M. 1999. Inhibitory effects of cadmium on carbonic anhydrase activity and ionic regulation of the estuarine crab *Chasmagnathus granulata* (Decapoda, Grapsidae). Comparative Biochemistry and Physiology, 122: 121-129.

Wang B L, Liu C Q, Wu Y Y. 2005. Effect of heavy metals on the activity of external carbonic anhydrase of microalga *Chlamydomonas reinhardtii* and microalgae from karst lakes. Bulletin of Environmental Contamination and Toxicology, 74: 227-233.

Weihrauch D, McNamara J C, Towle D W. 2004. Ion-motive ATPases and active, transbranchial NaCl uptake in the red freshwater crab, *Dilocarcinus pagei* (Decapoda, Trichodactylidae). Experimental Biology, 207: 4623-4631.

Wieth J O. 1979. Bicarbonate exchange through the human red cell membrane determined with ^{14}C-bicarbonate. Journal of Physiology, 294: 521-539.

Williams T G, Flanagan L B, Coleman J R. 1996. Photosynthetic gas exchange and discrimination against $^{13}CO_2$ and $C^{18}O^{16}O$ tobacco plants modified by an antisense construct to have low chloroplastic carbonic anhydrase. Plant Physiology, 112: 319-326.

Wu Y Y, Li P P, Zhao X Z. 2007. Effect of fluoride on the activity of carbonic anhydrase and photosynthetic oxygen evolution of *Chlamydomonas reinhardtii*. Fluoride, 40: 51-54.

Wu Y Y, Vreugdenhil D, Liu C, Fu W. 2012. Expression of carbonic anhydrase genes under dehydration and osmotic stress in *Arabidopsis Thaliana* leaves. Advanced Science Letters, 17(1): 261-265.

Wu Y Y, Zhao X Z, Li P P, Huang H K. 2007. Impact of Zn, Cu and Fe on the activity of carbonic anhydrase of erythrocytes in ducks. Biological Trace Element Research, 118: 227-232.

Wu Y Y, Zhao X Z, Li P P, Wang B L, Liu C Q. 2006. A study on the activities of carbonic anhydrase of two species of bryophytes, *Tortula sinensis* (Mull. Hal.) Broth. and Barbula convoluta Hedw. Cryptogamie Bryologie, 27(3): 349-355.

Wu Y Y, Zhao X Z, Li P P. 2010. Impact of estradiol benzoate and progesterone on the activity of carbonic anhydrase of erythrocytes in ducks (*Anatis domesticae* Caro). EPPH, IEEE Press.

Yang M T, Chen S L, Lin C Y, Chen Y M. 2005. Chilling stress suppresses chloroplast development and nuclear gene expression in leaves of mung bean seedlings. Planta, 221: 374-385.

Ying R R, Qiu R L, Tang Y T, et al. 2010. Cadmium tolerance of carbon assimilation enzymes and chloroplast in Zn/Cd hyperaccumulator *Picris divaricata*. Journal of Plant Physiology, 167: 81-87.

Yu S, Zhang X, Guan Q, Takano T, Liu S. 2007. Expression of a carbonic anhydrase gene is induced by environmental stresses in Rice (*Oryza sativa* L.). Biotechnology Letter, 29: 89-94.

第七章　岩溶湖泊中微藻碳酸酐酶的生物地球化学作用

摘　　要

生物化学反应所诱导的生物地球化学循环是地球环境中的重要"驱动力"。碳的生物地球化学循环是其中最为重要的元素循环之一，是其他元素循环的"龙头"，在很大程度上影响着氮、磷、硫等元素的生物地球化学循环。碳酸酐酶在碳的生物地球化学循环过程中起到重要的作用。其不仅强烈的影响着微藻等光合生物对水体可溶性无机碳(DIC)的利用，促进无机碳向有机碳的转化；而且还控制着水体碳酸钙的沉淀和碳酸盐岩的溶解，是水体碳循环及氮磷循环的重要调控因子。

阿哈湖、百花湖和红枫湖具有喀斯特湖泊的典型水化学特征，高 pH、高钙、高 HCO_3^-，溶质主要来源于碳酸盐岩的风化。湖泊里的不同藻类的生长发育受不同的环境因子和营养元素控制。蓝藻受温度控制明显，绿藻受钠离子浓度控制明显，硅藻受温度和钾离子浓度控制明显。同时，对不同岩溶湖泊的叶绿素 a 和光合放氧速率的研究表明，阿哈湖、红枫湖和百花湖可以分成两种类型，红枫湖和百花湖受到总氮的控制，阿哈湖受总磷的控制。

不同湖泊浮游植物的组成明显的不同，它们随着时间的变化而变化。而碳酸酐酶活力变化大且没有规律，并且与水体的营养元素没有较好相关性。但是，其与水体微藻的组成和种类具有一定的相关关系。阿哈湖中，总藻密度与蓝藻、绿藻显著相关，碳酸酐酶活力与总藻和蓝藻的密度成显著的正相关；百花湖中，总藻密度同样与蓝藻、绿藻显著相关，碳酸酐酶活力与绿藻的密度成显著的线性相关；红枫湖中，总藻密度与蓝藻、绿藻、硅藻显著相关，碳酸酐酶活力也与绿藻的密度成显著的线性相关。此外，湖泊微藻的碳酸酐酶的活力与该湖的微藻的物种丰富度有关。

水体无机碳和微藻胞外碳酸酐酶活力是影响岩溶湖泊微藻稳定碳同位素组成变化的决定性因素。不同季节岩溶湖泊微藻的碳汇能力和碳汇机制都不一样。微藻的间接碳汇和直接碳汇能力都在夏季处于较高水平；而在冬季处于较低水平。胞外碳酸酐酶主要促进微藻对大气碳源的利用。因此，微藻的胞外碳酸酐酶主要贡献于直接碳汇能力。同时，温度与水体微藻的直接和间接碳汇能力都有较好的正相关关系，其可能是岩溶湖泊水体微藻碳汇能力的主要限制性因素。这暗示我们，微藻将对由大气 CO_2 浓度升高而导致的全球温度升高，有一个相应的反馈机制，而碳酸酐酶在其中起到重要的作用。

Chapter 7 Biogeochemical action of microalgal carbonic anhydrase in karst lakes

Abstract

The biogeochemical cycle induced by biochemical reaction is an important "driver force" in the earth's environment. The biogeochemical cycle of carbon is one of the most important elements cycle; and it is the "leader" of other elements cycle and influences the biogeochemical cycle of nitrogen, phosphorus, sulfur and other elements to a great extent. The carbonic anhydrase plays an important role in the carbon biogeochemical cycle. It not only affects strongly the dissolved inorganic carbon (DIC) utilization by microalgae and other photosynthetic organism, and promotes the transformation from inorganic carbon in to organic carbon, but also controls the precipitation of calcium carbonate and the dissolution of carbonate rocks, and is an important regulatory factor in carbon, nitrogen and phosphorus cycle.

Aha, Baihua and Hongfeng Lakes have typical water chemical characteristics with high pH, high calcium, and high bicarbonate in Karst lakes, and their main source of the solute come from the weathering of carbonate rocks. The growth of the different algal species is controlled by various environmental factors and nutrient elements. *Cyanophyta* is controlled by temperature, *Chlorophyta* by sodium ion, and *Bacillariophyta* by temperature and the concentration of potassium ion, respctively. In addition, algal chlorophyll-*a* and photosynthetic rate in the different Karst lakes shows that Aha, Baihua and Hongfeng Lakes can be divided into two types. Baihua and Hongfeng Lakes are controlled by the total nitrogen and Aha Lake is under the control of the total phosphorus.

The compositions of phytoplankton were significantly different and changed with time in different lakes. The variations of carbonic anhydrase activity were great and fluctuate, and carbonic anhydrase had not a good correlation with the nutrient elements in water bodies. Carbonic anhydrase had a certain correlation with microalgal composition and species in the lake. The total algae density was significant correlation with *Cyanobacteria* and *Chlorophyta*, and carbonic anhydrase activity had a significant positive correlation with the density of algae and *Cyanobacteria* in Aha Lake. The total algae density was also significant correlation with *Cyanobacteria* and *Chlorophyta*, and carbonic anhydrase activity had a significantly linear correlated with the density of *Chlorophyta* in Baihua Lake. The total algae density was significant correlation with *Cyanobacteria*, *Chlorophyta* and *Bacillariophyta*, and carbonic anhydrase activity also had a significant linear correlation with the density of *Chlorophyta* in Hongfeng Lake. In addition, the carbonic anhydrase activity was related to the microalgal species richness in the lake.

Both the DIC and the activity of extracellular carbonic anhydrase (CAex) were the key factors influencing the stable carbon isotopic composition in microalgae. The carbon sink ability and carbon sequestration mechanism of microalgae in Karst lakes are different among various seasons. The direct and indirect carbon sink ability remains at a high level in summer, while a low level in winter. CAex mainly promote the utilization of inorganic carbon from atmosphere by microalgae. Therefore, the CAex of microalgae has a major contribution to the direct carbon sink. Meanwhile, temperature, which may be the main restrictive factors of the microalgal carbon sink ability in Karst lakes, has a good positive correlation with the direct and indirect carbon sink of unit water. It suggests that microalgae have a corresponding feedback mechanism in response to global temperature increase caused by the elevated of atmospheric CO_2 concentration and carbonic anhydrase plays an important role in it.

第一节　碳酸酐酶与岩溶水环境中的碳循环

地球环境的重要"驱动力"是调节所有的生物化学反应的生物地球化学循环。碳的循环是其中最为重要的元素循环之一。全球碳循环的通量反应始于光合作用。光合作用所固定的碳每年有99.95%被呼吸作用氧化，只有0.05%未被氧化而埋藏在沉积物中。光合作用是驱动碳循环的主动力。光合作用是绿色植物将水和二氧化碳同化成糖释放氧气的过程。二氧化碳是全球碳循环的纽带。浮游植物是水环境的主要生产者(刘建康，1999)。海洋中的生物量尽管只占地球上的1%，但它们的光合作用约占全球的50%。在水体中，碳以有机态和无机态两种形式存在。无机碳有CO_2、HCO_3^-、CO_3^{2-}等形态，这些无机碳形态在特定的生态环境下建立了特定的平衡。有机碳包括糖类、脂类、蛋白质、木质素等。水体中的自养生物利用无机碳通过光合作用同化成有机碳。异养生物通过食物链直接或间接地利用这些有机碳，使有机碳破坏、分解乃至矿化成无机碳。这种由自养生物（主要为绿色植物）把无机碳合成为有机碳，又由异养生物把有机碳分解成无机碳的过程就构成了水环境中的碳循环。

浮游植物的光合作用是水体中碳循环的关键过程。行使植物的光合作用的核酮糖-1,5-二磷酸羧化酶/加氧酶（Rubsico）需要无电荷的 CO_2（可以自由地出入细胞膜）。而在水体中，只有1%的无机碳是以 CO_2 形式存在，而99%的无机碳是以 HCO_3^- 形式存在。少量的 CO_2 形式的无机碳不足以满足植物的光合作用，因此，植物进化了一种将 HCO_3^- 转变成到细胞内 CO_2 的功能，这种功能就是碳酸酐酶（CA）的作用。碳酸氢根转化成水和二氧化碳的可逆反应，在无催化剂的条件下，达到反应平衡需要一分钟，这显然与光合作用的快速需要 CO_2 是不适应的。但有碳酸酐酶的作用下，这个反应达到平衡只需 10^{-6} s。水体浮游植物就是靠碳酸酐酶的作用来解决水体 CO_2 的限制问题，实现碳的同化和固定。

碳酸酐酶是光合作用的重要酶。该酶是一种诱导酶，在无机碳缺乏时，该酶被诱导出以适应水生环境（Ramazanov et al.，1993）。该酶受到环境的 pH、光质、无机和有机离子影响。它一方面因受环境因子对植物的生长发育及种群变化的间接影响而发生酶活力和同工酶的变化，另一方面也受到环境因子的直接影响。无机和有机阴离子能强烈地抑制碳酸酐酶活力（Hatch，1991）；碳酸酐酶活力因光照增加100倍，低光照度具有低活力的碳酸酐酶（Maribel et al.，1989a，b）；氮的缺乏会降低碳酸酐酶的活力，同时，可通过增加硝酸盐而慢慢回复到原活力（Burnell et al.，1990）；在酸性条件下生长的椭圆小球藻的碳酸酐酶的活力明显比在碱性条件下生长的椭圆小球藻的碳酸酐酶的活力要小（Rotatore and Colman，1991）；金属离子如 Zn、Cu、Cd、Mn、Co、Fe 也影响碳酸酐酶的活力（Agarwala et al.，1995；Cullen et al.，1999；Lane and Morel，2000；Lyszcz and Ruszkowska，1992；Yee and Morel，1996；Sas et al.，2006；Soyut et al.，2012；Wang et al.，2005；Wu et al.，2007；Sharma et al.，2009）。HCO_3^-、醋酸盐能调节该酶的活力（见第三章以及 Fett and Coleman，1994；Laurent et al.，2000；Merrett et al.，1996；Nimer et al.，1997）。氟、氯、溴离子以及磷酸盐等很多阴离子也对碳酸酐酶起着"激素"作用（见第六章以及 Baird et al.，1997；Wu et al.，2007；Li et al.，2013）。甘氨酸、葡萄糖和其他有机碳对碳酸酐酶的诱导有抑制作用（Shiraiwa and Umino，1991；Sugiharto et al.，1992b；Umino and Shiraiwa，1991）。植株缺 Na，叶片中的碳酸酐酶的活力是对照的两倍（Brownell et al.，1992）；植物激素也严重地影响酶的诱导和酶的活力，尤其是细胞分裂素可明显地促进碳酸酐酶的 mRNA 的表达（Sugiharto et al.，1992a）。温度和氧化-还原电位等对该酶的活力和诱导也有影响（Moubarak and Stemler，1994；Chaki et al.，2013）。这些情况表明水体中的环境强烈地影响碳酸酐酶的活力，水体环境因子通过影响碳酸酐酶来影响微藻的光合作用，从而影响水体碳循环。

碳酸酐酶不仅受水体环境的影响，也影响着水体环境。它的活力大小以及同工酶类型不仅间接地影响浮游生物的种群生态关系,而且直接地影响水体的pH、无机碳的组成和含量、阴阳离子的含量和组成、氧化-还原电位以及钙化作用(Cang and Roberts, 1992; Gao et al., 1999; Merrett et al., 1996; Moroney and Somanchi, 1999; Moubarak and Stemler, 1994; Rotatore and Colman, 1991; Shiraiwa et al., 1993; Beardall et al., 1998; Yin et al., 1993; Morel and Price, 2003; Raven et al., 2011)。所以,碳酸酐酶是水体碳循环的一个重要因子,是生物地球化学作用的一个重要标志,也是水体环境的"指示剂"。

水环境中 $CO_2+H_2O \leftrightarrow H^+ + HCO_3^-$ 反应是碳酸盐岩风化的第一步。岩溶水环境与碳酸酐酶的相互关系还体现在碳酸盐风化与碳酸酐酶的关系上。水体环境中的 $CO_2+H_2O \leftrightarrow H^+ + HCO_3^-$ 反应受到温度、P_{CO_2}、pH、Eh、HCO_3^- 等的影响,同时也影响到这些值的变化。外界环境对碳酸盐风化有显著影响(Blum et al., 1998; Roy et al., 1999; 刘再华, 2000, 2001; 刘丛强等, 2008; Li et al., 2010; Coogan and Gillis, 2013)。碳酸酐酶也能影响碳酸盐的风化(Liu and Dregbroat, 2001; Xie and Wu, 2014),碳酸酐酶通过影响碳酸盐风化促进 $CO_2+H_2O \leftrightarrow H^+ + HCO_3^-$ 反应速度,由此影响到水体环境中的 P_{CO_2}、pH、Eh、HCO_3^- 等指标。碳酸盐岩中的碳约占全球总碳量的99.55%,是地球上最主要的碳库。碳酸盐岩的风化不仅直接关系到水体的碳供给量而影响水体中碳通量的平衡,而且还可间接回收大气圈中的二氧化碳而影响温室气体的变化(Berner et al., 1983; Sarmiento and Sundquist, 1992; Liu and He, 1998; 刘再华等, 2000; 严国安等, 2001; 翁金桃, 1995; Taylor et al., 2012)。

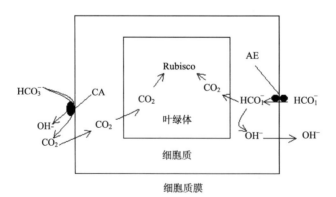

图 7-1 藻类无机碳利用方式的示意图

仿 Axelsson et al., 1999

Fig. 7-1 The schematic on the way of inorganic carbon utilization by Algae

Modified from Axelsson et al., 1999

水体生物具有广泛的多样性。浮游藻类众多(刘建康,1999),各种藻类对无机碳的利用方式不一样(Axelsson et al.,1999),这种多样性的营养方式带来了对环境的适应性。*Chlorella saccharophila* 和 *Chlorella ellipsoidea* 同属绿藻属植物,但对CO_2浓度的反应不同(Matsuda and Colman,1996)。*Chlorococcum littorale* 可以忍受高浓度CO_2浓度(Pesheva et al.,1994)。这种无机碳利用方式不同,一定程度上与有无碳酸酐酶,或碳酸酐酶的类型有关。在浮游藻类中,胞外碳酸酐酶在无机碳的转换中起着重要的作用。HCO_3^-带负电荷,在脂质中溶解性差,而CO_2可自由的溶解在脂质和水相中,对于脂质的细胞膜,CO_2进出自由,HCO_3^-转移较差。对于高效光合作用的藻类,进化了胞外碳酸酐酶(细胞质膜外),无疑可以起到CO_2浓缩作用。所以,对于藻类来说,无机碳可以以HCO_3^-直接进入细胞膜,也可以以CO_2方式直接进入细胞膜内,还可以通过胞外碳酸酐酶将HCO_3^-转换成CO_2进入细胞膜内。有无胞外碳酸酐酶就成为藻类利用环境中无机碳的限制因子之一(图7-1)。

HCO_3^-与藻类钙化作用紧密相关,从而制约着碳酸盐的沉淀。外源HCO_3^-是藻细胞钙化作用的无机碳来源(Dong et al.,1993,Leclercq et al.,2000;Budenbender et al.,2011)。水体中含有的Ca^{2+}对钙化作用和光合作用的CO_2利用都有显著的影响(McConnaughey and Falk,1991;缪晓玲,1998;Gattuso et al.,1998)。水体中含有过饱和的$CaCO_3$,当胞间CO_2浓度因生物光合作用而减少时,引起$CaCO_3$聚晶沉淀。其化学方程式:

$$2HCO_3^- + Ca^{2+} = CaCO_3\downarrow + CO_2\uparrow + H_2O$$

除了HCO_3^-、Ca^{2+}与CO_2对藻类钙化也有影响外,pH对藻体的钙化也有影响(McConnaughey and Falk,1991;Gao and Zheng,2010;Semesi et al.,2009)。在 *Halimeda tuna* 中,光合作用固定碳元素的速度超过一定量时就会发生$CaCO_3$沉淀。超过这个量,光合作用固定碳元素的速度与钙化作用速度呈线性关系。每固定$4\sim 8mol\ CO_2$,就沉淀$1mol\ CaCO_3$。介质中移去使介质碱化,促进了$CaCO_3$的沉淀。当介质中pH=6.3时,光合作用速度高而钙化作用低;pH=6.5~8.2时,钙化作用稳定;pH再增加时,光合作用速度低但钙化作用速度增加。碳酸盐岩的形成也与生物的钙化作用有关(Knoll et al.,1993;Riding,2000;Altermann et al.,2006)。

水体碳循环对其他元素的循环有着很大的影响。水体中存在NO_3^-、NH_4^+、$H_2PO_4^-$、PO_4^{3-}、SO_4^{2-}、SO_3^{2-}、S^{2-}等,这些是水生植物和细菌的营养。氮的循环中最重要的是生物固氮和硝酸盐还原以及亚硝酸盐还原作用。碳循环尤其是光合作用对它们的影响很大。光合作用通过提供能量如$NADPH_2$、$NADH_2$、ATP和中间代谢物如α-酮戊二酸、谷氨酸、铁氧还蛋白还原等来影响氮循环(曹宗巽和吴相钰,1978)。无机碳的利用显著地影响硝酸盐的还原和铵的代谢(Larsson et al.,

1985)。蓝藻类很多种都具有融固氮作用和 CO_2 浓缩机制于一身(Badger et al.，2006;Kranz et al.，2011)。生物对硫循环具有重大影响,由于它们对硫或其化合物的转换可以影响到水环境中的 pH、Eh、碳酸盐含量、氮磷元素含量,从而影响了整个水体生态系统的代谢。硫酸盐还原在水体硫循环中起着极其重要的作用。光合作用对硫酸盐还原影响极大,首先体现在硫酸盐还原需要在叶绿体中进行,另外还需要光合作用产生的 ATP 和铁氧还蛋白还原等。水体磷的循环主要由两个环节控制(Carpenter et al.，1992;阎葆瑞和张锡根,2000;Shilton et al.，2012),一是生物对磷的利用,它包括植物利用磷酸盐的磷变成有机磷。动物依靠植物获得它们所需要的磷;另一个环节就是水体环境下磷酸盐的溶解度。光合作用影响植物磷的利用,细胞的微环境 pH、Ca^{2+} 对磷酸盐的溶解度有显著影响。水体碳循环是其他元素循环的"龙头"。碳酸酐酶通过影响碳循环来影响其他元素的循环。

水体是水生生物的生存环境。水体营养元素对水生生态系统有严重影响。碳循环是"龙头",影响碳循环的碳酸酐酶也因此影响水生生态系统。水体中的营养物质(如氮、磷)是浮游植物生长、发育、繁殖阶段所必需的。但是,如果无机营养物尤其是氮、磷等大量进入湖泊、海湾等相对封闭、水流缓慢的水体,引起藻类和其他水生植物大量繁殖,水体溶解氧下降,水质恶化,其他水生生物大量死亡的现象,造成赤潮、水华现象。在适宜的光照、温度、pH 和充足营养物质的条件下,天然水体中藻类通过光合作用合成自身的原生质,其基本反应式为:$106CO_2 + 16NO_3^- + HPO_4^{2-} + 122H_2O + 18H^+ \rightarrow C_{104}H_{263}N_{16}P_1 + 138O_2$。反应式中的微量元素是指镁、锌、钼、钒、硼等元素的化合物。从反应式可看出,在藻类繁殖所需要的各种成分中,碳、氮、磷是比较重要的营养元素。藻类可以利用水中溶解的二氧化碳和有机物分解产生的二氧化碳作为自身生长所需要的碳源,而氮和磷就成为限制性因素。所以藻类的生长繁殖主要决定于水体中这两种成分的含量。在一般情况下,水体中藻类可利用的氮远比可利用的磷多,因此,磷的含量通常被作为富营养化的标志。蓝藻过多繁殖是许多水体造成富营养化的一个重要因子。而蓝藻中有一些种类如 Anabaena sp 含有 CO_2 浓缩机制(Price and Badger,1991;Beardall and Giordano,2002;Price et al.，2008;Ramanan et al.，2012),增加 CO_2 的利用,同时也增加了氮、磷的利用。调节蓝藻的 CO_2 浓缩能力(调节碳酸酐酶活力),可以协调藻类对氮、磷的利用,从而干预水体的富营养化过程。通过调节碳酸酐酶活力,还可调节光合作用,从而协调植物的生长发育,协调水体植物群落的发育和水体生态环境,也可以达到干预水体富营养化进程的作用。

第二节 岩溶湖泊水环境特征

西南喀斯特地区面积约有 55 万平方公里,是世界三大著名的喀斯特集中连片分布的地区之一(Han and Liu, 2004;刘丛强,2007)。贵州省是西南喀斯特的中心,贵阳周边著名的"两湖一库"是学者研究喀斯特岩溶湖泊的热点区域(Wan et al., 2008;Wang et al., 2005;Wang et al., 2012;Wu et al., 2008a;Wu et al., 2008b)。喀斯特岩溶地区的岩性主要是是石灰岩和白云岩,其主要成分是 $CaCO_3$ 或 $MgCa(CO_3)_2$ 等,它们都属于易受侵蚀的类型,相应地,喀斯特岩溶地区的无机碳迁移转化活动强烈,因此,它在全球气候变化方面也具有重要意义。

岩溶湖泊水体 pH 和 HCO_3^- 浓度较高,湖泊中的微藻 CAex 活力存在着较大的时间和空间差异(Wu et al., 2008a),因此,微藻是研究微藻 CAex 与藻体碳同位素组成的天然好材料。本研究从湖泊水体中的碳酸氢根离子含量、微藻碳酸酐酶胞外酶活力及藻体稳定碳同位素组成等方面出发,来寻找影响微藻碳同位素组成的关键因子,探讨微藻碳同位素组成变化与环境的相关性。通过双向同位素示踪技术,研究岩溶湖泊微藻的碳源利用策略及其对碳汇的影响,评价各湖泊的碳汇能力,探讨湖泊微藻碳汇能力的影响因素,揭示岩溶湖泊微藻碳酸酐酶的生物地球化学机制。

一、岩溶湖泊水文及物理特征

红枫湖和百花湖位于云贵高原中部,是两个相连的梯级多功能人工湖泊(水库)(图 7-2)。红枫湖是贵州最大的人工河道水库之一,属于长江流域的乌江水系,百花湖是红枫湖的下一级水库,以红枫湖的下泻水为主要补给;两湖相距 10km,具有相同的地质背景和气候条件。所在区域以三叠纪白云岩为主,岩溶作用发育,石灰土和黄壤广布,土层浅薄;具有石漠化景观。该流域属亚热带湿润季风气候,全年气候温和,雨量充沛。三湖水文特征见表 7-1。

表 7-1 红枫湖、百花湖和阿哈湖水文特征

Tab. 7-1 Hydrological characteristics of Hongfeng, Baihua and Aha Lake

湖泊名称	水位 /m	汇水面积 /km²	水面面积 /km²	补给系数	平均水深 /m	最大水深 /m	库容 /亿 m²	寄宿时间 /a	平均温度 /℃
阿哈湖	1108	190	3.4	55.9	13.2	24	0.45	0.44	15.3
红枫湖	1240	1578	32.2	49	9.25	50	2.98	0.32	14.1
百花湖	1189	1915	14.5	132	7.72	40	1.12	0.10	13.8

阿哈湖位于贵阳西南(图 7-2),是一个底层滞水带季节性缺氧的人工水库;为贵阳市饮用水源,兼有蓄水和防洪功能。阿哈湖汇水区域主要分布有二叠系灰岩及煤系地层,其上发育硅铝质和硅铁质黄壤;部分地区有三叠系碳酸盐岩及泥页岩出露,并发育着黑色、黄褐色石灰土。湖周围为疏林植被,灌木丛、灌草丛较多。

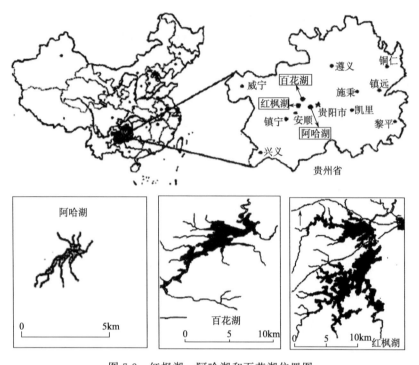

图 7-2 红枫湖、阿哈湖和百花湖位置图

Fig. 7-2 Location map of Hongfeng, Baihua and Aha Lake

二、岩溶湖泊水化学特征

(一) 岩溶湖泊水温随时间变化的情况

湖泊水温变化直接影响着湖泊水生生物的生长和繁殖，引起湖泊水体对流，造成物质迁移流场的改变。湖泊水温的高低是外界自然气候综合效应的结果，水温的变化反映出水体对太阳辐射吸收的强弱。太阳辐射强度表现出季节性差异，湖泊水温也表现出季节性的变化。图 7-3 显示的是 2010 年 5 月至 2011 年 4 月阿哈湖、百花湖和红枫湖水温的变化情况。三个湖泊水温最高(26～27℃)的月份都出现在 2010 年 7～8 月，之后的月份中水温都在下降。在 2011 年 2 月三个湖泊的水温都出现了最低值(7～8℃)，之后水温开始回升。

(二) 岩溶湖泊湖水 pH 随时间变化的情况

湖泊 pH 对水体中各类物质的迁移和转化过程具有控制作用。湖泊中元素的溶解沉淀、吸附解析、迁移转化都受到 pH 的影响。pH 对湖泊初级生产力也有

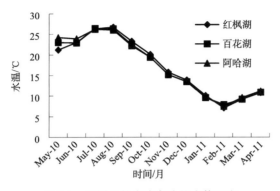

图 7-3 不同月份岩溶湖泊的水体温度

Fig. 7-3 Temperature of the karst lakes at different months

影响。图 7-4 显示的是 2010 年 5 月至 2011 年 4 月阿哈湖、百花湖和红枫湖 pH 的变化情况。除了 2010 年 9 月和 10 月(7.2~7.3),其他时间红枫湖水体的 pH 范围在 8.2~8.8 内,pH 波动不大。而百花湖和阿哈湖水体的 pH 最大值(8.5~9.2)出现在 2010 年 5 月和 6 月。阿哈湖水体的 pH 在 2011 年 4 月份 pH 也出现较大值(8.6)。三个湖泊水体 pH 最小值(7.2)均出现在 2010 年 9 月和 10 月,之后整体上三个湖泊水体的 pH 变化不大(8~8.7)。

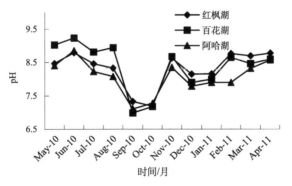

图 7-4 不同月份岩溶湖泊的水体 pH

Fig. 7-4 pH of the karst lakes at different months

(三) 岩溶湖泊湖水电导率随时间变化的情况

电导率与温度具有很大相关性。金属的电导率随着温度的增高而降低。半导体的电导率随着温度的增高而增高。水溶液的电导率高低取决于其内含溶质盐的浓度,或其他会分解为电解质的化学杂质含量。水样本的电导率是测量水的含盐成分、含离子成分、含杂质成分等的重要指标。水越纯净，电导率越低(电阻率越高)。在一段温度值域内，电导率可以被近似为与温度成正比。图 7-5 显示的是阿哈湖、百花湖和红枫湖电导率的变化情况。红枫湖和百花湖电导率的最小值($33\sim36\text{mS/m}$)出现在 2010 年 7 月份和 8 月份，最大值($42\sim46\text{mS/m}$)出现在 2011 年 3 月份和 4 月份。而阿哈湖的电导率要明显高于红枫湖和百花湖。阿哈湖电导率的最小值($53\sim56\text{mS/m}$)出现在 2010 年 6 月份和 7 月份，最大值(67.6mS/m)出现在 2011 年 3 月份。三个湖泊水温的最高值均出现在 2010 年 7 月和 8 月，而此时的其电导率处在最小值。当三个湖泊的电导率最大值出现在 3 月份时，其水温较低。说明三个湖泊水温和水体的电导率存在反比例关系。

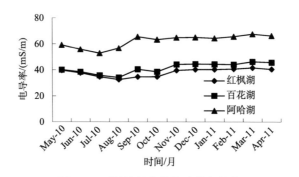

图 7-5 不同月份岩溶湖泊的电导率

Fig. 7-5 Conductivity of the karst lakes in different months

（四）岩溶湖泊阴阳离子年际变化

我们收集和测定了多年间红枫湖、阿哈湖和百花湖的主要阴阳离子浓度数据（各个数值为平均值 $n>10$），进行统计分析，结果见表 7-2 和 7-3。

表 7-2 岩溶湖泊各阴离子年际变化

Tab. 7-2 Interannual variability of anion concentration in karst lakes

（单位：mg/L）

湖泊	年份	HCO_3^-	SO_4^{2-}	NO_3^-	F^-
红枫湖	1996 年前	143.7±74.38	46.1±34.48	1.72±1.32	0.76±0.32
	2001~2002	140.3±87.65	65.43±38.52	1.01±1.43	0.29±0.14
	2012~2014	136.03±19.48	81.18±7.99	2.49±1.41	0.18±0.03
百花湖	1996 年前	139.3±78.56	97.3±16.37	2.2±2.42	0.76±0.41
	2001~2002	73.69±65.43	73.72±12.37	1.23±1.32	0.14±0.08
阿哈湖	2012~2013	133.59±22.14	147.57±15.65	3.37±1.52	0.23±0.04

表 7-3 岩溶湖泊各阳离子年际变化
Tab. 7-3 Interannual variability of cation concentration in karst lakes

(单位：mg/L)

湖泊	年份	Ca^{2+}	Mg^{2+}	K^+	Na^+
红枫湖	1996 年前	55.6±38.34	8.16±2.12	1.66±0.02	1.73±0.02
	2001~2002	57.24±44.02	11.12±2.43	2.9±0.03	1.86±0.02
	2012~2014	51.81±7.45	16.27±1.31	2.20±0.31	6.81±1.43
百花湖	1996 年前	48.24±27.56	12.79±0.87	0.64±0.04	4.42±0.06
	2001~2002	60.51±32.46	10.18±0.84	2.78±0.04	2.55±0.05
阿哈湖	2012~2013	67.64±7.00	19.94±3.12	4.17±1.71	9.41±1.26

从表 7-2 和表 7-3，可以看出不同年份岩溶湖泊水体的主要阴阳离子都有一定的变化。其中，红枫湖水体中的 F^- 随着时间的推移不断降低，而 Na^+、Mg^{2+} 和 SO_4^{2-} 离子不断增加。同时，我们也可以发现，在 2001~2002 年，红枫湖和百花湖水体中的 NO_3^- 明显降低，这主要是由于在 1999 年当地政府实施了以削减氮、磷入湖排放量为目的的"两湖"综合治理一期工程（吴沿友等，2004）。

（五）红枫湖和阿哈湖溶质来源示踪

阿列金提出按天然水中主要阴离子和阳离子的类别以及离子间毫克当量比例的差异为原则进行分类。按主要阴离子分成三类：重碳酸盐-碳酸盐水、硫酸盐水及氯化物水；每一类再按主要阳离子又可分成 Ca 组、Na 组和 Mg 组；每组又按离子间的毫克当量分成四个水型：Ⅰ $[HCO_3^-]>[Ca^{2+}]+[Mg^{2+}]$；Ⅱ $[HCO_3^-]<[Ca^{2+}]+[Mg^{2+}]<[HCO_3^-]+[SO_4^{2-}]$；Ⅲ $[HCO_3^-]+[SO_4^{2-}]<[Ca^{2+}]+[Mg^{2+}]$ 或 $[Cl^-]>[Na^+]$；Ⅳ $[HCO_3^-]=0$（刘建康，1999；朱颜明和何岩，2002）。阿哈湖（AH）和红枫湖（HF）湖水的主要离子组成变化见图 7-6。从图中可以看出，阿哈湖和红枫湖的主要阳离子为钙离子；而红枫湖的主要阴离子为 HCO_3^-，阿哈湖的主要阴离子为 HCO_3^- 和 SO_4^{2-}。红枫湖主要溶质的质量浓度顺序分别为：$[Ca^{2+}]>[Mg^{2+}]>[K^+]>[Na^+]$，$[HCO_3^-]>[SO_4^{2-}]>[Cl^-]>[NO_3^-]$；阿哈湖主要溶质的质量浓度顺序分别为 $[Ca^{2+}]>[Mg^{2+}]>[K^+]>[Na^+]$，$[SO_4^{2-}]>[HCO_3^-]>[Cl^-]>[NO_3^-]$，但是在冬季随着 HCO_3^- 浓度的增加，出现了 $[HCO_3^-]>[SO_4^{2-}]>[Cl^-]>[NO_3^-]$。按阿列金天然水分类原则，红枫湖属于重碳酸盐类，钙组，三型水；阿哈湖属于硫酸盐类，钙组，三型水。

湖水的化学组成由湖水的来源（河流输入和大气降水等）、湖水区域的水/岩相互作用、人为活动的影响共同决定。由图 7-6 可知，湖水化学组成主要受河流输入影响。通常可用溶质元素比率和同位素比率的相关性，或是溶质的离子平衡来鉴定

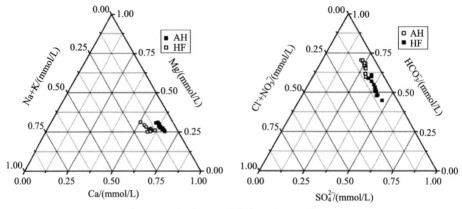

图 7-6　红枫湖和阿哈湖主要离子组成变化

Fig. 7-6　Main ion composition changes of Hongfeng Lake and Aha Lake

溶质的来源(Xu and Liu, 2007)。在此利用 Na^+/Ca^{2+} 对 Mg^{2+}/Ca^{2+}(物质的量浓度比)作图(图 7-7),可以发现,红枫湖的 Na^+/Ca^{2+} 和 Mg^{2+}/Ca^{2+} 分别为:0.169~0.312 和 0.422~0.616;阿哈湖的 Na^+/Ca^{2+} 和 Mg^{2+}/Ca^{2+} 分别为:0.201~0.314 和 0.372~0.515。然而,沿途主要出露的灰岩、白云岩和硅酸盐岩的 Na^+/Ca^{2+} 和 Mg^{2+}/Ca^{2+} 分别约为:0.005 与 0.02、0.01 与 1、0.8 与 5。因此,红枫湖和阿哈湖的溶质至少存在三个风化来源:石灰岩、白云岩和硅酸盐岩。

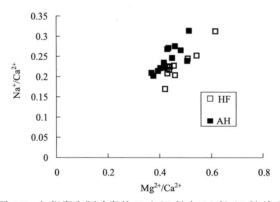

图 7-7　红枫湖和阿哈湖的 Na^+/Ca^{2+} 与 Mg^{2+}/Ca^{2+} 关系

Fig. 7-7　Relation of Na^+/Ca^{2+} and Mg^{2+}/Ca^{2+} in Hongfeng Lake and Aha Lake

为了进一步区分碳酸盐岩和硅酸盐各自对湖泊溶质的贡献,通过 Ca^{2+}、Mg^{2+}、Na^+、K^+、HCO_3^-、SO_4^{2-}、Cl^- 之间的比值关系来进行判定。假设来源于蒸发岩溶解或大气输入或硫化物氧化的 SO_4^{2-} 仅仅只用于 $Ca^{2+}+Mg^{2+}$ 平衡,那么 $[Ca^{2+}+Mg^{2+}]*([Ca^{2+}+Mg^{2+}]*=[Ca^{2+}+Mg^{2+}]-[SO_4^{2-}])$ 就来源于碳酸

盐岩或硅酸盐岩风化。因此，$[Ca^{2+}+Mg^{2+}]*/[HCO_3^-]$比值代表了碳酸盐岩和硅酸盐岩风化的$Ca^{2+}$和$Mg^{2+}$的相对含量，其值应小于1。同样地，水样中的$[Na^++K^+]*$（$[Na^++K^+]*=[Na^++K^+]-Cl^-$）应该等于碳酸盐岩和硅酸盐岩的风化。水样中的$[Na^++K^+]*/[HCO_3^-]$与$[Ca^{2+}+Mg^{2+}]*/[HCO_3^-]$的变化反映了碳酸盐岩和硅酸盐岩风化对河水溶质化学组成的相对贡献(Han and Liu, 2004)。如图 7-8 所示，大多数的样品点都落在$[Ca^{2+}+Mg^{2+}]*/[HCO_3^-]=1$和$[Na^++K^+]*/[HCO_3^-]=0$比值线的附近。因此，碳酸盐岩溶解阿哈湖和红枫湖的主要溶质来源。而这两个湖泊相比，红枫湖受硅酸盐岩溶蚀的影响较大。

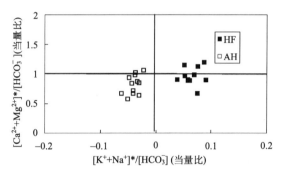

图 7-8　湖泊$[Ca^{2+}+Mg^{2+}]*/[HCO_3^-]$与$[Na^++K^+]*/[HCO_3^-]$的关系

Fig. 7-8　Relation of $[Ca^{2+}+Mg^{2+}]*/[HCO_3^-]$ and $[Na^++K^+]*/[HCO_3^-]$ in Hongfeng Lake and Aha Lake

三、岩溶湖泊微藻的生长特征

（一）岩溶湖泊水环境对水体微藻的生长影响

岩溶湖泊水环境的多样性。岩溶湖泊水环境具有多样性，水环境中的元素含量和气候因子对藻类生长有着显著的影响。以红枫湖、百花湖、阿哈湖为研究对象，借助 SPSS 统计分析软件对水环境中的元素含量、气候因子、藻类进行相关分析和曲线拟合，来研究岩溶湖泊水环境的受制因素和水环境对藻类生长的影响。结果表明，NO_3^-与水环境的其他元素无相关关系，其他元素两者都具有显著地相关，说明湖泊水体的化学组成受流域的岩石和土壤侵蚀所控制。蓝藻随着温度的升高，生长速度加快，绿藻的生长为钠离子所抑制，硅藻的生长受到温度和钾离子的促进，总藻随温度的升高密度加大。不同的藻类的生长发育受不同的环境因子和营养元素控制(吴沿友等，2003)。

湖泊中化学组成的相关性。把湖泊中的水体中的元素含量用 SPSS 软件进行相关性分析，除了NO_3^-含量与其他元素不相关外，其他元素两者都具有显著地相关，说明湖泊水体的化学组成受SO_4^{2-}、Cl^-、HCO_3^-、K、Na、Ca、Mg 的控制,也就是说,岩溶湖泊水环境主要受流域岩石侵蚀和酸雨等因素控制,而在一定范围内,不受人为的 NO_3^- 的控制(吴沿友等,2003)。

岩溶湖泊水环境对藻类生长的影响。以各元素、温度和降雨量为自变量,藻的密度为因变量进行逐步回归,结果表明,水体藻的密度与温度极显著相关,随着温度的升高,藻的生长速度加快。蓝藻的密度与温度极显著相关,随着温度的升高,蓝藻的生长速度加快。绿藻的密度与钠离子浓度极显著相关,随着钠离子浓度升高,绿藻的生长速度减慢。硅藻的密度与温度和钾离子浓度极显著相关,随着温度和钾离子浓度的升高,硅藻的生长速度加快。从以上结果可以看出,水环境的各因子对不同的藻类影响的程度不同。水体整个生态系统的藻类生长受外界环境协同控制,了解不同的藻类生长的不同条件,有利于富营养化、"水华"的防治。在岩溶湖泊,藻类生长不受流域岩石的溶蚀和土壤的侵蚀影响,仅仅受到钠、钾和温度的影响,这对岩溶湖泊水环境的保护具有指导作用,同时对于其他水体环境的保护有借鉴作用。

(二)岩溶湖泊叶绿素 a 和光合放氧能力

对岩溶湖泊的叶绿素 a 和光合放氧速率的研究表明,阿哈湖、红枫湖和百花湖可以分成两种类型,红枫湖和百花湖受到总氮的控制,阿哈湖受总磷的控制。在阿哈湖中,叶绿素 a 与总磷有显著的正相关,净光合氧的释放速率随着总磷的变化而变化(表 7-4)。在红枫湖和百花湖中,总氮、叶绿素 a 与净光合氧的释放速率具有显著的相关性(表 7-5,表 7-6)。叶绿素 a 与净光合氧的释放速率随着总氮的变化而变化。水体的营养状态特别是总氮和总磷决定着湖泊微藻的生物量。

表 7-4 阿哈湖中总氮(TN)、总磷(TP)与微藻的叶绿素 a、
净光合速率的相关系数($n=24$)

Tab. 7-4 Correlation coefficient of TN, TP, microalgal chl a,
net photosynthetic rate in Aha Lake($n=24$)

变量	TN/(mg/L)	TP/(mg/L)	Chl$a$/(μg/L)	$Pn1$[②]/(nmolO_2 h^{-1}μgchla^{-1})
TP	−0.16			
Chl$a$	−0.24	0.74[①]		
$Pn1$[②]	−0.18	0.30	0.37	
$Pn2$[②]	−0.18	0.11	0.08	0.92[①]

① 0.01 水平下的显著性(2-tailed)。

②$Pn1$:150μmol/(m²·s)光下的净光合速率;$Pn2$:500 μmol/(m²·s)光下的净光合速率。

①Correlation is significant at the 0.01 level (2-tailed).

②$Pn1$: the net photosynthetic O_2 evolution rate under a photon flux density of 150μmol/(m²·s); $Pn2$: the net photosynthetic O_2 evolution rate under a photon flux density of 500μmol/(m²·s).

表 7-5 百花湖中总氮(TN)、总磷(TP)与微藻的叶绿素 a、
净光合速率的相关系数($n=24$)

Tab. 7-5 Correlation coefficient of TN, TP, microalgal chl a, net photosynthetic rate and compositions in Baihua Lake($n=24$)

变量	TN/(mg/L)	TP/(mg/L)	Chl a/(μg/L)	$Pn1$/(nmolO$_2$ h^{-1} μgchl a^{-1})
TP	-0.09			
Chl a	0.83**	-0.08		
$Pn1$	0.84**	-0.20	0.82**	
$Pn2$	0.76**	-0.35	0.71**	0.76**

** 0.01水平下的显著性(2-tailed)。

注：$Pn1$：150μmol/(m^2·s)光下的净光合速率；$Pn2$：500 μmol/(m^2·s)光下的净光合速率。

** Correlation is significant at the 0.01 level (2-tailed).

Note：$Pn1$：the net photosynthetic O$_2$ evolution rate under a photon flux density of 150μmol/(m^2·s); $Pn2$：the net photosynthetic O$_2$ evolution rate under a photon flux density of 500μmol/(m^2·s).

表 7-6 红枫湖中总氮(TN)、总磷(TP)与微藻的叶绿素 a、
净光合速率的相关系数($n=24$)

Tab. 7-6 Correlation coefficient of TN, TP, microalgal chl$a$, net photosynthetic rate and compositions in Hongfeng Lake($n=24$)

变量	TN/(mg/L)	TP/(mg/L)	Chl a/(μg/L)	$Pn1$/(nmolO$_2$ h^{-1} μgchl a^{-1})
TP	-0.06			
Chl a	0.71**	-0.06		
$Pn1$	0.61**	0.04	0.61**	
$Pn2$	0.62**	-0.11	0.65**	0.87**

** 0.01水平下的显著性(2-tailed)。

注：$Pn1$：150μmol/(m^2·s)光下的净光合速率；$Pn2$：500 μmol/(m^2·s)光下的净光合速率。

** Correlation is significant at the 0.01 level (2-tailed).

Note：$Pn1$：the net photosynthetic O$_2$ evolution rate under a photon flux density of 150μmol/(m^2·s); $Pn2$：the net photosynthetic O$_2$ evolution rate under a photon flux density of 500μmol/(m^2·s).

湖泊 N:P 的比例随时间的变化而变化，阿哈湖的 N:P 的比例不同于百花湖与红枫湖的 N:P 的比例。在阿哈湖中，N:P 的比例为 9.38~77.29(平均 44.02)，在百花湖中，N:P 的比例为 43.40~349.12(平均 150.22)，在红枫湖中，N:P 的比例为 51.26~473.33(平均 180.26)。造成 N:P 比例不同的原因是百花湖和红枫湖经过了综合治理，磷减少的速度快于按比例的氮的减少速度。磷的浓度接近于富营养化的临界值，氮成为湖泊营养的控制因素。而在阿哈湖中，很少点源污染，进入水体的氮、磷都很少，因此磷成为浮游植物生长的限制因子。营养的波动使藻类不至于过度生长，从而使水体不至于形成

"水华"现象。

第三节 岩溶湖泊微藻种类组成与碳酸酐酶
——以 2002～2003 年为例

一、微藻种群组成和碳酸酐酶变化特征

对红枫湖、百花湖和阿哈湖的藻类的组成和碳酸酐酶胞外酶活力进行了系统的研究。研究结果表明，不同湖泊浮游植物的组成明显的不同，它们随着时间的变化而变化。在红枫湖中，表层水的蓝藻密度为 1.26×10^5～12.17×10^5 cells/L（平均 4.60×10^5 cells/L）（图 7-9）；在百花湖中，表层水的蓝藻密度为 0.50×10^5～11.64×10^5 cells/L（平均 4.93×10^5 cells/L）（图 7-10）；在阿哈湖中，表层水的蓝藻密度为 1.77×10^5～11.10×10^5 cells/L（平均 4.98×10^5 cells/L）（图 7-11）。在红枫湖中，表层水的绿藻密度为 0.76×10^5～5.20×10^5 cells/L（平均 2.38×10^5 cells/L）（图 7-9）；在百花湖中，表层水的绿藻密度为 0.61×10^5～13.70×10^5 cells/L（平均 3.32×10^5 cells/L）（图 7-10）；在阿哈湖中，表层水的绿藻密度为 0.47×10^5～15.00×10^5 cells/L（平均 2.58×10^5 cells/L）（图 7-11）。在红枫湖中，表层水的硅藻密度为 0.11×10^5～6.10×10^5 cells/L（平均 1.08×10^5 cells/L）（图 7-9）；在百花湖中，表层水的硅藻密度为 0.01×10^5～17.30×10^5 cells/L（平均 2.32×10^5 cells/L）（图 7-10）；在阿哈湖中，表层水的硅藻密度为 0.12×10^5～16.00×10^5 cells/L（平均 3.97×10^5 cells/L）（图 7-11）。在红枫湖中，表层水的总藻密度

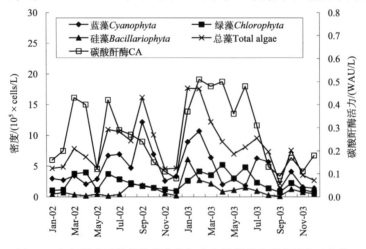

图 7-9 红枫湖中浮游植物群落的组成和微藻碳酸酐酶活力日变化

Fig. 7-9 Variation of phytoplankton communities' composition and microalgae's CA activity in Hongfeng Lake

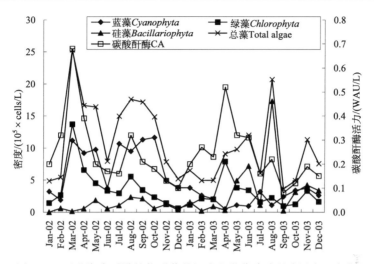

图 7-10 百花湖中浮游植物群落的组成和微藻碳酸酐酶活力日变化

Fig. 7-10 Variation of phytoplankton communities' composition and microalgae's CA activity in Baihua Lake

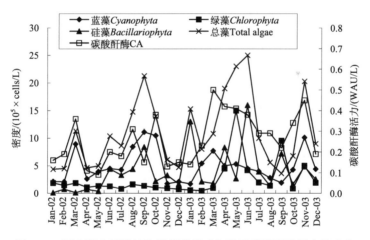

图 7-11 阿哈湖中浮游植物群落的组成和微藻碳酸酐酶活力日变化

Fig. 7-11 Variation of phytoplankton communities' composition and microalgae's CA activity in Aha Lake

为 $2.14\times10^5 \sim 17.70\times10^5$ cells/L(平均 8.29×10^5 cells/L)（图 7-9）；在百花湖中，表层水的总藻密度为 $3.62\times10^5 \sim 25.30\times10^5$ cells/L(平均 10.65×10^5 cells/L)（图 7-10）；在阿哈湖中，表层水的总藻密度为 $3.62\times10^5 \sim 25.00\times10^5$ cells/L(平均 11.08×10^5 cells/L)（图 7-11）。碳酸酐酶活力变化大且没有规律。在红枫

湖中，微藻的碳酸酐酶活力为 $0.08\sim0.51$ WAU/L（平均 0.27 WAU/L）（图 7-9）；在百花湖中，微藻的碳酸酐酶活力为 $0.08\sim0.68$ WAU/L（平均 0.24 WAU/L）（图 7-10）。在阿哈湖中，微藻的碳酸酐酶活力为 $0.09\sim0.50$ WAU/L（平均 0.26 WAU/L）（图 7-11）。在所有湖泊中微藻的碳酸酐酶胞外酶活力与水体的营养元素没有相关性，这可能与水体中的元素含量较低有关。

二、碳酸酐酶活力与微藻的组成和种类的相关性

湖泊微藻的碳酸酐酶活力与水体微藻的组成和种类具有一定的相关性。在阿哈湖中，总藻的密度与蓝藻、绿藻显著相关（表 7-7），碳酸酐酶活力与总藻的密度成显著的正相关（$Y(\text{WUA/L})=-0.00+0.12\ln X$ (10^5 cells/L)（$R^2=0.34$，$n=24$，$P<0.01$）），与蓝藻的密度也成显著的正相关（$Y(\text{WUA/L})=0.12+0.08\ln X$ (10^5 cells/L)（$R^2=0.37$，$n=24$，$P<0.01$）。

表 7-7 阿哈湖中浮游植物组成和碳酸酐酶活力的相关系数（$n=24$）
Tab. 7-7 Correlation coefficient of phytoplankton composition and microalgae's CA activity in Aha Lake（$n=24$）

变量	Cyanophyta	Chlorophyta	Bacillariophyta	Total algae
Chlorophyta	0.00			
Bacillariophyta	0.02	0.18		
Total algae	0.58**	0.41*	0.65**	
CA	0.55**	0.37	0.12	0.56**

** 0.01 水平下的显著性（2-tailed）；* 0.05 水平下的显著性（2-tailed）。

** Correlation is significant at the 0.01 level (2-tailed); * Correlation is significant at the 0.05 level (2-tailed).

表 7-8 百花湖中浮游植物组成和碳酸酐酶活力的相关系数（$n=24$）
Tab. 7-8 Correlation coefficient of phytoplankton composition and microalgae's CA activity in Baihua Lake（$n=24$）

变量	Cyanophyta	Chlorophyta	Bacillariophyta	Total algae
Chlorophyta	0.41*			
Bacillariophyta	−0.30	−0.12		
Total algae	0.67**	0.69**	0.35	
CA	0.15	0.92**	−0.06	0.52**

* 0.05 水平下的显著性（2-tailed）；** 0.01 水平下的显著性（2-tailed）。

* Correlation is significant at the 0.05 level (2-tailed); ** Correlation is significant at the 0.01 level (2-tailed).

在百花湖中，总藻的密度同样与蓝藻、绿藻显著相关（表 7-8），并且蓝藻与绿藻之间也具有相关性。碳酸酐酶活力与绿藻的密度成显著的线性相关（Y

$(WUA/L)=0.04+0.10X(10^5 \text{ cells/L})(R^2=0.84, n=24, P<0.01))$。

在红枫湖中，总藻的密度与蓝藻、绿藻、硅藻显著相关（表 7-9），碳酸酐酶活力也与绿藻的密度成显著的线性相关$(Y(WUA/L)=0.03+0.10X(10^5 \text{ cells/L}), (R^2=0.89, n=24, P<0.01))$。

表 7-9 红枫湖中浮游植物组成和碳酸酐酶活力的相关系数（$n=24$）

Tab. 7-9 Correlation coefficient of phytoplankton composition and microalgae's CA activity in Hongfeng Lake ($n=24$)

变量	Cyanophyta	Chlorophyta	Bacillariophyta	Total algae
Chlorophyta	−0.03			
Bacillariophyta	0.42*	0.20		
Total algae	0.65**	0.50*	0.74**	
CA	0.05	0.94**	0.35	0.60**

* 0.05 水平下的显著性（2-tailed）；** 0.01 水平下的显著性（2-tailed）。

* Correlation is significant at the 0.05 level (2-tailed); ** Correlation is significant at the 0.01 level (2-tailed).

三、岩溶湖泊微藻种类与碳酸酐酶协同变化特征

湖泊微藻的碳酸酐酶的活力与该湖的微藻的物种丰富度有关。红枫湖、百花湖、阿哈湖三个湖泊的物种丰富度是不同的。在 2002～2003 年的调查中，阿哈湖有 13 个优势种，百花湖有 21 个优势种，红枫湖有 19 个优势种。从表 7-10 中可以看到，11 种绿藻、8 种硅藻、2 种甲藻和 2 种蓝藻有可能的胞外碳酸酐酶。对生物量贡献最大的绿藻是球衣藻（Chlamydomonas globosa）、小球藻（Chlorella vulgaris）、湖生卵囊藻（Oocystis lacustis），双对栅藻（Scenedesmus bijugattus）、二形栅藻（Scenedesmus dimorphus）和微小四角藻（Tetraedron minimum），其中的 5 种有胞外碳酸酐酶活力。对生物量贡献最大的硅藻是中型脆杆藻（Fragilaria intermedia）和尖针杆藻（Synedra acus），而这两种藻并未发现有胞外碳酸酐酶的存在。对生物量贡献最大的蓝藻是针晶拟指球藻（Dactylococcopsis rhaphidioides）、水生集胞藻（Synechocystis aquatilis）和伪双点颤藻（Oscillatoria pseudogeminata），而只有在水生集胞藻中发现有胞外碳酸酐酶的存在。在三个湖中，蓝藻的比例最大，3 种优势种中 1 种具有碳酸酐酶胞外酶活力的蓝藻是丰富的，6 种中的 5 种具有碳酸酐酶胞外酶活力的绿藻是丰富的。在阿哈湖中，具有碳酸酐酶胞外酶活力的双对栅藻（Scenedesmus bijugattus）并不丰富，而另一种具有碳酸酐酶胞外酶活力的微小四角藻（Tetraedron minimum）

表 7-10 2002～2003 年阿哈湖、百花湖、红枫湖中的浮游植物的组成和丰富度

Tab. 7-10 Phytoplankton taxonomic composition and occurrence in Aha(AH), Baihua(BH) and Hongfeng (HF) Lakes during 2002～2003

		AH	BH	HF	References (related to external CA)
蓝藻门	Cyanophyta				
水华鱼腥藻	Anabaena flos-aguae (Lyngh.) Breb.	+	+	++	Ogawa and Kaplan, 2003; Smith and Ferry, 1986; Aizawa and Miyachi, 2000; Badger and Price, 1992; Sültemeyer et al., 2003
水华束丝藻	Aphanizomenon flos-aquae Ralf	+	+	++	
微小隐球藻	Aphanocapsa delicatissma W. et G. S. West	+	++	+++	
细小隐球藻	Aphanocapsa elachista W. et G. S. West	+	+	++	
针晶拟指球藻	Dactylococcopsis rhaphidioides Hansg.	++++	++	+++	
针晶拟指球藻长形变型	Dactylococcopsis rhaphidioides F. longior Geitler	++	+++	+++	
水华微囊藻	Microcystis flos-aquae (Wittr) Kirch.	+	+	+++	
点形裂面藻	Merismopedia punctata.	++			
水生集胞藻	Synechocystis aquatilis Sauvageau, Algues recolt.	+++	+++	+++	Ogawa and Kaplan, 2003; Smith and Ferry, 1986; Aizawa and Miyachi, 2000; Badger and Price, 1992; Sültemeyer et al., 2003
伪双点颗藻	Oscillatoria pseudogeminata G. Schm.	++++	+++	++	
伪双点颗藻颗粒变种	Oscillatoria pseudogeminata var. unigranulata Biswa.	++	+	++	
金藻门	Chrysophyta				
分歧锥囊藻	Dinobryon divergens Imh	++	+++	++	
甲藻门	Pyrrophyta				

续表

		AH	BH	HF	References (related to external CA)
二角多甲藻	*Peridinium bipes* Stein.	+++	++	+	Berman-Frank et al., 1995, 1998
微小多甲藻	*Peridinium peridinium* (Pen.) Lemm.	++	+	+	Berman-Frank et al., 1995, 1998
硅藻门	Bacillariophyta				
易变胞藻	*Astasia mutata* Shi.	+	++		
广缘小环藻	*Cyclotella bodanica* Eul.	++	+++	+++	Colman and Rotatore, 1988, 1995
科曼小环藻	*Cyclotella comensis* Grun.	++	+++	+++	Colman and Rotatore, 1988, 1995
梅尼小环藻	*Cyclotella meneghiniana* Kutz.	++	+++	+++	Colman and Rotatore, 1988, 1995
小桥弯藻	*Cynbella laevis* Nag.	+++	+++	+++	
中型脆杆藻	*Fragilaria intermedia* Grun	++ + ++	+ ++ ++	++++	
颗粒直链藻最窄变种	*Melosira grarulata* var. *angustissima* Mull.	+++	+++	+++	
筒单舟形藻	*Navicula anglica* Ralfa.	++	++++	++++	Rotatore and Colman, 1992
尖针杆藻	*Synedra acus* Kutz	+++++	+++++	+++++	
绿藻门	Chlorophyta				
镰形纤维藻奇异变种	*Ankistrodesmus falcatus* var. *mirabilis* G. S. West.	++	++	++	
球衣藻	*Chlamydomonas globosa* Snow.	+++	++++	+++	Aizawaand and Miyachi 1986; Palmqvist et al., 1994; Amoroso et al., 1998; Badger and Price, 1992; Ghoshal et al., 2002; Hunnik et al., 2000, 2001
小球藻	*Chlorella vulgaris* Bqij.	+ ++ ++	++++ ++++	+ + ++	Kozlowska-Szerenos et al., 2004; Umino and Shiraiwa and Umino, 1991

续表

		AH	BH	HF	References (related to external CA)
土生绿球藻	Chlorococcum humicola (Nag.) Rab.	++	+	+	
极毛顶棘藻	Chodatella cilliata (Lag) Lemm.	+	+	++	
盐生顶棘藻	Chodatella subsalsa Lemm.	+	++	++	
纤细新月藻	Closterium gracile Breb.	++	+	+++	
拟新月藻	Closteropsis longissima Lemm	+	+	++	
小空星藻	Coelastrum microporum Nag.	++	++	++	
网状空星藻	Coelastrum reticulatum (Dang.) Senn.		++	+	
四角十字藻	Crucigenia terapedia (Kirch) W. et G. S. West.	++	+++		
胶网藻	Dictyosphaerium dhrenbergianum Nag.	+	++	++	
美丽胶网藻	Dictyosphaerium pulchellum Wood.	+	++	++	
空球藻	Eudorina elegans Ehr.		+	++	
多芒藻	Golenkinia radiata Chod.	++	+	++	
肥壮蹄形藻	Kirchneriella obesa (West) Schm.	+	+++	+	
波吉卵囊藻	Oocystis borgei Snow.	+++++	+++++	+++++	
湖生卵囊藻	Oocystis lacustis Chod.	++	++	+	
小形卵囊藻	Oocystis parva W. et G. S West.	+	+	++	
实球藻	Pandorina morum (Muell.) Bory.	++	+++	+	
单角盘星藻具孔变种	Pediastrum simplex var. duodenrium (Bail.) Rabenh.				
井联藻	Quadrigula chodatii G. M. Smith	+	+++	++	

续表

		AH	BH	HF	References (related to external CA)
阿库栅藻	*Scenedesmus acunae* Coma.	+	++	+	Palmqvist et al., 1994; Findenegg, 1976; Thielmann et al., 1990.
弯曲栅藻	*Scenedismus arcuatus* Lemm		++	+	Palmqvist et al., 1994; Findenegg, 1976; Thielmann et al., 1990.
双对栅藻	*Scenedesmus bijugattus* (Turp) Lag.	++	+++ ++	++++	Palmqvist et al., 1994; Findenegg, 1976; Thielmann et al., 1990.
龙骨栅藻	*Scenedesmus carinatus* (Lemm.) Chod.	++	+	+	Palmqvist et al., 1994; Findenegg, 1976; Thielmann et al., 1990.
二形栅藻	*Scenedesmus dimorphus* (Turp.) Kutz	+++	++	+++	Palmqvist et al., 1994; Findenegg, 1976; Thielmann et al., 1990.
爪哇栅藻	*Scenedesmus javaensis* Chod.	+	+	+	Palmqvist et al., 1994; Findenegg, 1976; Thielmann et al., 1990.
四尾栅藻	*Scenedesmus quadricauda* (Turp) Breb.	++	+		Palmqvist et al., 1994; Findenegg, 1976; Thielmann et al., 1990.
四棘栅藻	*Scenedesmus quadrispina* Chod.	++	++	+	Palmqvist et al., 1994; Findenegg, 1976; Thielmann et al., 1990.
钝齿角星鼓藻	*Staurastrum crenulatum* (Naeg.) Delp.	+	+	++	Palmqvist et al., 1994; Findenegg, 1976; Thielmann et al., 1990.
纤细角星鼓藻	*Staurastrum gracile* Ralfs.	+	++		Palmqvist et al., 1994; Findenegg, 1976; Thielmann et al., 1990.
具齿角星鼓藻	*Staurastrum indentatum* W. et G. S. West		+	++	Palmqvist et al., 1994; Findenegg, 1976; Thielmann et al., 1990.
微小四角藻	*Tetraedron minimum* (A. Br.) Hansg	+++	+++++	+++++	Hunnik et al., 2000, 2001
膨胀四角藻	*Tetraedron tumidulum* (Br) Hansg	++	++	++	Hunnik et al., 2000, 2001

注：+++++：最丰富 the most abundant，++++：非常丰富 very abundant，+++：丰富 abundant，++：零星存在 sporadical presence，+：稀少 rare

Note：+++++：the most abundant，++++：very abundant，+++：abundant，++：sporadical presence，+：rare.

的丰富度也显著低于在百花湖和红枫湖中的微小四角藻（Tetraedron minimum）的丰富度。也就是说，在阿哈湖中，具有碳酸酐酶胞外酶活力的绿藻占总藻的比例较小，而相应的，具有碳酸酐酶胞外酶活力的蓝藻占总藻的比例较大，这样就表现出碳酸酐酶胞外酶活力与蓝藻成正相关。在百花湖和红枫湖中，具有碳酸酐酶胞外酶活力的双对栅藻（Scenedesmus bijugattus）和微小四角藻（Tetraedron minimum）是非常丰富的。因此，在百花湖和红枫湖中，碳酸酐酶活力与绿藻显著相关，而与蓝藻密度无关。在阿哈湖中，硅藻的密度占总藻的密度的36%，但没有1种具有碳酸酐酶胞外酶活力的硅藻是丰富的藻类，因此，在阿哈湖中，碳酸酐酶活力与硅藻没有显著相关性。在百花湖和红枫湖中，硅藻的密度占总藻的密度分别为13%和22%。具有碳酸酐酶胞外酶活力的硅藻的比例极小，所以，在百花湖和红枫湖中，碳酸酐酶活力与硅藻也没有显著相关性。岩溶湖泊水化学成分尤其是HCO_3^-和CO_2变化较大。碳酸酐酶胞外酶能把外界环境中的HCO_3^-转化成CO_2补充进入藻细胞。微藻中碳酸酐酶胞外酶的CO_2浓缩机制是其适应水化学成分变化的重要机制。在红枫湖和百花湖中，具有碳酸酐酶胞外酶活力的绿藻敏感地响应外界CO_2的变化，在阿哈湖中，具有碳酸酐酶胞外酶活力的蓝藻敏感地响应外界CO_2的变化。碳酸酐酶胞外酶与藻类的净光合放氧具有显著的关系（Wu et al.，2008a）。较高的碳酸酐酶胞外酶活力能够增强藻类的净光合放氧能力。因此，碳酸酐酶胞外酶能够影响藻类的生长和组成。调节碳酸酐酶活力是调节藻类组成和丰富度的又一个重要途径，有可能在防止湖泊"水华"上起重要作用。

第四节　岩溶湖泊无机碳的变化和微藻碳酸酐酶

在自然水体中，由于不同地域等因素造成各种环境的差异，不同湖泊中的微藻种群差异较大，相应地，它的胞外碳酸酐酶活性也存在较大差异。在同一个湖泊中，由于不同季节的光照、温度和营养环境等因素的差异，微藻种群也存在季节性变化，最终也会造成不同季节的湖泊微藻胞外碳酸酐酶活性变化（Wu et al.，2008a）。湖泊微藻的胞外碳酸酐酶活性存在较大的地域差异性和季节差异性，最终都将影响湖泊无机碳变化。

一、岩溶湖泊无机碳浓度及碳同位素组成的变化特征

水中重要的溶解性无机碳HCO_3^-，是水体CO_2的重要储存库，也是微藻的重要无机碳源。降水和温度对三个湖中的HCO_3^-影响较大；降水对HCO_3^-具有稀释效应；温度升高，水中溶解的CO_2量下降，使得方程$CO_2+H_2O \rightleftharpoons HCO_3^-+H^+$向

CO_2 方向移动，进而减少了 HCO_3^- 的量。图 7-12 描述的是阿哈湖、百花湖和红枫湖湖水 HCO_3^- 浓度的变化情况。三个湖泊 HCO_3^- 浓度的最小值（0.86～1.23mmol/L）出现在 2010 年 6 月份。之后的几个月水温升高。从 2010 年 9 月份开始，三个湖泊的 HCO_3^- 浓度增大，最大值（2.2～3.2mmol/L）出现在 2011 年 3 月份和 4 月份。贵州处于东南季风和西南季风的交汇地带，全年气候温和，雨量充沛，降雨多集中于 5～10 月份，降水量占年总水量 75% 以上；11 月至次年 4 月降水量明显减少，冬季（12 月至次年 2 月）降雨量最少，为年降水量的 5% 以下。丰水期一般是 5～10 月，枯水期为 12 月至次年 3 月。阿哈湖、百花湖和红枫湖的 HCO_3^- 水平相当，在夏季 5、6 月份浓度最低，冬季浓度最高。

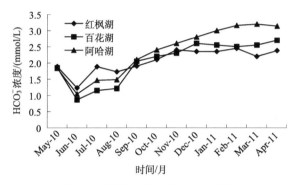

图 7-12　不同月份 HCO_3^- 含量变化

Fig. 7-12　Variation in the content of HCO_3^- at different months

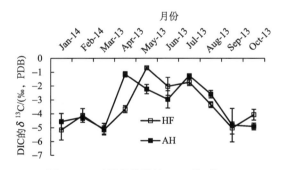

图 7-13　不同月份湖泊 DIC 的 $\delta^{13}C_{DIC}$

Fig. 7-13　Variation of $\delta^{13}C_{DIC}$ in different months

水中 DIC 的碳同位素组成能较好地反映河流侵蚀和水生生物对 DIC 的利用情况（李军等，2010）。如图 7-13 所示，湖泊中可溶性无机碳同位素组成（$\delta^{13}C_{DIC}$）呈规律性变化。从总体来看，$\delta^{13}C_{DIC}$ 具有夏季偏正，冬季偏负的特征。

一方面，由于夏季雨水充沛，流域侵蚀增加，碳酸盐岩溶蚀所产生的 DIC 大量注入河水，造成富含 ^{13}C 的无机碳进入水体的量增加；另一方面气温升高，CO_2 的溶解度降低，排除水体，使得 HCO_3^-/CO_2 比例增加，两者共同作用造成水体中的 $\delta^{13}C_{DIC}$ 偏正。此外，在 6 月份的时候，$\delta^{13}C_{DIC}$ 出现了一个下降，这主要是由于生物量在 5～6 月份已经达到最大值，随后便是生物量下降，这时水生生物的呼吸和有机质的分解作用排放的 DIC 量大于生物的光合作用吸收 DIC 的量，从而使 $\delta^{13}C_{DIC}$ 下降，这一点也可以从水体的 HCO_3^- 含量得到证明（图 7-12），其从 6 月份开始逐渐上升。

二、湖泊水体中的碳酸氢根离子浓度与藻体 $\delta^{13}C$ 的关系

三个湖泊中的碳酸氢根离子含量与藻体稳定碳同位素组成之间都呈现负的相关关系（图 7-14），其中在两个湖泊中，相关性很好，分别是：阿哈水库（$y=-0.21x-4.03$，$R^2=0.88$，$P<0.01$，$n=12$）；百花湖（$y=-0.19x-3.63$，$R^2=0.88$，$P<0.01$，$n=12$）。而在红枫湖中，碳酸氢根离子含量与其对应的藻体稳定碳同位素组成之间虽存在负相关关系，但是，相关性不显著（$y=-0.05x+0.65$，$R^2=0.17$，$P=0.19>0.05$，$n=12$）。

三个湖泊水体中的碳酸氢根离子含量相近，且都具有相似的季节性变化规律，碳酸氢根离子最高浓度都出现在冬季，最低浓度都出现在夏季，这可能主要受温度和当地降雨量的影响（张乃星，2012）。冬季水温低，由大气 CO_2 进入水体中的可溶性无机碳增加，再加上冬季降水稀少稀释作用弱，最终造成冬季水体中的碳酸氢根离子含量升高。随着夏季气温和水温的升高，水体可溶性无机碳大量释放到大气中，再加上大量降水所带来的稀释作用，最终造成水体中的碳酸氢根离子含量下降，同时，夏季微藻生长旺盛，光合作用大量消耗无机碳，进一步加剧了表层水体中可溶性无机碳含量的下降。

湖泊中碳酸氢根离子浓度的变化带来水体中可溶性无机碳同位素组成（$\delta^{13}C_{DIC}$）的变化，$\delta^{13}C_{DIC}$ 具有夏季偏正，冬季偏负的特征。一方面，由于夏季气温升高，流域侵蚀增加，造成富含 ^{13}C 的无机碳进入水体的量增加；另一方面，气温升高，溶于水中的大气 CO_2 减少，两者共同作用造成水体中的 $\delta^{13}C_{DIC}$ 偏正。相应地，藻类吸收利用偏正的 $\delta^{13}C_{DIC}$，导致藻体 $\delta^{13}C$ 偏正。反之，冬季藻体 $\delta^{13}C$ 偏负。因而，水体中碳酸氢根离子浓度与藻体稳定碳同位素组成之间呈现出负的相关关系。

研究认为浮游生物的 $\delta^{13}C$ 与环境中的 CO_2 浓度的倒数呈负相关关系（Laws et al.，1995），而本研究发现微藻 $\delta^{13}C$ 与湖水中 DIC 浓度呈负相关关系。两者之间存在差异，这主要是因为前者是从大尺度上建立起来的浮游生物的 $\delta^{13}C$ 与环境中的 CO_2 浓度之间的相关性，忽略了短时间内无机碳的浓度变化；而本研

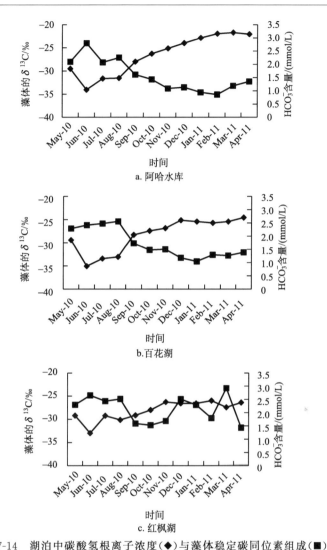

图 7-14 湖泊中碳酸氢根离子浓度(◆)与藻体稳定碳同位素组成(■)

Fig. 7-14 Concentration of bicarbonate ion(◆) and microalgal stable carbon isotope composition(■) in lakes

究是从季节变化的小尺度上建立的微藻 $\delta^{13}C$ 与湖水中 DIC 浓度关系,综合了温度等环境变化所带来的水体中无机碳含量的月变化信息。

三个湖泊的微藻胞外碳酸酐酶活力都具有典型的季节性变化规律。夏季(7、8 月份)的微藻胞外碳酸酐酶活力较大,冬季(1、2 月份)的微藻胞外碳酸酐酶活力很低(图 7-15)。同时,不同湖泊的微藻胞外碳酸酐酶活力也具有一定的差异性,以红枫湖的微藻胞外碳酸酐酶活力最大,这可能跟红枫湖的微藻种群以绿藻为主,且绿藻的胞外碳酸酐酶活力较大有关(Wu et al.,2008a)。在阿哈水库,6月份的胞外碳酸酐酶活力出现剧烈变化,这可能是由于季节变化所引起的水体中微藻种群发生剧烈变化,正在由胞外碳酸酐酶活力较弱的种群向胞外碳酸酐酶活力较强的种群转变,故在此时出现了一个较大波动。

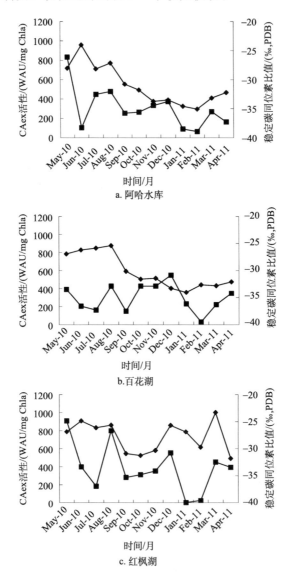

图 7-15 湖泊中的微藻胞外碳酸酐酶活力(■)与藻体稳定碳同位素组成(◆)

Fig. 7-15 Activity of extracellular carbonic anhydrase(■) and algal stable carbon isotopecomposition(◆) in lakes

由图 7-15 可以看出，湖泊藻体 $\delta^{13}C$ 与微藻的胞外碳酸酐酶活力都具有相似的季节变化规律，湖泊微藻的胞外碳酸酐酶活力较高的夏季，对应的藻体 $\delta^{13}C$ 偏正；湖泊微藻的胞外碳酸酐酶活力较低的冬季，对应的藻体 $\delta^{13}C$ 偏负。因为胞外碳酸酐酶能够催化无机碳进入细胞的过程，在有足够胞外碳酸酐酶快速催化的过程中，只存在约 1.1‰的稳定碳同位素分馏；而在无胞外碳酸酐酶催化的情况下，将产生约 10‰的稳定碳同位素分馏(Marlier and O'Leary，1984；Mook et al.，1974)。此外，胞外碳酸酐酶活力大小也会影响到微藻吸收无机碳过程中的稳定碳同位素分馏程度(Wu et al.，2011)。

本研究发现：水体无机碳和碳酸酐酶胞外酶活力是影响藻体稳定碳同位素组成变化的两个重要因素，最终带来自然水体中藻体 $\delta^{13}C$ 的变化范围较大(−23.30‰～−35.08‰，PDB)。在阿哈水库和百花湖中，碳酸氢根离子浓度与藻体稳定碳同位素组成之间的相关性显著，说明水体无机碳对藻体 $\delta^{13}C$ 的影响超过了碳酸酐酶胞外酶对藻体 $\delta^{13}C$ 的影响；而红枫湖的碳酸氢根离子浓度与藻体稳定碳同位素组成之间的相关性不显著，说明碳酸酐酶胞外酶对藻体 $\delta^{13}C$ 的影响超过了水体无机碳对藻体 $\delta^{13}C$ 的影响。

第五节 岩溶湖泊微藻碳汇能力

一、岩溶湖泊微藻的生物量的变化

与室内培养的微藻叶绿素 a 相比，三个湖泊表层水中生长的微藻的叶绿素 a 含量都较低。阿哈湖、百花湖和红枫湖的微藻种类以绿藻、蓝藻和硅藻为主(王宝利，2005)。图 7-16 描述的是阿哈湖、百花湖和红枫湖微藻叶绿素 a 含量的变化情况。三个湖泊微藻叶绿素 a 含量的最小值(0.044～0.098μg/(mL/L))均出现在 2010 年 11 月份，阿哈湖和百花湖的最大值(0.328～0.44μg/(mL/L))出现在 2010 年 8 月份，而红枫湖的最大值(0.162μg/(mL/L))出现在 2010 年 7 月份。红枫湖微藻的叶绿素 a 含量相对较小。图 7-16 同样也描述了阿哈湖、百花湖和红枫湖微藻蛋白质含量的变化情况。与叶绿素 a 含量的情况不同，三个湖泊微藻蛋白质含量的最小值(0.03～0.25μg/(mL/L))均出现在 2010 年 5 月，三个湖泊微藻蛋白质含量的最大值(2.44～3.28μg/(mL/L))出现在 2011 年 3 月和 4 月。阿哈湖、百花湖和红枫湖的微藻生物量和种类随着季节的更替而发生变化，在冬季微藻生物量很低。温带地区许多湖泊浮游植物群落种类组成变化的周期极其相似。冬季低水温和光照，尽管营养物充足，浮游植物生物量还是很低；春天温度仍然很低，但光照增强，促进浮游植物大量增殖，从而出现春季水华；随着温度升高，水体分层，浮游动物增加，由于捕食压力，浮游植物生物量下降；由于夏末水体分层结束后发生混匀，常在秋季出现较小的硅藻增殖(刘建康，1999)。

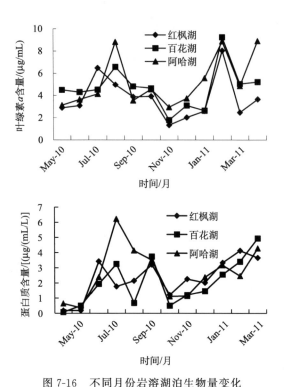

图 7-16　不同月份岩溶湖泊生物量变化

Fig. 7-16　Variation of biomass in karst lakes at different months

二、岩溶湖泊微藻无机碳碳源份额和无机碳利用途径

(一) 实验处理

碳酸氢钠浓度梯度：1.0mmol/L、2.5mmol/L、5.0mmol/L、20.0mmol/L，AZ 浓度梯度：0、1.0mmol/L、10.0mmol/L。正交设计实验。同时，分别添加两种标记的碳酸氢钠，它们的 $\delta^{13}C$ 值分别为 $-17.4‰$ 和 $-28.4‰$。初始 pH 调到 8.10 附近，并控制微藻接种量。

(二) 微藻碳同位素组成变化

从处理后的微藻碳同位素组成来看，藻体 $\delta^{13}C$ 值随添加碳酸氢钠浓度的增加而偏负（表 7-11）。尤其是添加 $\delta^{13}C$ 值偏负的碳酸氢钠对藻体 $\delta^{13}C$ 值所带来的影响更加明显。

表 7-11 不同碳酸氢根浓度及 AZ 处理下微藻碳同位素组成
Tab. 7-11 Value of microalgae's $\delta^{13}C$ in different HCO_3^- and AZ concentration treatments

aNaHCO$_3$ /(mmol/L)	0mmol/L AZ		1.0mmol/L AZ		10.0mmol/L AZ	
	δ_{T1}	δ_{T2}	δ_{T1}	δ_{T2}	δ_{T1}	δ_{T2}
1	30.4±0.3	31.0±0.1	31.7±0.2	32.4±0.3	34.8±0.1	35.2±0.3
2.5	30.7±0.1	31.6±0.1	32.2±0.3	33.1±0.4	35.1±0.2	36.0±0.2
5	32.7±0.2	33.7±0.3	33.2±0.2	35.1±0.2	34.6±0.4	36.0±0.3
20	34.5±0.3	36.7±0.3	35.9±0.4	39.8±0.3	37.4±0.2	41.7±0.3

注：a：培养液中添加的 NaHCO$_3$ 浓度；δ_{T1}：添加 $\delta^{13}C$ 为 $-17.4‰$ 的 NaHCO$_3$ 的培养液所培养出的微藻 $\delta^{13}C$ 值；δ_{T2}：添加 $\delta^{13}C$ 为 $-28.4‰$ 的 NaHCO$_3$ 的培养液所培养出的微藻 $\delta^{13}C$ 值。

Note：a：concentration of NaHCO$_3$ added in medium；δ_{T1}：The value of microalgae's $\delta^{13}C$ cultured in the medium added NaHCO$_3$ with an $\delta^{13}C$ value of $-17.4‰$；δ_{T2}：The value of microalgae's $\delta^{13}C$ cultured in the medium added NaHCO$_3$ with an $\delta^{13}C$ value of $-28.4‰$.

添加 AZ 对藻体 $\delta^{13}C$ 值所带来的影响显著，且随添加 AZ 浓度的增加，藻体 $\delta^{13}C$ 值越来越偏负（表 7-11）。研究发现：添加 10.0 mmol/L AZ，对纯培养的莱茵衣藻带来约 9‰ 的碳同位素分馏（Wu et al.，2012），而同样条件下培养的野外混合微藻，添加 10.0 mmol/L AZ，所带来的碳同位素分馏较小。这主要是因为野外的微藻种群较复杂，既有碳酸酐酶胞外酶活性强的藻类，也有碳酸酐酶胞外酶活性低的藻类，综合起来，野外微藻相比于纯培养的莱茵衣藻的碳酸酐酶胞外酶活性较低，所以造成添加 10.0 mmol/L AZ 对野外微藻所带来的碳同位素组成的影响较小。

从添加两种标记的不同 $\delta^{13}C$ 值的碳酸氢钠来看，所添加的碳酸氢钠 $\delta^{13}C$ 值越负，对应藻体的 $\delta^{13}C$ 值也越负。在相同条件下，由两种碳酸氢钠所造成的微藻碳同位素组成的差值也随添加碳酸氢钠浓度的增加而不断扩大。

（三）微藻吸收利用不同无机碳碳源的份额及不同形态无机碳的途径份额

微藻利用添加的碳酸氢根占所利用的总无机碳的份额（f_B）随添加碳酸氢钠浓度的变化而变化（表 7-12）。f_B 随添加碳酸氢钠浓度的增加而变大；添加 AZ，抑制了碳酸酐酶胞外酶的活性（Moroney et al.，1985），最终也会带来 f_B 增大，而且，f_B 随添加 AZ 浓度的增加也具有不断变大的趋势。在添加 20.0 mmol/L 碳酸氢钠和 10.0 mmol/L AZ 的条件下，f_B 的最大值达到了 0.4。

微藻利用碳酸氢根离子途径所占的份额（f_b）也随添加碳酸氢钠和 AZ 的浓度变化而变化（表 7-12）。一般情况下，f_b 随添加碳酸氢钠浓度的增加而下降，而且，f_b 也随添加 AZ 浓度的增加而下降。在添加 20.0 mmol/L 碳酸氢钠和 10.0 mmol/L AZ 的共同作用下，造成 f_b 为 0，说明此时微藻吸收利用无机碳的碳酸氢根离子途径受到了完全抑制。

在未添加 AZ 的前提下，野外微藻在较低浓度碳酸氢钠（1.0～2.5mmol/L）条件下，碳酸氢根离子的利用途径占主导（$f_b > 0.5$），但在添加过高浓度碳酸氢钠条件下，碳酸氢根离子的利用途径急剧下降。添加 AZ，抑制了碳酸酐酶胞外酶的活性（Moroney et al.，1985），造成微藻的碳酸氢根离子的利用途径下降，尤其是在添加 10.0mmol/L AZ 的条件下，微藻利用碳酸氢根离子的途径受到了极大抑制，造成此时的 f_b 都接近于 0（表 7-12）。

表 7-12 不同处理下微藻的增殖倍数、碳源利用份额及不同形态无机碳的途径
Tab. 7-12 Algal growth and variation in the propotion of carbon sources and carbon utilization pathway with different treatments

$NaHCO_3$ /(mmol/L)	0mmol/L AZ			1.0mmol/L AZ			10.0mmol/L AZ		
	P	f_B	f_b	P	f_B	f_b	P	f_B	f_b
1	4.89±0.07	0.06	0.54	3.75±0.25	0.06	0.39	1.59±0.24	0.03	0.02
2.5	5.27±0.52	0.08	0.52	3.67±0.39	0.08	0.35	1.93±0.30	0.09	0.04
5	6.28±0.44	0.09	0.31	3.93±0.19	0.17	0.3	2.16±0.19	0.12	0.12
20	6.55±0.32	0.2	0.18	4.46±0.28	0.35	0.14	2.53±0.47	0.4	0

注：P 为处理后的微藻生物量相对于接种时的增殖倍数；f_B 为微藻利用添加的无机碳占总碳源的份额；f_b 为微藻利用碳酸氢根离子途径所占的份额。

Note: P: the proliferating multiple; f_B: the proportion of the utilization of DIC from the added HCO_3^- in the whole carbon sources by the microalgae; f_b: the proportion of the HCO_3^- pathway.

三、乙酰唑胺与碳酸氢钠共同作用下的岩溶湖泊微藻碳汇

（一）实验处理

分别于每个月将采回来的微藻进行室内模拟实验，并于每季度中期进行一次大的模拟实验，具体实验设计见表 7-13。

表 7-13 实验处理
Tab. 7-13 Experimental treatments

编号	$NaHCO_3$/(mmol/L)	AZ/(mmol/L)	月份
0-0		0	1、2、3、4、5、6、7、8、9、10、11、12
0-1	0	1	1、4、7、10
0-10		10	1、2、3、4、5、6、7、8、9、10、11、12
1-0		0	1、4、7、10
1-1	1	1	1、4、7、10
1-10		10	1、4、7、10

续表

编号	NaHCO$_3$/(mmol/L)	AZ/(mmol/L)	月份
2-0		0	1、2、3、4、5、6、7、8、9、10、11、12
2-1	2	1	1、4、7、10
2-10		10	1、2、3、4、5、6、7、8、9、10、11、12
20-0		0	1、2、3、4、5、6、7、8、9、10、11、12
20-1	20	1	1、4、7、10
20-10		10	1、2、3、4、5、6、7、8、9、10、11、12

(二) 不同月份微藻在 AZ 和 NaHCO$_3$ 梯度处理下的利用份额变化

不同季节温度不同，湖水的 DIC 组成不同，微藻的生理机能也不同，从而导致微藻的碳汇能力不同。从不同季节的微藻在不同处理下对添加 NaHCO$_3$ 的利用情况来看，在夏季，微藻对添加 NaHCO$_3$ 的利用能力最强；而在冬季，微藻对添加 NaHCO$_3$ 的利用能力相对较弱（图 7-17）。一般情况下，微藻对添加 NaHCO$_3$ 的利用能力随着 AZ 浓度的增加而增加（图 7-18）。此外，红枫湖和阿哈湖的微藻在不同处理下对 NaHCO$_3$ 的利用情况有所不同。对于红枫湖的微藻来说，在春季，其对添加 NaHCO$_3$ 的利用能力有下降的趋势，随后便上升；而阿哈湖的微藻则主要呈现从冬季到夏季逐渐升高的趋势。这可能与不同湖泊微藻的利用碳源的途径不同有关。

为了进一步研究不同湖泊微藻的碳源利用途径差异及其主要影响因素，我们对全年每月份的微藻进行不同的处理。如图 7-18 所示，红枫湖的微藻在相同 NaHCO$_3$

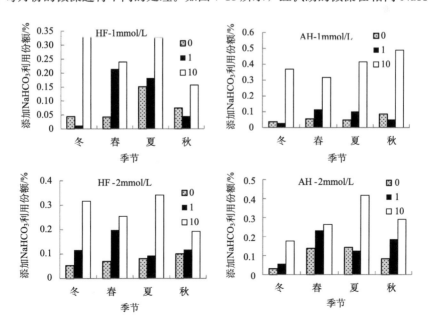

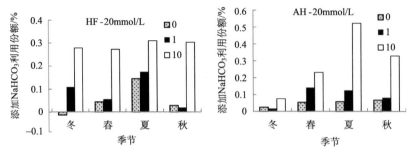

图 7-17 不同季节湖泊微藻在不同处理下对添加 $NaHCO_3$ 的利用份额

HF：红枫湖；AH：阿哈湖

Fig. 7-17 The propotion of the utilization on DIC from the added HCO_3^- with different treatments in different seasons by microalgae

HF：Hongfeng Lake；AH：Aha Lake

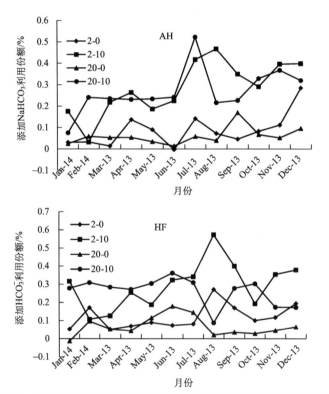

图 7-18 不同月份湖泊微藻在不同处理下对添加 $NaHCO_3$ 的利用份额

HF：红枫湖；AH：阿哈湖

Fig. 7-18 The propotion of the utilization on DIC from the added HCO_3^- with different treatments in different months by microalgae

HF：Hongfeng Lake；AH：Aha Lake

浓度处理下有相似的变化规律；而阿哈湖则在相同 AZ 浓度处理下有相似的变化规律。因此，DIC 的浓度可能是红枫湖微藻碳汇能力的主要影响因素，阿哈湖的微藻碳汇能力则主要受胞外碳酸酐酶的影响。

同时，阿哈湖的微藻在 10mmol/L AZ 处理的利用份额与 0mmol/L AZ 处理下利用份额的差异也明显大于红枫湖的微藻，这一点也进一步说明阿哈湖的胞外碳酸酐酶对微藻的碳汇能力影响大于红枫湖。

四、岩溶湖泊微藻碳汇能力的评价

微藻一方面能够利用水体中固有的 DIC；另一方面能够利用大气中 CO_2。其中，我们将微藻对水体 DIC 的利用称之为间接碳汇；而对大气 CO_2 的利用称之为直接碳汇。

(一) 单位质量叶绿素 a 微藻的碳汇能力

从图 7-19 中可以看出，在夏季，无论是微藻的间接碳汇还是直接碳汇都处于最高水平；而冬季的时候，微藻的碳汇能力较低。同时，微藻的直接碳汇能力明显大于间接碳汇能力。此外，通过对有无 AZ 处理下微藻的直接碳汇能力和间接碳汇能力进行对比，可以发现，无 AZ 处理下微藻的直接碳汇能力普遍大于有 AZ 的处理；而有 AZ 处理下微藻的间接碳汇能力普遍大于无 AZ 处理。这主要是由于微藻胞外碳酸酐酶主要通过促进微藻对大气 CO_2 的利用来增强微藻的碳汇能力(见第五章)。其中，对于 20mmol/L $NaHCO_3$ 处理下的 6~10 月的红枫湖的微藻来说，无 AZ 处理下微藻的直接碳汇能力却小于有 AZ 的处理。这主要是由于，在 6~10 月，微藻的碳汇能力表现的较强，而前面提到 DIC 的浓度可能是红枫湖微藻碳汇能力的主要影响因子，在高 $NaHCO_3$ 处理下微藻的胞外碳酸酐酶将受到抑制，而 HCO_3^- 直接利用通道可能受到诱导，从而导致以 HCO_3^- 直接利用通道为主要无机碳利用途径的处理(20 mmol/L $NaHCO_3$)下的微藻的间接碳汇能力较强。总体来说，微藻胞外碳酸酐酶主要贡献于湖泊微藻的直接碳汇能力，特别是受胞外碳酸酐酶主要控制的阿哈湖，表现最为明显。

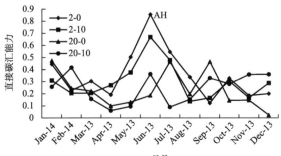

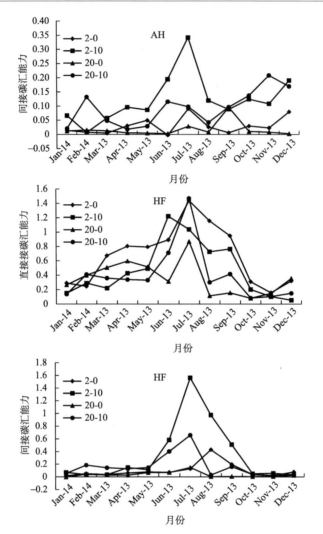

图 7-19 不同月份单位叶绿素 a 湖泊微藻的直接碳汇和间接碳汇能力

HF：红枫湖；AH：阿哈湖

Fig. 7-19 Direct and indirect carbon sink in unit Chl-a of microalgae with different treaments in different months

HF：Hongfeng Lake；AH：Aha Lake

（二）岩溶湖泊单位水体微藻的碳汇能力

微藻作为湖泊等水体的主要初级生产力，是湖泊碳汇的主要来源，被广泛的称为"微型生物泵"（Stone，2010）。因此，不同月份下单位水体微藻的碳汇能力在很大程度上能较好地反映全年湖泊的碳汇能力。从图 7-20 可以看出，无论是红枫湖还是阿哈湖，其单位水体微藻的直接碳汇能力和间接碳汇能力都在春季和夏季处于较高水平，而在冬季和秋季处于较低水平。

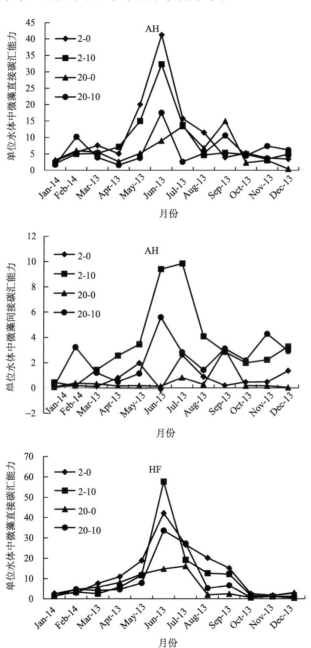

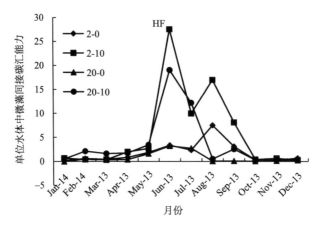

图 7-20　不同月份单位水体湖泊微藻的直接碳汇和间接碳汇能力
HF：红枫湖；AH：阿哈湖
Fig. 7-20　Direct and indirect carbon sink in unit volume lake
water with different treatments in different months
HF：Hongfeng Lake；AH：Aha Lake

究其原因，可能是不同季节的气候条件和物源输入差异导致水体物理化学条件发生变化。然而，多种水化学和水环境生物学参数可能与岩溶湖泊水体微藻的碳汇能力产生相互影响，比如：叶绿素、HCO_3^-、pH、NO_3^- 浓度、温度、阳离子浓度、阴离子浓度和总离子浓度等。将各参数与不同处理下岩溶湖泊单位水体微藻的直接碳汇能力和间接碳汇能力进行相关性分析。我们可以发现不同处理下湖泊微藻各碳汇能力的影响因素，见表 7-14 和表 7-15。

表 7-14　不同 AZ 处理下阿哈湖单位水体微藻直接和间接碳汇与各参数的相关性($n=12$)
Tab. 7-14　The correlation between direct and indirect carbon sink in unit volume lake water and different parameters with different AZ treatments in Aha Lake($n=12$)

处理	碳汇类型	叶绿素	HCO_3^-	pH	NO_3^-	温度	DIC 同位素	阳离子	阴离子	总离子
2-0	直接碳汇	0.81**	−0.62*	0.47	0.24	0.62*	0.73**	0.27	0.03	0.11
2-0	间接碳汇	0.13	−0.40	0.16	−0.03	0.42	0.55	0.05	−0.08	−0.05
2-10	直接碳汇	0.78**	−0.62*	0.43	0.27	0.59*	0.71*	0.25	0.03	0.10
2-10	间接碳汇	0.63*	−0.66*	0.28	0.34	0.75**	0.64*	−0.03	−0.13	−0.11

* 在 0.05 水平(双侧)上显著相关；** 在 0.01 水平(双侧)上显著相关。

* significant correlation at the 0.05 level (bilateral)；** Significant correlation at the 0.01 level (bilateral).

表 7-15 不同 AZ 处理下红枫湖单位水体微藻直接和间接碳汇与各参数的相关性($n=12$)

Tab. 7-15 The correlation between direct and indirect carbon sink of unit volume lake water and different parameters with different AZ treatments in Hongfeng Lake($n=12$)

处理	碳汇类型	叶绿素	HCO_3^-	pH	NO_3^-	温度	DIC 同位素	阳离子	阴离子	总离子
2-0	直接碳汇	0.92**	−0.87**	0.45	0.29	0.80**	0.62*	−0.64*	−0.30	−0.60*
2-0	间接碳汇	0.44	−0.61*	0.31	−0.14	0.64*	0.42	−0.61*	−0.36	−0.61*
2-10	直接碳汇	0.97**	−0.81**	0.25	0.27	0.61*	0.41	−0.52	−0.24	−0.49
2-10	间接碳汇	0.87**	−0.85**	0.24	0.16	0.66*	0.41	−0.59*	−0.30	−0.56

* 在 0.05 水平(双侧)上显著相关；** 在 0.01 水平(双侧)上显著相关。

* significant correlation at the 0.05 level (bilateral); ** Significant correlation at the 0.01 level (bilateral).

从表 7-14 中可以看出,对于阿哈湖来说,除了 2-0 处理下的微藻的间接碳汇,其他各处理下的单位水体的微藻各碳汇都与叶绿素、HCO_3^- 浓度、温度和 $\delta^{13}C_{DIC}$ 有较好的相关关系。而由于 2-0 处理下的微藻间接碳汇能力很小,各月份间变化趋势不明显,从而导致相关性较差。此外,如表 7-5 所示,各处理下红枫湖的单位水体的微藻的各碳汇能力与叶绿素 a、HCO_3^- 浓度、温度和阳离子浓度有较好的相关关系。总体来看,温度可能是是岩溶湖泊水体微藻碳汇能力的主要限制性因素。它与各处理下单位水体微藻的直接和间接碳汇能力都有较好的正相关关系。同时,以阳离子作为指示的风化物源物质的输入也可能是另一个限制性因素,特别是红枫湖。其中,温度升高对水体微藻碳汇能力的影响又将诱导叶绿素 a 浓度升高,并由此导致微藻对水体 HCO_3^- 等形式的 DIC 利用增强,从而导致水体 HCO_3^- 浓度降低。这也暗示我们,微藻将对由大气 CO_2 浓度升高而导致的全球温度升高,有一个相应的反馈机制来抵抗"温室效应",而碳酸酐酶在其中起到重要的作用。

参 考 文 献

曹宗巽,吴相钰. 1978. 植物生理学(上册). 北京:高等教育出版社,1-252.

李军,刘丛强,李龙波,李思亮,王宝利,Chetelat B. 2010. 硫酸侵蚀碳酸盐岩对长江河水 DIC 循环的影响. 地球化学,39(4):305-313.

刘丛强,蒋颖魁,陶发祥,郎赟超,李思亮. 2008. 西南喀斯特流域碳酸盐岩的硫酸侵蚀与碳循环. 地球化学,37(4):404-414.

刘丛强. 2007. 生物地球化学过程与地表物质循环:西南喀斯特流域侵蚀与生源要素循环. 北京:科学出版社.

刘建康. 1999. 高级水生生物学. 北京:科学出版社.

刘再华,Dreybrodt W. 2001. 不同 CO_2 分压条件下的白云岩溶解动力学机理. 中国科学(D 辑:

地球科学),31(4): 377-384.

刘再华,袁道先,何师意,张美良,张加桂. 2000. 地热 CO_2-水-碳酸盐岩系统的地球化学特征及其 CO_2 来源-以黄龙沟、康定和云南中甸下给为例. 中国科学(D辑:地球科学),30(2): 209-214.

刘再华. 2000. 碳酸酐酶对碳酸盐岩溶解的催化作用及其在大气 CO_2 沉降中的意义. 地球学报,22(5): 477-480.

缪晓玲. 1998. 海洋浮游藻(Emiliania huxleyi)无机碳利用机理. 上海师范大学学报(自然科学版),27(4): 44-48.

王宝利. 2005. 高原湖泊微藻和微量金属元素的相互作用过程及实验模拟研究. 中国科学院.

翁金桃. 1995. 碳酸盐岩在全球碳循环过程中的作用. 地球科学进展,10(2): 154-158.

吴沿友,李萍萍,刘丛强. 2003. 无机元素和气候因子对岩溶湖泊水体微藻生长的影响. 南京林业大学学报,27(增刊): 67-70.

吴沿友,李萍萍,王宝利,刘丛强,何梅,陈椽. 2004. 红枫湖百花湖水质及浮游植物的变化. 农业环境科学学报,23(4): 745-747.

严国安,刘永定. 2001. 水生生态系统的碳循环及对大气 CO_2 汇. 生态学报,21(5): 827-823.

阎葆瑞,张锡根. 2000. 微生物成矿学. 北京: 科学出版社.

张乃星,吴凤丛,任荣珠,等. 2012. 渤海海峡冬季表层海水中溶解无机碳分布特征分析. 海洋科学,36(2): 56-61.

朱颜明,何岩. 2002. 环境地理学导论. 北京: 科学出版社.

Agarwala S C, Nautiyal B D, Chatterjee C, Nautiyal N. 1995. Variations in copper and zinc supply influence growth and activities of some enzymes in maize. Soil Science and Plant Nutrition, 41(2): 329-335.

Aizawa K, Miyachi S. 1986. Carbonic anhydrase and CO_2 concentrating mechanisms in microalgae and cyanobacteria. Fems Microbiology Letters, 39(3): 215-233.

Altermann W, Kazmierczak J, Oren A, Wright D T. 2006. Cyanobacterial calcification and its rock-building potential during 3.5 billion years of Earth history. Geobiology, 4(3): 147-166.

Amoroso G, Sültemeyer D, Thyssen C, Fock H P. 1998. Uptake of HCO_3^- and CO_2 in cells and chloroplasts from the microalgae Chlamydomonas reihardtii and Dunaliella tertiolecta. Plant Physiol, 116(1): 193-201.

Axelsson L, Larsson C, Ryberg H. 1999. Affinity, capacity and oxygen sensitivity of two different mechanisms for bicarbonate utilization in Ulva lactuca L. (Chlorophyta). Plant, Cell and Envirnment, 22: 969-978.

Badger M R, Price G D, Lon B M, Woodger F J. 2006. The environmental plasticity and ecological genomics of the cyanobacterial CO_2 concentrating mechanism. Journal of experimental botany, 57(2): 249-265.

Badger M R, Price G D. 1992. The CO_2 concentrating mechanism in cyanobacteria and microalgae. Physiol. Plantarum 84(4): 606-615.

Baird T T, Waheed A, Okuyama T, Sly W S, Fierke C A. 1997. Catalysis and inhibition of human carbonic anhydrase IV. Biochemistry, 36(9): 2669-2678.

Beardall J, Giordano M. 2002. Ecological implications of microalgal and cyanobacterial CO_2 concentrating mechanisms and their regulation. Functional Plant Biology, 29(3): 335-347.

Beardall J, Johnston A, Raven J. 1998. Environmental regulation of CO_2-concentrating mechanisms in microalgae. Canadian Journal of Botany, 76(6): 1010-1017.

Berman-Frank I, Erez J, Kaplan A. 1998. Changes in inorganic carbon uptake during the progression of a dinoflagellate bloom in a lake ecosystem. Canadian Journal of Botany-revue Canadienne de botanique, 76(6): 1043-1051.

Berman-Frank I, Kaplan A, Zohary T, Dubinsky Z. 1995. Carbonic anhydrase activity in the bloom-forming dinoflagellate Peridinium gatunense. Journal of Phycology, 31(6): 906-913.

Berner R A, Lasaga A C, Garrels R M. 1983. The carbonate-silicate geochemical cycle and its effect on atmospheric carbon dioxide over the past 100 million years. American Journal of Science, 283(7): 641-683.

Blum J D, Gazis C A, Jacobson A D, Chamberlain C P. 1998. Carbonate versus silicate weathering in the Raikhot watershed within the High Himalayan Crystalline Series. Geology, 26(5): 411-414.

Brownell P F, Bielig L M, Grof C. 1992. Increased carbonic anhydrase activity in leaves of sodium deficient C4 plants. Australian Journal of Plant Physiology, 18(6): 589-592.

Budenbender J, Riebesell U, Form A. 2011. Calcification of the Arctic coralline red algae Lithothamnion glaciale in response to elevated CO_2. Marine Ecology Progress Series, 441: 79-87.

Burnell J N, Suzuki I, Sugiyama T. 1990. Light induction and the effect of nitrogen status upon the activity of Carbonic anhydrase in maize leaves. Plant Physiology, 94(1): 384-387.

Cang K, Roberts J. 1992. Quantitation of rates of transport, metabolic fluxes, and cytoplasmic levels of inorganic carbon in maize root tips during potassium ion uptake. Plant Physiology, 99(1): 291-297.

Carpenter S R, Kraft C E, Wright R, He X, Soranno P A, Hodgson J R. 1992. Resilience and resistance of a lake phosphorus cycle before and after food web manipulation. American Naturalist, 140(5): 781-798.

Chaki M, Carreras A, López-Jaramillo J, et al. 2013. Tyrosine nitration provokes inhibition of sunflower carbonic anhydrase (β-CA) activity under high temperature stress. Nitric Oxide, 29: 30-33.

Colman B, Rotatore C. 1988. Uptake and accumulation of inorganic carbon by a freshwater diatom. Journal of Experimental Botany, 39(205): 1025-1032.

Colman B, Rotatore C. 1995. Photosynthetic inorganic carbon uptake and accumulation in two marine diatoms. Plant Cell and Environment, 18(8): 919-924.

Coogan L A, Gillis K M. 2013. Evidence that low-temperature oceanic hydrothermal systems play an important role in the silicate-carbonate weathering cycle and long-term climate regulation. Geochemistry Geophysics Geosystems, 14(6): 1771-1786.

Cullen J T, Lane T W, Morel F M, Sherrell R M. 1999. Modulation of cadmium uptake in phytoplankton by seawater CO_2 concentration. Nature, 402 (6758): 165-167.

Dong L F, Nimer N A, Okus E, Merrett M J. 1993. Dissolved inorganic carbon utilization in relation to calcite production in Emiliania huxleyi (lohmann) kamptne. New Phytologist, 123 (4): 679-684.

Fett J P, Coleman J R. 1994. Regulation of periplasmic carbonic anhydrase expression in *Chlamydomonas reinhardtii* by acetate and pH. Plant Physiology, 106(1): 103-108.

Findenegg G R. 1976. Correlations between accessibility of carbonic anhydrase for external substrate and regulation of photosynthetic use of CO_2 and HCO_3^- by Scenedesmus obliquus. Zeitschrift fur Pflanzenphysiologie, 79(5): 428-437.

Gao K, Zheng Y. 2010. Combined effects of ocean acidification and solar UV radiation on photosynthesis, growth, pigmentation and calcification of the coralline alga *Corallina sessilis* (*Rhodophyta*). Global Change Biology, 16(8): 2388-2398.

Gao Y Z, Zhao Z L, Guo M L, Wang Z. 1999. The role of carbonic anhydrase in regulating photosynthetic CO_2 fixation in high plants. Current Research in photosynthesis (Vols 1-4), Netherland: Kluwer Academic Publishers, 497-500.

Gattuso J P, Frankignoulle M, Bourge I, Romaine S, Buddemeier R W. 1998. Effect of calcium carbonate saturation of seawater on coral calcification. Global and Planetary Change, 18(1): 37-46.

Ghoshal D, Husic H D, Goyal A. 2002. Dissolved inorganic carbon concentration mechanism in *Chlamydomonas moewusii*. Plant PhysiolLgy and Biochemistry, 40(4): 299-305.

Han G L, Liu C Q. 2004. Water geochemistry controlled by carbonate dissolution: a study of the river waters draining karst-dominated terrain, Guizhou Province, China. Chemical Geology, 204(1-2): 1-21.

Hatch M D. 1991. Carbonic anhydrase assay: Strong inhibition of the leaf enzyme by carbon dioxide in certain buffers. Analytical Biochemistry, 192(1): 85-89.

Hunnik E, Ende H, Timmermans K R, Laan P, Leeuw J W. 2000. A comparison of CO_2 uptake by the green alga *Tetraedron minimum* and *Chlamydomonas noctigama*. Plant Biology, 2(6): 624-627.

Hunnik E, Livine A, Pogenberg V, Spijkerman E, Ende H, Mendoza E G, Sültemeyer D, Leeuw J W. 2001. Identification and localization of a thylakoid bound carbonic anhydrase from the green algae *Tetraedron minimum* (*Chlorophyta*) and *Chlamydomonas noctigama* (*Chlorophyta*). Planta, 212(3): 454-459.

Knoll A H, Fairchild I J, Swett K. 1993. Calcified microbes in Neoproterozoic carbonates - implications for our understanding of the Proterozoic/Cambrian transition. Palaios, 8(6): 512-525.

Kozłowska-Szerenos B, Bialuk I, Maleszewski S. 2004. Enhancement of photosynthetic O_2 evolution in Chlorella vulgaris under high light and increased CO_2 concentration as a sign of acclimation to phosphate deficiency. Plant Physiology and Biochemistry, 42(5): 403-409.

Kranz S A, Eichner M, Rost B. 2011. Interactions between CCM and N_2 fixation in Trichodesmium. Photosynthesis research, 109(1-3): 73-84.

Lane T W, Morel F M M. 2000. Regulation of carbonic anhydrase expression by zinc, cobalt, and carbon dioxide in the marine diatom *Thalassiosira weissflogii*. Plant Physiology, 123(1): 345-352.

Larsson M, Olsson T, Larsson C M. 1985. Distribution of reducing power between photosynthetic carbon and nitrogen assimilation in *Scenedesmus*. Planta, 164(2): 246-253.

Laurent B, Norihide K, Miyachi S. 2000. Effect of external pH on inorganic carbon assimilation in unicellular marine green algae. Phycological Research, 48(1): 47-54.

Laws E A, Popp B N, Bidigare R R, Kennicutt M C, Macko S A. 1995. Dependence of phytoplankton carbon isotopic composition on growth-rate and $[CO_2]$aq: Theoretical considerations and experimental results. Geochimica Et Cosmochimica Acta, 59(6): 1131-1138.

Leclercq N C, Gattuso J P, Jaubert J E A N. 2000. CO_2 partial pressure controls the calcification rate of a coral community. Global Change Biology, 6(3): 329-334.

Li Q, Wu Y Y, Wu Y D. 2013. Effects of fluoride and chloride on the growth of *Chlorella pyrenoidosa*. Water Science and Technology, 68(3): 722-727.

Li S L, Liu C Q, Li J, Lang Y C, Ding H, Li L. 2010. Geochemistry of dissolved inorganic carbon and carbonate weathering in a small typical karstic catchment of Southwest China: Isotopic and chemical constraints. Chemical Geology, 277(3): 301-309.

Liu Z H, He D. 1998. Special speleo thems in cenment-grouting tunels and their implications of the atmospheric CO_2 sink. Environmental Geology, 35(4): 258-262.

Lyszcz S, Ruszkowska M. 1992. Effect of zinc excess on carbonic anhydrase activity of crops. Acta Physiologiae Plantarium, 14(1): 35-39.

Maribel L D, Tsuzuki M, Miyachi S. 1989a. Blue light induction of carbonic anhydrase activity in *chlamydomonas reihardtii*. Plant cell physiology, 30(2): 215-219.

Maribel L D, Tsuzuki M, Miyachi S. 1989b. Light requirement for carbonic anhydrase induction in *chlamydomonas reihardtii*. Plant cell physiology, 30(2): 207-213.

Marlier J F, O'Leary M H. 1984. Carbon kinetic isotope effects on the hydration of carbon dioxide and the dehydration of bicarbonate ion. Journal of the American Chemical Society, 106(18): 5054-5057.

Matsuda Y, Colman B. 1996. Active uptake of inorganic carbon by *Chlorella saccharophila* is not repressed by growth in high CO_2. Journal of Experimental Botany, 47(305): 1951-1956.

McConnaughey T A, Falk R H. 1991. Calcium-proton exchange during algal calcification. Biological Bulletin, 180(1): 185-195.

Merrett M J, Nimer N A, Dong L F. 1996. The utilization of bicarbonate ions by the marine microalga *Nannochloropis oculata* (*Droop*) Hibberd. Plant Cell and Environment, 19(4): 478-484.

Mook W G, Bommerson J C, Staverman W H. 1974. Carbon isotope fractionation between dissolved bicarbonate and gaseous carbon dioxide. Earth and Planetary Science Letters, 22(2): 169-176.

Morel F M M, Price N M. 2003. The biogeochemical cycles of trace metals in the oceans. Science, 300(5621): 944-947.

Moroney J V, Husic H D, Tolbert N E. 1985. Effect of carbonic-anhydrase inhibitors on inorganic carbon accumulation by *chlamydomons reinhardtii*. Plant physiology, 79(1): 177-183.

Moroney J V, Somanchi A. 1999. How do algae concentrate CO_2 to increase the efficiency of photosynthetic carbon fixation? Plant Physiology, 119(1): 9-16.

Moubarak M M, Stemler A. 1994. Oxidation-reduction potential dependence of photosystem II carbonic anhydrase in maize thylakoids. Biochemistry, 33(14): 4432-4438.

Nimer N A, Iglesias R M D, Merrett M J. 1997. Bicarbonate utilization by marine phytoplankton species. Journal of Phycology, 33(4): 625-631.

Ogawa T, Kaplan A. 2003. Inorganic carbon acquisition systems in *cyanobacteria*. Photosynthesis Research, 77(2-3): 105-115.

Palmqvist K, Yu J W, Badger M R. 1994. Carbonic anhydrase activity and inorganic carbon fluxes in low and high-Ci cells of *Chlamydomonas reinhardtii* and *Scenedesmus obliquus*. Physiol Plantarum, 90(3): 537-547.

Pesheva I, Kodama M, Dionisio-Sese M L, Miyachi S. 1994. Changes in photosynthetic characteristics induced by transferring air-grown cells of *Chlorococcum littorale* to high-CO_2 conditions. Plant Cell Physiology, 35: 379-387.

Price G D, Badger M R, Woodger F J, Long B M. 2008. Advances in understanding the cyanobacterial CO_2-concentrating-mechanism (CCM): functional components, Ci transporters, diversity, genetic regulation and prospects for engineering into plants. Journal of Experimental Botany, 59(7): 1441-1461.

Price G D, Badger M R. 1991. Evidence for the role of carboxysomes in the *cyanobacterial* CO_2-concentration mechanism. Canadian Journal of Botany revue Canadienne de Botanique, 69(5): 963-973.

Ramanan R, Vinayagamoorthy N, Sivanesan S D, Kannan K, Chakrabarti T. 2012. Influence of CO_2 concentration on carbon concentrating mechanisms in cyanobacteria and green algae: A proteomic approach. Algae, 27(4): 295-301.

Ramazanov Z, Mason C B, Geraghty A M, Spalding M H, Moroney J V. 1993. The low carbon dioxide inducible 36-kilodalton protein is localized to the chloroplast envelope of *Chlamydomonas reinhardtii*. Plant Physiology, 101(4): 1195-1199.

Raven J A, Giordano M, Beardall J, Maberly S C. 2011. Algal and aquatic plant carbon concentrating mechanisms in relation to environmental change. Photosynthesis Research, 109(1-3): 281-296.

Riding R. 2000. Microbial carbonates: the geological record of calcified bacterial-algal mats and biofilms. Sedimentology, 47(1): 179-214.

Rotatore C, Colman B. 1991. The acquisition and accumulation of inorganic carbon by the unicellular green alga *Chlorella ellipsoidea*. Plant Cell and Environment, 14(4): 377-382.

Rotatore C, Colman B. 1992. Active uptake of CO_2 by the diatom *Navicula pelliculosa*. Journal of Expermental Botany, 43(249): 571-576.

Roy S, Gaillardet J, Allegre C J. 1999. Geochemistry of dissolved and suspended loads of the Seine river, France: anthropogenic impact, carbonate and silicate weathering. Geochimica Et Cosmochimica Acta, 63(9): 1277-1292.

Sarmiento J L, Sundquist E T. 1992. Revised budget for the oceanic uptake of anthropogenic carbon dioxide. Nature, 356(6370): 589-593.

Sas K N, Kovacs L, Zsiros O, Gombos Z, Garab G, Hemmingsen L, Danielsen E. 2006. Fast cadmium inhibition of photosynthesis in cyanobacteria in vivo and in vitro studies using perturbed angular correlation of gamma-rays. Journal of Biological Inorganic Chemistry, 11(6): 725-734.

Semesi I S, Kangwe J, Bjork M. 2009. Alterations in seawater pH and CO_2 affect calcification and photosynthesis in the tropical coralline alga, *Hydrolithon* sp (*Rhodophyta*). Estuarine, Coastal and Shelf Science, 84(3): 337-341.

Sharma A, Bhattacharya A, Singh S. 2009. Purification and characterization of an extracellular carbonic anhydrase from *Pseudomonas fragi*. Process Biochemistry, 44(11): 1293-1297.

Shilton A N, Powell N, Guieysse B. 2012. Plant based phosphorus recovery from waste water via algae and macrophytes. Current Opinion in Biotechnology, 23(6): 884-889.

Shiraiwa Y, Goyal A, Tolbert N E. 1993. Alkalization of the medium by unicellular green algae during uptake of dissolved inorganic carbon. Plant and cell Physiology, 34(5): 649-657.

Shiraiwa Y, Umino Y. 1991. Effect of glucose on the induction of the carbonic anhydrase and the change in $K_{1/2}$ (CO_2) of photosynthesis in *Chlorella vulgaris* 11h. Plant and cell Physiology, 32(2): 311-314.

Smith K S, Ferry J G. 2000. Prokaryotic carbonic anhydrases. FEMS Microbiology Reviews, 24(4): 335-366.

Soyut H, Beydemir S, Ceyhun S B, Erdogan O, Kaya E D. 2012. Changes in carbonic anhydrase activity and gene expression of Hsp70 in rainbow trout (*Oncorhynchus mykiss*) muscle after exposure to some metals. Turkish Journal of Veterinary & Animal Sciences, 36(5): 499-508.

Stone R. 2010. The invisible hand behind a vast carbon reservoir. Science, 328(5985): 1476-1477.

Sugiharto B, Burnell J N, Sugiyama T. 1992a. Cytokinin is required to induce the nitrogen-dependent accumulation of mRNAs for phosphoenolpyruvate carboxylase and carbonic anhydrase in detached maize leaves. Plant Physiology, 100(1): 153-156.

Sugiharto B, Suzuki I, Burnell J N, Sugiyama T. 1992b. Glutamine induces the nitrogen-dependent accumulation of mRNA's encoding phosphoenolpyruvate carboxylase and carbonic anhydrase in detached maize leaf tissue. Plant Physiology, 100(4): 2066-2070.

Sültemeyer D, Schmidt C, Fock H P. 1993. Carbonic anhydrase in higher plants and aquatic microorganisms. Plantarum, 88(1): 179-190.

Taylor L L, Banwart S A, Valdes P J, Leake J R, Beerling D J. 2012. Evaluating the effects of terrestrial ecosystems, climate and carbon dioxide on weathering over geological time: a global-scale process-based approach. Philosophical Transactions of the Royal Society B: Biological Sciences, 367(588): 565-582.

Thielmann J, Tolbert N E, Goyal A, Senger H. 1990. Two systems for concentrating CO_2 and bicarbonate during photosynthesis by *Scenedesmus*. Plant Physiology, 92(3): 622-629.

Umino Y, Shiraiwa Y. 1991. Effect of metabolites on carbonic anhydrase induction in *Chlorella vulgaris*. Journal of Plant Physiology, 139(1): 41-44.

Wan G J, Lee H N, Wan E Y, Wang S L, Yang W, Wu F C, Chen J A, Wang C S. 2008. Analyses of Pb-210 concentrations in surface air and in rain water at the central Guizhou, China. Tellus Series B-Chemical and Physical Meteorology, 60(1): 32-41.

Wang B L, Liu C Q, Wu Y Y. 2005. Effect of heavy metals on the activity of external carbonic anhydrase of microalga *Chlamydomonas reinhardtii* and microalgae from karst lakes. Bulletin of environmental contamination and toxicology, 74(2): 227-233.

Wang S L, Yeager K M, Wan G J, Liu C Q, Wang Y C, Lu Y C. 2012. Carbon export and HCO_3^- fate in carbonate catchments: A case study in the karst plateau of southwestern China. Applied Geochemistry, 27(1): 64-72.

Wu Y Y, Li P P, Wang B L, Liu C Q, He M, Chen C. 2008b. Variation in trophic status of three Karst reservoirs in the Yunnan-Guizhou Plateau in China. Cryptogamie, Algologie, 29(1): 81-93.

Wu Y Y, Zhao X Z, Li P P, Huang H K. 2007. Impact of Zn, Cu, and Fe on the activity of carbonic anhydrase of erythrocytes in ducks. Biological Trace Element Research, 118(3): 227-232.

Wu Y Y, Li P P, Wang B L, Liu C Q, He M, Chen C. 2008a. Composition and activity of external carbonic anhydrase of microalgae from karst lakes in China. Phycological Research, 56(2): 76-82.

Wu Y Y, Xu Y, Li H T, Xing D K. 2011. Effect of acetazolamide on stable carbon isotope fractionation in *Chlamydomonas reinhardtii* and *Chlorella vulgaris*. Chinese Science Bulletin, 57(7): 786-789.

Xie T X, Wu Y Y. 2014. The role of microalgae and their carbonic anhydrase on the biological dissolution of limestone. Environmental Earth Sciences, 71(12): 5231-5239.

Xu Z H, Liu C Q. 2007. Chemical weathering in the upper reaches of Xijiang River draining the Yunnan-Guizhou Plateau, Southwest China. Chemical Geology, 239(1-2): 83-95.

Yee D, Morel F M M. 1996. In vivo substitution of zinc by cobalt in carbonic anhydrase of a marine diatom. Limnology and Oceanography, 41(3): 573-577.

Yin Z H, Heber U, Raghavendra A S. 1993. Light-induced pH changes in leaves of C4 plants: Comparison of cytosolic alkalization and vacuolar acidification with that of C3 plants. Planta, 189(2): 267-277.

第八章 展 望

碳酸酐酶不完全都是只含锌辅基，它们的生理作用和功能也具有多样化。碳酸酐酶不仅是碳代谢的关键调节点，也是其他代谢的桥梁和纽带。拓展碳酸酐酶的生物地球化学作用研究，将最终建立酶生物地球化学及大分子生物地球化学科学体系。因此，在未来的五年内，我们将在以下几个方面开展工作。

一、含镉及其他金属碳酸酐酶的生物地球化学作用

随着含 Cd、Co、Fe、Mn 等非 Zn 金属的碳酸酐酶的发现，人们对碳酸酐酶的生物地球化学作用范畴产生了遐想。与含 Zn 的碳酸酐酶的作用不同，含 Cd、Co、Fe、Mn 等的碳酸酐酶响应环境的方式和机制也可能不同，"激素"效应的剂量也不相同。由于 Co、Fe、Mn 是很多关键代谢酶的辅基，而碳酸酐酶是含量较多的一种酶。环境中 Cd、Co、Fe、Mn 等元素的在不同酶之间的分配对含 Cd、Co、Fe、Mn 等非 Zn 金属的碳酸酐酶的功能和作用至关重要，因此，研究含非 Zn 金属辅基的碳酸酐酶的生物地球化学作用将会增加我们对生物地球化学循环中的一些未知过程的理解。

二、微藻碳酸酐酶对有机碳的生物地球化学作用

有机物一方面能调节碳酸酐酶的活力，另一方面也能影响碳酸酐酶的基因表达和合成，表现出对环境有机物的高度响应。同时，碳酸酐酶也能改造有机物，突出地表现在它的脂酶和转甲基功能上；在生物的分解和合成中发挥作用。此外，碳酸酐酶还具有对碳酰基硫(COS)的分解能力，在消除温室气体的不利效应方面将发挥重要影响。微藻碳酸酐酶对有机碳的生物地球化学作用的研究将是一个崭新的课题。

三、微藻碳酸酐酶的生物多样性与生物地球化学作用的多样性

无论是自然水体还是人工水体，微藻都具有丰富的生物多样性。微藻的种类不同，碳酸酐酶类型和活性也显著不同。即使是同为绿藻门的莱茵衣藻和蛋白核小球藻，它们的碳酸酐酶对环境的响应也明显不同。微藻碳酸酐酶的生物多样性，带来了碳酸酐酶的生物地球化学作用的多样性。

四、碳酸酐酶在全球碳氮磷硫循环中的作用

微藻的胞外碳酸酐酶具有无机碳浓缩机制,使无机碳进入到细胞内,在胞内碳酸酐酶的协同作用下,实现了无机碳的还原。胞外碳酸酐酶是无机碳代谢的开关,而藻类碳代谢又是生态系统碳代谢的纽带。响应于环境的胞外碳酸酐酶的生物地球化学作用,使得全球碳循环能够快速应对环境的变化。碳代谢为氮、磷、硫代谢提供碳骨架和能量,因此碳酸酐酶的生物地球化学作用,也势必对全球碳氮磷硫循环产生深远影响。

五、碳酸酐酶对硅酸岩风化及碳汇的影响

硅酸盐指的是硅、氧与其他化学元素(主要是铝、铁、钙、镁、钾、钠等)结合而成的化合物的总称。它在地壳中分布极广,占地壳岩石质量的80%以上,是构成多数岩石和土壤的主要成分。在地质时间尺度上,硅酸盐岩一方面在水和二氧化碳的作用下,发生酸溶蚀作用,产生碳汇。另一方面在碱性环境下发生碱溶作用。碳酸酐酶因能够影响二氧化碳的通量和环境的质子,因此,很可能深刻地调节硅酸岩的溶蚀。而硅酸岩的溶蚀对土壤乃至整个生态系统的金属元素的生物地球化学循环有巨大影响,这必将导致生态系统的其他元素的生物地球化学循环的变化。因此,对碳酸酐酶对硅酸岩的风化的研究,将极大丰富碳酸酐酶的生物地球化学作用的研究内容,能更深刻地揭示生物地球化学作用规律。

六、生态系统中碳酸酐酶的调控和应用

水体中微藻碳酸酐酶的变化,影响水环境;反过来,水环境的变化也影响着水体中微藻碳酸酐酶对环境的响应模式。调控水体微藻的碳酸酐酶,一方面可以优化水环境,修复氮磷及其他元素如氟离子污染的环境,治理水华,另一方面还可以优化藻体的群落结构,为水产养殖提供优质饵料。此外,因碳酸酐酶能影响有机质转化中碳氢氧的同位素以及碳酸盐岩沉积过程中的碳氢氧同位素的分馏,考虑碳酸酐酶对同位素分馏作用将大大提高用于反演古气候变化而构建的模型精度。

七、酶生物地球化学作用及大分子的生物地球化学作用

碳酸酐酶是碳代谢的关键酶,它深刻地影响着稳定碳同位素分馏、改变着无机碳的利用方式和份额,驱动了碳酸盐岩的溶蚀,影响了岩溶碳汇和光合碳汇。碳酸酐酶可以作为酶生物地球化学作用及生物大分子的地球化学作用的模式蛋白。硝酸还原酶是硝酸盐代谢乃至氮代谢的关键酶,它决定着微藻无机氮的利用方式和利用份额,影响着水体无机氮的组成和微藻的种群分布,驱动水体元素生

物地球化学循环，也将影响着水体氮的通量，最终也将影响界面乃至整个生态系统氮的生物地球化学循环。腺苷酰硫酸还原酶是硫酸盐乃至硫循环的关键酶，同样地也影响水体硫的通量，影响岩石的溶蚀和风化，改变着水体元素的组成和群落的结构，最终也将影响界面硫的循环乃至整个生态系统硫的生物地球化学循环。总之，研究碳、氮、硫、磷循环的这些关键酶的生物地球化学作用，将可最终建立酶生物地球化学及大分子生物地球化学科学体系。

Chapter 8　Prospects

Carbonic anhydrase are not all Zn-containing metalloenzyme, their physiological action and functions have the diversification. Carbonic anhydrase is not only the key and regulation point to the carbon metabolism, but also the bridge and the link to other metabolic. Expanding the research on biogeochemical action of carbonic anhydrase will eventually establish the scientific system about the enzymatic and macromolecular biogeochemistry. Therefore, in the next five years, we will work on the following aspects.

1. Biogeochemistry action of Cd- and other metal-containing carbonic anhydrase

Carbonic anhydrase will be endowed the perfect outlook about biogeochemistry action with the discovery of non-zinc such as Cd, Co, Fe and Mn-containing metalloenzyme. The mode and mechanism of Cd, Co, Fe and Mn-containing metalloenzyme responded to environment, as well as their actions and dosage of "hormone" effect, differ from that of zinc-containing enzyme. Carbonic anhydrase had a lot of proportion among all of enzymes, and Co, Fe, and Mn are prothetic groups of many key metabolic enzymes. The distribution of Cd, Co, Fe, and Mn in different species of enzymes plays important role in the functions and actions of the non-zinc metal-containing carbonic anhydrase. Thus, biogeochemistry action of the non-zinc metal-containing carbonic anhydrase will help us deeply understand some unknown process in biogeochemistry cycles.

2. Effect of microalgal carbonic anhydrase on biogeochemistry action of organic carbon

Organic matter, on one hand, can regulate the activity, on the other hand, have an impact on the gene expression and synthesis of carbonic anhydrase, which shows highly response to organic matter in the environment. Meanwhile, carbonic anhydrase can also transform organic matter, especially highlight on the functions of its lipoidase and transmethylation, and plays a part in biological decompose and synthesis. In addition, carbonic anhydrase decomposes carbonyl sulfide (COS), which has a great role in eliminating the adverse effect of greenhouse gases. Biogeochemistry action of organic carbon on carbonic anhydrase in microalgae will be a brand topic.

3. Biodiversity versus the diversity of biogeochemical process of the microalgal carbonic anhydrase

Microalgae had rich biodiversity both in natural water and artificial water. The isoforms and activity of carbonic anhydrase in different algal species was also significantly different. Both *Chlamydomonas reinhardtii* and *Chlorella pyrenoidosa* are Chlorophyta, and their carbonic anhydrase response to the environment is obviously different. The biodiversity of microalgal carbonic anhydrase results in the diversity of its biogeochemical process.

4. The action of carbonic anhydrase on the global C, N, P and S cycles

Microalgae with extracellular carbonic anhydrase, which has inorganic carbon concentration mechanism to enrich inorganic carbon in the cells, achieved the inorganic carbon reduction under the synergistic action of intracellular carbonic anhydrase. Extracellular carbonic anhydrase behaves as a switch for the inorganic carbon metabolism, whereas the carbon metabolism of algae behaves as the ligament among ecosystem carbon metabolism. The biogeochemical action of extracellular carbonic anhydrase responded to the environment induced the global carbon cycle quickly responded to the change of the environment. The carbon metabolism provides the carbon skeleton and energy for the nitrogen, phosphorus and sulphurs metabolism. Therefore, the biogeochemical action of carbonic anhydrase will definitely have a profound effect on the global carbon, nitrogen, phosphorus and sulfur cycle.

5. The effect of carbonic anhydrase on the weathering process of silicate minerals and the carbon sink

Silicate is defined as a kind of compound which constitute silicon and oxygen and some other elements (mainlyaluminum, iron, calcium, magnesium, potassium, sodium, etc.). It is widespread in the earth's crust, account for more than 80% of amount of earth crustal rocks, and constitute the majority of the main composition of rock and soil. In the geological time scale, on the one hand, the acid dissolution of silicate minerals produce carbon sink through the action of water and carbon dioxide. On the other hand, the silicate minerals can be dissolved in the alkaline environment. Carbonic anhydrase, which can influence carbon dioxide flux and the protons in the environment, is likely to be profoundly adjust the dissolution of silicate rock. Silicate rock dissolution has great influence on biogeochemical cycles of metal elements in soil system and the whole ecosystem, which will lead to the change in biogeochemical cycles of other elements in the ecosystem. Therefore, the research on weathering process of silicate minerals induced by carbonic anhydrase could greatly enrich the contents of biogeochemical action from carbonic anhydrase and could more deeply reveal biogeochemical action law.

6. The application and regulation of carbonic anhydrase in ecosystem

The change ofmicroalgal carbonic anhydrase in aquatic media will influence the water environment; vice versa, the change of water environment will also affect the mode of microalgal carbonic anhydrase responding to the environment. The regulating the carbonic anhydrase of microalgae in water not only restore the environment of nitrogen, phosphorus and other elements such as fluoride pollution, solve the blooms, but also optimize the microalgae community structure with providing high quality bait for the aquaculture. In addition, carbonic anhydrase had an impact on the isotope discrimination of the carbon, hydrogen and oxygen in the conversion of organic matter and the deposition of carbonate. Therefore, the consideration about the effect of carbonic anhydrase on stable isotope fractionation will greatly improve the precision of model built for the inversion of ancient climate change.

7. The enzymatic and macromolecular biogeochemical action

The carbonic anhydrase, a key enzyme involved in carbon metabolism, affect the stable carbon isotope fractionation in depth, change the way and share of inorganic carbon utilization, drive the dissolution of carbonate, and influence the karst and photosynthetic carbon sink. Carbonic anhydrase plays a role as model protein in the enzymatic and macromolecular biogeochemical action. The nitrate reductase, a key enzyme involved in nitrate metabolism, even in nitrogen metabolism, determine the way and share of inorganic nitrogen utilization, affect the composition of inorganic nitrogen and the microalgal species distribution, drive the element biogeochemical cycle, influence the nitrogen flux in aquatic media, and finally change the biogeochemical cycle of nitrogen in the interfaces, even in whole ecosystem. Adenosine phosphosulfate reductase is a key enzyme involved in sulfate and sulfur cycle, which, similarly affect the sulfur flux, influence the corrosion and weathering of rocks, change the composition of elements and the algal species structure, and in the end, influence the biogeochemical cycle of sulfur in the interface, even in whole ecosystem. In a word, researching the biogeochemical action of these key enzymes involved in the cycles of carbon, nitrogen, sulfur, and phosphorus, will lead to building the scientific system about the enzymatic and macromolecular biogeochemistry.